Mechatronic Systems 2

Mechatronic Systems 2
Applications in Material Handling Processes and Robotics

Edited by

Leonid Polishchuk
Vinnytsia National Technical University, Vinnytsia, Ukraine

Orken Mamyrbayev
Institute of Information and Computational Technologies CS MES RK, Almaty, Kazakhstan

Konrad Gromaszek
Lublin University of Technology, Lublin, Poland

LONDON AND NEW YORK

Cover image: Andrzej Kotyra
First published 2021
by Routledge/Balkema
Schipholweg 107C, 2316 XC Leiden, The Netherlands
e-mail: enquiries@taylorandfrancis.com
www.routledge.com – www.taylorandfrancis.com

Routledge/Balkema is an imprint of the Taylor & Francis Group, an informa business

© 2021 selection and editorial matter, Leonid Polishchuk, Orken Mamyrbayev, Konrad Gromaszek; individual chapters, the contributors

The right of Leonid Polishchuk, Orken Mamyrbayev, Konrad Gromaszek to be identified as the authors of the editorial material, and of the authors for their individual chapters, has been asserted in accordance with sections 77 and 78 of the Copyright, Designs and Patents Act 1988.

All rights reserved. No part of this book may be reprinted or reproduced or utilized in any form or by any electronic, mechanical, or other means, now known or hereafter invented, including photocopying and recording, or in any information storage or retrieval system, without permission in writing from the publishers.

Although all care is taken to ensure integrity and the quality of this publication and the information herein, no responsibility is assumed by the publishers nor the author for any damage to the property or persons as a result of operation or use of this publication and/or the information contained herein.

Library of Congress Cataloging-in-Publication Data
A catalog record has been requested for this book

ISBN: 978-1-032-10585-7 (Hbk)
ISBN: 978-1-032-12621-0 (Pbk)
ISBN: 978-1-003-22544-7 (eBook)

DOI: 10.1201/9781003225447

Typeset in Times New Roman
by codeMantra

Contents

List of editors ix
List of contributors xi

1 **Development of perspective equipment for the regeneration of industrial filters** 1
  I. Sevostyanov, I. Zozulyak, Y. Ivanchuk, O. Polischuk, K. Koval, W. Wójcik, A. Kalizhanova, and A. Kozbakova

2 **Intelligent implants in dentistry: Realities and prospects** 15
  S. Zlepko, S. Tymchyk, O. Hrushko, I. Vishtak, Z. Omiotek, S. Amirgaliyeva, and A. Tuleshov

3 **Modeling of the exhaustion and regeneration of the resource regularities of objects with different natures** 27
  V. Mykhalevych, V. Kraievskyi, O. Mykhalevych, O. Hrushko, A. Kotyra, P. Droździel, O. Mamyrbaev, and S. Orazalieva

4 **Increase in durability and reliability of drill column casing pipes by the surface strengthening** 39
  I. Aftanaziv, L. Shevchuk, L. Strutynska, I. Koval, I. Svidrak, P. Komada, G. Yerkeldessova, and K. Nurseitova

5 **Experimental research of forming machine with a spatial character of motion** 51
  I. Nazarenko, O. Dedov, M. Ruchynskyi, A. Sviderskyi, O. Diachenko, P. Komada, M. Junisbekov, and A. Oralbekova

6 **Research of ANSYS Autodyn capabilities in evaluating the landmine blast resistance of specialized armored vehicles** 61
  S. Shlyk, A. Smolarz, S. Rakhmetullina, and A. Ormanbekova

7	**Phenomenological aspects in modern mechanics of deformable solids** V. Ogorodnikov, T. Arkhipova, M.O. Mokliuk, P. Komada, A. Tuleshov, U. Zhunissova, and M. Kozhamberdiyeva	77
8	**The determination of deformation velocity effect on cold backward extrusion processes with expansion in the movable die of axisymmetric hollow parts** I. Aliiev, V. Levchenko, L. Aliieva, V. Kaliuzhnyi, P. Kisała, B. Yeraliyeva, and Y. Kulakova	87
9	**Stress state of a workpiece under double bending by pulse loading** V. Dragobetskii, V. Zagoryanskii, D. Moloshtan, S. Shlyk, A. Shapoval, O. Naumova, A. Kotyra, M. Mussabekov, G. Yusupova, and Y. Kulakova	101
10	**Tensor models of accumulation of damage in material billets during roll forging process in several stages** V. Matviichuk, I. Bubnovska, V. Mykhalevych, M. Kovalchuk, W. Wójcik, A. Tuleshov, S. Smailova, and B. Imanbek	111
11	**Synergetic aspects of growth in machining of metal materials** E. Posviatenko, N. Posvyatenko, O. Mozghovyi, R. Budyak, A. Smolarz, A. Tuleshov, G. Yusupova, and A. Shortanbayeva	121
12	**Theoretical and experimental studies to determine the contact pressures when drawing an axisymmetric workpiece without a blank flange collet** R. Puzyr, R. Argat, A. Chernish, R. Vakylenko, V. Chukhlib, Z. Omiotek, M. Mussabekov, G. Borankulova, and B. Yeraliyeva	131
13	**Modification of surfaces of steel details using graphite electrode plasma** V. Savulyak, V. Shenfeld, M. Dmytriiev, T. Molodetska, V. M. Tverdomed, P. Komada, A. Ormanbekova, and Y. Turgynbekov	141

14 **Complex dynamic processes in elastic bodies and the methods of their research**   151
B. I. Sokil, A. P. Senyk, M. B. Sokil, A. I. Andrukhiv,
O. O. Koval, A. Kotyra, P. Drożdziel, M. Kalimoldayev,
and Y. Amirgaliyev

15 **Analysis of the character of change of the profilogram of micro profile of the processed surface**   165
N. Veselovska, S. Shargorodsky, V. Rutkevych,
R. Iskovych-Lototsky, Z. Omiotek, O. Mamyrbaev,
and U. Zhunissova

16 **Investigation of interaction of a tool with a part in the process of deforming stretching with ultrasound**   175
N. Weselowska, V. Turych, V. Rutkevych, G. Ogorodnichuk,
P. Kisała, B. Yeraliyeva, and G. Yusupova

17 **Robotic complex for the production of products special forms with filling inside made from dough**   185
R. Grudetskyi, O. Zabolotnyi, P. Golubkov, V. Yehorov,
A. Kotyra, A. Kozbakova, and S. Amirgaliyeva

18 **Theoretical preconditions of circuit design development for the manipulator systems of actuators of special-purpose mobile robots**   197
S. Strutynskyi, W. Wójcik, A. Kalizhanova, and M. Kozhamberdiyeva

19 **Analysis of random factors in the primary motion drive of grinding machines**   213
V. Tikhenko, O. Deribo, Z. Dusaniuk, O. Serdiuk, A. Kotyra,
S. Smailova, and Y. Amirgaliyev

20 **Dynamic characteristics of "tool-workpiece" elastic system in the low stiffness parts milling process**   225
Y. Danylchenko, A. Petryshyn, S. Repinskyi, V. Bandura,
M. Kalimoldayev, K. Gromaszek, and B. Imanbek

21 **Modeling of contact interaction of microroughnesses of treated surfaces during finishing anti-friction non-abrasive treatment FANT**   237
I. Shepelenko, Y. Nemyrovskyi, Y. Tsekhanov, E. Posviatenko,
Z. Omiotek, M. Kozhamberdiyeva, and A. Shortanbayeva

22 Practices of modernization of metal-cutting machine tool CNC systems   247
V. Sychuk, O. Zabolotnyi, P. Harchuk, D. Somov, A. Slabkyi, Z. Omiotek, S. Rakhmetullina, and G. Yusupova

23 Improving the precision of the methods for vibration acceleration measurement using micromechanical capacitive accelerometers   257
V. F. Hraniak, V. V. Kukharchuk, Z. Omiotek, P. Drożdziel, O. Mamyrbaev, and B. Imanbek

24 Modeling of the technological objects movement in metal processing on machine tools   267
G. S. Tymchyk, V. I. Skytsiouk, T. R. Klotchko, P. Komada, S. Smailova, and A. Kozbakova

25 Physical bases of aggression of abstract objects existence   279
G. S. Tymchyk, V. I. Skytsiouk, T. R. Klotchko, W. Wójcik, Y. Amirgaliyev, and M. Kalimoldayev

26 Development and investigation of changes in the form of metal when obtaining the crankshaft's crankpin using free forging   291
V. Chukhlib, A. Okun, S. Gubskyi, Y. Klemeshov, R. Puzyr, P. Komada, M. Mussabekov, D. Baitussupov, and G. Duskazaev

27 Approaches to automation of strength and durability analysis of crane metal structures   303
S. Gubskyi, V. Chukhlib, A. Okun, Y. Basova, S. Pavlov, K. Gromaszek, A. Tuleshov, and A. Toigozhinova

List of editors

Konrad Gromaszek was born in Poland in 1978. He was professor of the Lublin University of Technology at the Faculty of Electrical Engineering and Computer Sciences. After obtaining a doctorate in December 2006, he was employed at the Department of Electronics and Information Technology. In 2019, he obtained DSc degree and now works as a university professor.

In the years 2007–2008, he was the manager of the FP6 project, related to the development of the Regional Innovation Strategy for the Lubelskie Voivodeship. He participated in a total of about 15 courses and training in data management and processing. He has received two awards from the Rector of the Lublin University of Technology (second and third degree). Konrad Gromaszek is the author and coauthor of over 53 publications (including three monographs, two manuals, and scripts). He was the tutor of 89 diploma theses. He tries to combine research and teaching with organizational activities. He participated in the preparation of applications for seven research projects and was the contractor in four. He is also active in non-university projects as an expert. He belongs to the following organizations: IEEE, Polish Association of Measurements for Automation and Robotics (POLSPAR), Polish Society of Theoretical and Applied Electrical Engineering (PTETiS), Polish Information Technology Society (PTI), and Lubelskie Towarzystwo Naukowe (LTN).

Orken Mamyrbayev was born in Kazakhstan, 1979. He was Deputy General Director in science and head of the laboratory of computer engineering of intelligent systems at the Institute of Information and Computational Technologies. In 2014, he obtained his Ph.D. in Information Systems at the Kazakh National Technical University named after K. I. Satbayev and was Associate Professor in 2019 at the Institute of Information and Computational Technologies. He is a member of the dissertation council "Information Systems" at L.N. Gumilyov Eurasian National University in the specialties Computer Sciences and Information Systems. His main research field of activity is related to machine learning, deep learning, and speech technologies. In total, he has published more than five books, over 120 papers, and authored several patents and copyright certificates for an intellectual property object in software. Currently, he manages two scientific projects: the development of an end-to-end automatic speech recognition system for agglutinative languages and information model and software tools for the system of automatic search and analysis of multilingual illegal web content based on the ontological approach.

Leonid Polishchuk was born in Ukraine in 1954. He is the Head of the Department of Industrial Engineering at Vinnitsia National Technical University, Ukraine. Leonid is Doctor of Technical Sciences, professor, academician at the Ukraine Academy of Hoisting-and-Transport Sciences, member of the editorial board of two scientific and technical publications in Ukraine, and member of the specialized scientific council for the defense of doctoral dissertations. In 1994, he defended his thesis "Dynamic Load of the Mechanical System of the Belt Conveyor with the Built-in Drive" in the concentration of "Dynamics, Strength of Machines, Devices and Equipment". In 2017, he defended the specialized council doctoral thesis "Dynamics of Drive System and Boom Construction of Belt Conveyors on Mobile Machinery" in the concentration of "Dynamics and Strengths of Machines." The scientific focus is the dynamics of drive systems with devices and control systems with variable operating modes and diagnostics of metal structures of hoisting-and-transport and technological machines. He has more than 200 scientific publications of which two are monographs, 123 are of scientific and 18 are of educational and methodological nature, 8 are of scientific nature in publications such as Scopus and WoS, and 33 are patents.

List of contributors

I. Aftanaziv, Descriptive Geometry and Engineering Graphics Department, Institute of Applied Mathematics and Basic Sciences, National University "Lvivska Polytechnica", Lviv, Ukraine

I. Aliiev, Automation of Mechanical Engineering and Information Technologies, Processing of Metal Forming Department, Donbass State Engineering Academy, Kramatorsk, Ukraine

L. Aliieva, Automation of Mechanical Engineering and Information Technologies Processing of Metal Forming Department, Donbass State Engineering Academy, Kramatorsk, Ukraine

Y. Amirgaliyev, Institute of Information and Computational Technologies CS MES RK, Almaty, Kazakhstan

S. Amirgaliyeva, Institute of Information and Computational Technologies CS MES RK, Almaty, Kazakhstan, IT Department, Academy of Logistics & Transport, Almaty, Kazakhstan

A. I. Andrukhiv, Transport Technologies Department, Lviv Polytechnic National University, Lviv, Ukraine

R. Argat, Mechanical Engineering Technology Department, Institute of Mechanics and Transport, Kremenchuk Mykhailo Ostrohradskyi National University, Kremenchuk, Ukraine

T. Arkhipova, Department of Material Resistance and Applied Mechanics, Vinnytsia National Technical University, Vinnytsia, Ukraine

D. Baitussupov, IT Department, Academy of Logistics & Transport, Almaty, Kazakhstan

V. Bandura, Agricultural Engineering and Technical Service Department, Vinnytsia National Agrarian University, Vinnytsia, Ukraine

Y. Basova, Processing of Metal Forming Department, National Technical University "Kharkiv Polytechnic Institute", Kharkiv, Ukraine

G. Borankulova, Faculty of Information Technology, Automation and Telecommunications, M. Kh. Dulaty Taraz Regional University, Taraz, Kazakhstan

I. Bubnovska, Electrical Systems, Technologies and Automation in Agriculture Department, Vinnytsia National Agrarian University, Vinnytsia, Ukraine

R. Budyak, Machine-Tractor Fleet Operation and Maintenance Department, Vinnytsia National Agrarian University, Vinnytsia, Ukraine

A. Chernish, Mechanical Engineering Technology Department, Institute of Mechanics and Transport, Kremenchuk Mykhailo Ostrohradskyi National University, Kremenchuk, Ukraine

V. Chukhlib, Processing of Metal Forming Department, Kharkiv Polytechnic Institute, National Technical University "Kharkiv Polytechnic Institute", Kharkiv, Ukraine

Y. Danylchenko, Machine Design Department, National Technical University of Ukraine "Igor Sikorsky Kyiv Polytechnic Institute", Kyiv, Ukraine

O. Dedov, Machines and Equipment of Technological Processes Department, Kyiv National University of Construction and Architecture, Kyiv, Ukraine

O. Deribo, Machine-Building Technology and Automation Department, Vinnytsia National Technical University, Vinnytsia, Ukraine

O. Diachenko, Machines and Equipment of Technological Processes Department, Kyiv National University of Construction and Architecture, Kyiv, Ukraine

M. Dmytriiev, Industrial Engineering Department, Vinnitsia National Technical University, Vinnytsia, Ukraine

V. Dragobetskii, Mechanical Engineering Technology Department, Institute of Mechanics and Transport, Kremenchuk Mykhailo Ostrohradskyi National University, Kremenchuk, Ukraine

P. Droździel, Faculty of Mechanical Engineering, Lublin University of Technology, Lublin, Poland

Z. Dusaniuk, Machine-Building Technology and Automation Department, Vinnytsia National Technical University, Vinnytsia, Ukraine

G. Duskazaev, IT Department, Academy of Logistics & Transport, Almaty, Kazakhstan

P. Golubkov, Automation of Technological Processes and Robotic Systems Department, Odessa National Academy of Food Technologies, Odessa, Ukraine

K. Gromaszek, Faculty of Electrical Engineering and Computer Science, Lublin University of Technology, Lublin, Poland

R. Grudetskyi Automation and Computer-Integrated Technologies Department, Lutsk National Technical University, Lutsk, Ukraine

S. Gubskyi, Processing of Metal Forming Department, National Technical University "Kharkiv Polytechnic Institute", Kharkiv, Ukraine

P. Harchuk, Automation and Computer-Integrated Technologies Department, Lutsk National Technical University, Lutsk, Ukraine

V. F. Hraniak, Theoretical Electrical Engineering and Electrical Measurements Department, Vinnytsia National Technical University, Vinnytsia, Ukraine

O. Hrushko, Material Resistance and Applied Mechanics Department, Vinnytsia National Technical University, Vinnytsia, Ukraine

B. Imanbek, Faculty of Information Technology, Al-Farabi Kazakh National University, Almaty, Kazakhstan

R. Iskovych-Lototsky, Industrial Engineering Department, Vinnytsia National Technical University, Vinnytsia, Ukraine

Y. Ivanchuk, Computer Science Department, Vinnytsia National Technical University, Vinnytsia, Ukraine

M. Junisbekov, Automation and Telecommunications, M. Kh. Dulaty Taraz Regional University, Taraz, Kazakhstan

M. Kalimoldayev, Institute of Information and Computational Technologies CS MES RK, Almaty, Kazakhstan

V. Kaliuzhnyi, Aircraft Production Technologies Department, National Technical University of Ukraine 'Igor Sikorsky Kyiv Polytechnic Institute', Kyiv, Ukraine

A. Kalizhanova, Institute of Information and Computational Technologies CS MES RK, Almaty, Kazakhstan, Faculty of Information Technology, Al-Farabi Kazakh National University, Almaty, Kazakhstan

P. Kisała, Faculty of Electrical Engineering and Computer Science, Lublin University of Technology, Lublin, Poland

Y. Klemeshov, Metal Forming Department, Z. I. Nekrasov Iron & Steel Institute of NAS of Ukraine, Kyiv, Ukraine

T. R. Klotchko, Devices Production Department, National Technical University of Ukraine "Sikorsky Kyiv Polytechnic Institute", Kyiv, Ukraine

P. Komada, Faculty of Electrical Engineering and Computer Science, Lublin University of Technology, Lublin, Poland

A. Kotyra, Faculty of Electrical Engineering and Computer Science, Lublin University of Technology, Lublin, Poland

I. Koval, Integration of Education with Enterprises Department, National University "Lvivska Polytechnica", Lviv, Ukraine

K. Koval, Integration of Education with Enterprises Department, Vinnytsia National Technical University, Vinnytsia, Ukraine

O. O. Koval, Industrial Engineering Department, Vinnytsia National Technical University, Vinnytsia, Ukraine

M. Kovalchuk, Higher Mathematics Department, Vinnytsia National Technical University, Vinnytsia, Ukraine

A. Kozbakova, Institute of Information and Computational Technologies CS MES RK, Almaty, Kazakhstan, Faculty of Engineering and IT, Almaty Technological University, Almaty, Kazakhstan

M. Kozhamberdiyeva, Faculty of Information Technology, Al-Farabi Kazakh National University, Almaty, Kazakhstan

V. Kraievskyi, Higher Mathematics Department, Vinnytsia National Technical University, Vinnytsia, Ukraine

V. V. Kukharchuk, Theoretical Electrical Engineering and Electrical Measurements Department, Vinnytsia National Technical University, Vinnytsia, Ukraine

Y. Kulakova, Institute of Automation and Information Technologies, Satbayev University, Almaty, Kazakhstan

V. Levchenko, Automation of Mechanical Engineering and Information Technologies, Processing of Metal Forming Department, Donbass State Engineering Academy, Kramatorsk, Ukraine

O. Mamyrbaev, Institute of Information and Computational Technologies CS MES RK, Almaty, Kazakhstan

V. Matviichuk, Electrical Systems, Technologies and Automation in Agriculture Department, Vinnytsia National Agrarian University, Vinnytsia, Ukraine

M. O. Mokliuk, Department of Material Resistance and Applied Mechanics, Vinnytsia State Pedagogical University named after M. Kotsiubynsky, Vinnytsia, Ukraine

T. Molodetska, Industrial Engineering Department, Vinnitsia National Technical University, Vinnitsia, Ukraine

D. Moloshtan, Mechanical Engineering Technology Department, Institute of Mechanics and Transport, Kremenchuk Mykhailo Ostrohradskyi National University, Kremenchuk, Ukraine

O. Mozghovyi, Physics and Methods of Teaching Physics, Astronomy Department, Vinnytsia Mykhailo Kotsiubynskyi State Pedagogical University, Vinnytsia, Ukraine

M. Mussabekov, IT Department, Academy of Logistics & Transport, Almaty, Kazakhstan

O. Mykhalevych, Higher Mathematics Department, Vinnytsia National Technical University, Vinnytsia, Ukraine

V. Mykhalevych, Higher Mathematics Department, Vinnytsia National Technical University, Vinnytsia, Ukraine

O. Naumova, Mechanical Engineering Technology Department, Institute of Mechanics and Transport, Kremenchuk Mykhailo Ostrohradskyi National University, Kremenchuk, Ukraine

I. Nazarenko, Machines and Equipment of Technological Processes Department, Kyiv National University of Construction and Architecture, Kyiv, Ukraine

Y. Nemyrovskyi, Exploitation and Repairing Machine Department, Central Ukrainian National Technical University, Kirovohrad, Ukraine

K. Nurseitova, Department of Information Technology, D. Serikbayev East Kazakhstan State Technical University, Ust-Kamenogorsk, Kazakhstan

G. Ogorodnichuk, Industrial Engineering Department, Vinnytsia National Agrarian University, Vinnytsia, Ukraine

V. Ogorodnikov, Department of Material Resistance and Applied Mechanics, Vinnytsia National Technical University, Vinnytsia, Ukraine

A. Okun, Processing of Metal Forming Department, National Technical University "Kharkiv Polytechnic Institute", Kharkiv, Ukraine

Z. Omiotek, Faculty of Electrical Engineering and Computer Science, Lublin University of Technology, Lublin, Poland

A. Oralbekova, IT Department, Kazakh University Ways of Communications, Almaty, Kazakhstan

S. Orazalieva, Faculty of Engineering and IT, Almaty Technological University, Almaty, Kazakhstan

A. Ormanbekova, Faculty of Information Technology, Al-Farabi Kazakh National University, Almaty, Kazakhstan

S. Pavlov, Biomedical Engineering Department, Vinnytsia National Technical University, Vinnytsia, Ukraine

A. Petryshyn, Machine Design Department, National Technical University of Ukraine "Igor Sikorsky Kyiv Polytechnic Institute", Kyiv, Ukraine

O. Polischuk, Safety of Life and Security Pedagogy Department, Vinnytsia National Technical University, Vinnytsia, Ukraine

N. Posvyatenko, Manufacturing, Repair and Materials Engineering Department, National Transport University, Kyiv, Ukraine

E. Posviatenko, Manufacturing, Repair and Materials Engineering Department, National Transport University, Kyiv, Ukraine

R. Puzyr, Mechanical Engineering Technology Department, Institute of Mechanics and Transport, Kremenchuk Mykhailo Ostrohradskyi National University, Kremenchuk, Ukraine

S. Rakhmetullina, East Kazakhstan State Technical University named after D.Serikbayev, Ust-Kamenogorsk, Kazakhstan

S. Repinskyi, Machine-Building Technology and Automation Department, Vinnytsia National Technical University, Vinnytsia, Ukraine

M. Ruchynskyi, Machines and Equipment of Technological Processes Department, Kyiv National University of Construction and Architecture, Kyiv, Ukraine

V. Rutkevych, Machinery and Equipment of Agricultural Production Department, Vinnytsia National Agrarian University, Vinnytsia, Ukraine

V. Savulyak, Industrial Engineering Department, Vinnitsia National Technical University, Vinnytsia, Ukraine

A. P. Senyk, Transport Technologies Department, Lviv Polytechnic National University, Lviv, Ukraine

O. Serdiuk, Machine-Building Technology and Automation Department, Vinnytsia National Technical University, Vinnytsia, Ukraine,

I. Sevostyanov, Technological Processes and Equipment of Processing and Food Production Department, Vinnytsia National Agrarian University, Vinnytsia, Ukraine

A. Shapoval, Mechanical Engineering Technology Department, Institute of Mechanics and Transport, Kremenchuk Mykhailo Ostrohradskyi National University, Kremenchuk, Ukraine

S. Shargorodsky, Machinery and Equipment of Agricultural Production Department, Vinnitsia National Technical University, Vinnitsia, Ukraine

V. Shenfeld, Industrial Engineering Department, Vinnitsia National Technical University, Vinnitsia, Ukraine

I. Shepelenko, Exploitation and Repairing Machine Department, Central Ukrainian National Technical University, Kirovohrad, Ukraine

L. Shevchuk, Descriptive Geometry and Engineering Graphics Department, Institute of Applied Mathematics and Basic Sciences, National University "Lvivska Polytechnica", Lviv, Ukraine

S. Shlyk, Mechanical Engineering Technology Department, Institute of Mechanics and Transport, Kremenchuk Mykhailo Ostrohradskyi National University, Kremenchuk, Ukraine

A. Shortanbayeva, Faculty of Information Technology, Al-Farabi Kazakh National University, Almaty, Kazakhstan

V. I. Skytsiouk, Devices Production Department, National Technical University of Ukraine "Sikorsky Kyiv Polytechnic Institute", Kyiv, Ukraine

A. Slabkyi, Industrial Engineering Department, Lutsk National Technical University, Lutsk, Ukraine

S. Smailova, Department of Information Technology, D. Serikbayev East Kazakhstan State Technical University, Ust-Kamenogorsk, Kazakhstan

A. Smolarz, Faculty of Electrical Engineering and Computer Science, Lublin University of Technology, Lublin, Poland

D. Somov, Automation and Computer-Integrated Technologies Department, Lutsk National Technical University, Lutsk, Ukraine

B. I. Sokil, Engineering Mechanics Department, Hetman Petro Sahaidachnyi National Army Academy, Lviv, Ukraine

M. B. Sokil, Transport Technologies Department, Lviv Polytechnic National University, Lviv, Ukraine, D. Somov, Automation and Computer-Integrated Technologies Department, Lutsk National Technical University, Lutsk, Ukraine

L. Strutynska, Descriptive Geometry and Engineering Graphics Department, Institute of Applied Mathematics and Basic Sciences, National University "Lvivska Polytechnica", Lviv, Ukraine

S. Strutynskyi, Applied Hydroaeromechanics and Mechanotronics Department, Igor Sikorsky Kyiv Politechnic Institute, Kyiv, Ukraine

A. Sviderskyi, Machines and Equipment of Technological Processes Department, Kyiv National University of Construction and Architecture, Kyiv, Ukraine

I. Svidrak, Descriptive Geometry and Engineering Graphics Department, Institute of Applied Mathematics and Basic Sciences, National University "Lvivska Polytechnica", Lviv, Ukraine

V. Sychuk, Automation and Computer-Integrated Technologies Department, Lutsk National Technical University, Lutsk, Ukraine

V. Tikhenko, Metal-Cutting Machines, Metrology and Certification Department, Odessa National Politechnical University, Odessa, Ukraine

A. Toigozhinova, IT Department, Academy of Logistics and Transport, Almaty, Kazakhstan

Y. Tsekhanov, Department of Engineering and Computer Graphics, Voronezh State Technical University, Voronezh, Russia

A. Tuleshov, Institute of mechanics and engineering science CS MES RK, Almaty, Kazakhstan

Y. Turgynbekov, Faculty of Information Technology, Automation and Telecommunications, M. Kh. Dulaty Taraz Regional University, Taraz, Kazakhstan

V. Turych, Industrial Engineering Department, Vinnytsia National Agrarian University, Vinnytsia, Ukraine

V. M. Tverdomed, Faculty of Infrastructure and Railway Rolling Stock, State University of Infrastructure and Technology, Kyiv, Ukraine

G. S. Tymchyk, Devices Production Department, National Technical University of Ukraine "Sikorsky Kyiv Polytechnic Institute", Kyiv, Ukraine

S. Tymchyk, Biomedical Engineering Department, Vinnytsia National Technical University, Vinnytsia, Ukraine

R. Vakylenko, Mechanical Engineering Technology Department, Institute of Mechanics and Transport, Kremenchuk Mykhailo Ostrohradskyi National University, Kremenchuk, Ukraine

N. Veselovska, Machinery and Equipment of Agricultural Production Department, Vinnytsia National Technical University, Vinnytsia, Ukraine

I. Vishtak, Material Resistance and Applied Mechanics Department, Vinnytsia National Technical University, Vinnytsia, Ukraine

N. Weselowska, Industrial Engineering Department, Vinnytsia National Agrarian University, Vinnytsia, Ukraine

W. Wójcik, Faculty of Electrical Engineering and Computer Science, Lublin University of Technology, Lublin, Poland

V. Yehorov, Automation of Technological Processes and Robotic Systems Department, Odessa National Academy of Food Technologies, Odessa, Ukraine

B. Yeraliyeva, Faculty of Information Technology, Automation and Telecommunications, M. Kh. Dulaty Taraz Regional University, Taraz, Kazakhstan

G. Yerkeldessova, IT Department, Academy of Logistics & Transport, Almaty, Kazakhstan

G. Yusupova, Department of IT, Turan University, Almaty, Kazachstan

O. Zabolotnyi, Automation and Computer-Integrated Technologies Department, Lutsk National Technical University, Lutsk, Ukraine

V. Zagoryanskii, Mechanical Engineering Technology Department, Institute of Mechanics and Transport, Kremenchuk Mykhailo Ostrohradskyi National University, Kremenchuk, Ukraine

U. Zhunissova, Department of IT, Astana Medical University, Nur-Sultan City, Kazakhstan

S. Zlepko, Biomedical Engineering Department, Vinnytsia National Technical University, Vinnytsia, Ukraine

I. Zozulyak, Technological Processes and Equipment of Processing and Food Production Department, Vinnytsia National Agrarian University, Vinnytsia, Ukraine

Chapter 1

Development of perspective equipment for the regeneration of industrial filters

I. Sevostyanov, I. Zozulyak, Y. Ivanchuk, O. Polischuk,
K. Koval, W. Wójcik, A. Kalizhanova,
and A. Kozbakova

CONTENTS

1.1 Introduction 1
1.2 Analysis of literature data and problem statement 2
1.3 Materials and research methods 6
1.4 Development of equipment for the regeneration of ion-exchange resin 6
1.5 Conclusion 12
References 12

1.1 INTRODUCTION

The problem of the lack of clean drinking water is becoming urgent for an increasing number of regions in different countries, especially in large cities with high levels of environmental pollution and in industrial zones. Quality water is a determining factor for the food and chemical industry, pharmacology, microbiology, and microelectronics. In addition, in Europe, including Belarus, Russia, and the Ukraine, there are quite a few areas with high natural hardness of water. In this regard, industrial and household filters and filter systems are increasingly being used.

One of the main stages of high-quality water purification is its degreasing and softening. In most cases, they are implemented using ion-exchange resin cartridge filters as filter fillers. However, in the process of using such filters, when cleaning sufficiently hard water, they lose their performance within a year. It should be added that the majority of enterprises and private users of these filters are interested in their regeneration. The latter is carried out using a 10% solution of salt. The regenerated resin is soaked in the salt solution for 8–10 hours. At the same time, for sufficiently high-quality recovery, it is necessary to ensure periodic mixing of the resin in the salt solution with a velocity of $v_{min} = 0.01$ m/s. This is needed to penetrate into the lower layers of the portion of consumables. However, this is not always done, and therefore, regeneration is incomplete, and the life of such cartridges will be limited. Further, it is necessary to consider the viscosity of the mixed material. According to our experimental data, it is $\mu_m = 1.32–1.38$ Pa/s, as well as the true resin density of $\rho_m = 1.04 \cdot 10^{-3}$ kg/m^3.

Thus, the actual task is the selection or development of equipment for the effective mixing of ion-exchange resin in saline solution. This equipment should be compact, reliable, inexpensive, and convenient in operation.

DOI: 10.1201/9781003225447-1

1.2 ANALYSIS OF LITERATURE DATA AND PROBLEM STATEMENT

From the known equipment, the most suitable are the agitators and mixers for wet dispersed materials. They are used in construction and in the food industry. Mixers and mixers for food production are divided into (Dragilev & Drozdov, 1999; Saravacos, 2002) high-speed and low-speed, uninterrupted, and periodic action, with stationary fixed and nonstationary chambers with screw, vane, rotor, anchor, propeller, turbine, drum, and finger actuators, providing radial, axial, or radial–axial flows of a stirred medium. Taking into account the above conditions and requirements for equipment for mixing ion-exchange resin in saline solution, batch mixers with a stationary fixed chamber and a radial–axial flow of the mixed solution are most suitable. Then, in accordance with the diagram given in (Dragilev & Drozdov, 1999), it is necessary to use propeller turbines with flat blades, paddle, or frame mixers, and mixers for mixing materials with the above viscosity (Bartholomai, 1987, Hakansson et al., 2016a, b).

Figure 1.1 shows the scheme of the predeterminer PR-3 with a turbine agitator (Dragilev & Drozdov, 1999). It is used in sugar beet production, the executive elements of which are the blades (9, 12, and 13), mounted on a shaft (8), driven in rotation by an electric motor (7). In this case, the blades (9) are designed to remove foam in the duct (5), and mixing the product coming through the nozzle (4) provides the blades (12). The blades (13) serve to prevent congestion when removing the product for unloading through the nozzle at the bottom of the pre-defector. The mixing process is also facilitated by counter-patches (11), mounted on the inner surface of the housing (10). In our opinion, the mixing process is effectively carried out only in the lower central part of this unit, while the upper and peripheral layers are less affected. In addition, the circulation and return of the product to the zone of more intensive mixing are not ensured. With a shaft rotation frequency of 8 $n = 64 \min^{-1}$ and a case diameter of 10 $D = 2.4$ m (Dragilev & Drozdov, 1999), the average linear velocity of the product being mixed is shown in equation (1.1):

$$v_n = \frac{\pi \cdot n}{30} \frac{D}{2} = \frac{3.14 \cdot 64 \cdot 1.2}{30 \cdot 2} \approx 4 \, \text{m/s} \tag{1.1}$$

which is significantly higher than $v_{\min}$ and in accordance with the formula for determining the power of rotation of the shaft (8), shown in equation (1.2):

$$N_n = F_c v_n, \tag{1.2}$$

in which F_c is the strength of resistance to mixing. It depends mainly on the processed product and leads to a corresponding unnecessary increase in energy consumption for mixing (the nominal power of the electric motor of the predeterminer PR-3 is $N_{\text{nom}} = 13 \, \text{kW}$).

Figure 1.2 shows a homogenizer with a paddle stirrer (5) (Dragilev & Drozdov, 1999). It is driven from the electric motor (9), through the V-belt transmission (8) and the shaft (6). The product is fed through the nozzle (7), mixed in the hopper (4), and discharged through the nozzle (1). This device is designed for quick mixing of the product, which in continuous flow passes through the hopper (4). In this regard, the latter has a relatively small capacity. This circumstance does not allow for the use of

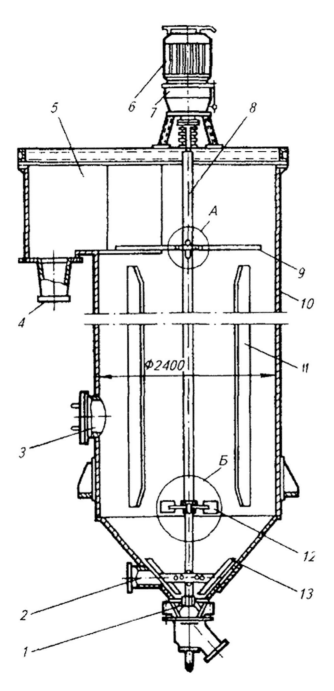

Figure 1.1 Scheme of predeterminer PR-3: (1) bearing; (2) inlet fitting; (3) access hatch; (4) outlet; (5) foam box; (6) electric motor; (7) gearbox; (8) vertical shaft; (9, 12, 13) mixers; (10) the case; (11) counterfeit.

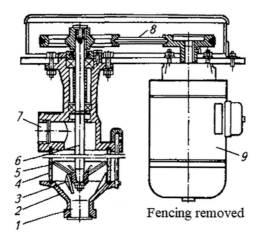

Figure 1.2 Scheme of the homogenizer: (1) outlet; (2) bottom; (3) ledge; (4) housing; (5) stirrer; (6) vertical shaft; (7) inlet; (8) V-belt transmission; (9) electric motor.

a homogenizer for long-term mixing of sufficiently large portions of the ion-exchange resin with a salt solution (Concidine, 2012).

The frequency of rotation of the shaft (6) of the homogenizer is $n = 950\,\text{min}^{-1}$, and the diameter of the body 4 is $D = 0.16\,\text{m}$ (Dragilev & Drozdov, 1999). Then, in accordance with the formula (1), the average linear velocity of the stirred product is:

$$v_n = \frac{\pi \cdot n}{30} \frac{D}{2} = \frac{3.14 \cdot 950 \cdot 0.08}{30 \cdot 2} = 3.97\,\text{m/s},$$

which also leads to unnecessary energy consumption (see formula (equation 1.2). The nominal power of the motor homogenizer is $N_{\text{nom}} = 0.28\,\text{kW}$ (Dragilev & Drozdov, 1999).

MT-250 machines for mixing and tempering various viscous masses, kneading machines with horizontal shafts, and a horizontal DÜC-C con-machine are also complicated, constructive, non-technological in manufacturing, expensive, and excessively powerful (Dragilev & Drozdov, 1999; Sevostyanov, 2013). The design of the kneading machine TM-63M (Dragilev & Drozdov, 1999) provides for the continuous passage of the processed material through it. In addition, the Z-shaped blades of the machine, performing the function of actuating elements, are rather non-technological.

The bubbler (Sevostyanov et al., 2015) shown in Figure 1.3 is a pneumatic-type agitator. Compressed air is fed into its tank, which is filled with the material being processed, through a tube or a system of tubes with small holes. Pop-up bubbles of the latter capture and mix material particles. However, as noted in (Geissler et al., 2014; Sevostyanov et al., 2015), pneumatic mixing is much less energy efficient than mechanical.

The most consistent with the above requirements and conditions is the mixer SMKN (Hakansson et al., 2016a, b) (Figure 1.4). It consists of two horizontal shafts (6), equipped with figured blades (5), deployed relative to the axes of the shafts at 60 degrees. This arrangement of the blades and the presence of holes in them ensure

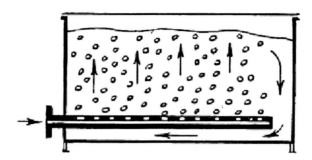

Figure 1.3 Diagram of the bubbler.

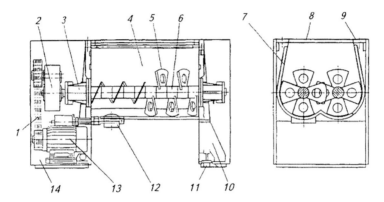

Figure 1.4 Mixer circuit SMKN: (1) V-belt transmission; (2) gearbox; (3) shaft support; (4) working chamber; (5) the blade; (6) shaft; (7) shirt; (8) lattice; (9) cover; (10) bed; (11) sensor; (12) inlet; (13) electric motor; (14) drive housing.

effective mixing of the product with its oncoming movement along the walls of the housing (4). The shaft (6) is driven from the electric motor (13) through the V-belt drive (1) and the gearbox (2). However, the mixer contains a shirt (7) that is not needed for regeneration of the resin. It is filled with paraffin oil with electric heaters. Taking into account the speed of rotation of the shafts (6) – $n = 38\,\text{min}^{-1}$ and the diameter of the blades (5) – $D = 0.9\,\text{m}$ by the formula (equation 1.1), we find that the average velocity of the particles of the mixed product provided by the mixer equals $0.89\,\text{m/s}$:

$$v_n = \frac{\pi \cdot n}{30} \frac{D}{2} = \frac{3.14 \cdot 38 \cdot 0.45}{30 \cdot 2} = 0.89\,\text{m/s},$$

This is noticeably more $v_{min} = 0.01\,\text{m/s}$ and allows us to conclude that the mixer has insufficient energy efficiency. The nominal power of the mixer motor is $N_{nom} = 55\,\text{kW}$ (Hakansson et al., 2016a, b).

In accordance with the results of the above analysis, it is obvious that the available serial machines for mixing dispersed masses do not satisfy the basic requirements for the regeneration equipment of ion-exchange resin.

During the operation, they must ensure uniform continuous circulation of ion-exchange resin with a dynamic viscosity $\mu_m = 1.38$ Pa/s in the working chamber with a speed of $v_{min} = 0.01$ m/s. In addition, they should have a simple and reliable design, created based on proven components of known equipment.

To achieve this goal, it is necessary to solve the following main tasks:

1. To eliminate the abovementioned disadvantages of the known equipment for mixing wet dispersed materials to optimize their design in terms of the power and speed of mixing.
2. Develop dependencies to determine the basic operating parameters of the proposed equipment, including the power and speed of mixing, necessary for the subsequent development of the methodology for its design calculation.

1.3 MATERIALS AND RESEARCH METHODS

As starting materials for solving the formulated tasks, we use the above schemes of the known equipment for mixing wet dispersed materials. We also use the identified deficiencies of the known equipment and formulas to take into account their specific requirements to determine their main operating parameters.

The dependencies of mechanics, hydraulics, and the theory of vibro-impact machines were used as a method for deriving dependencies to calculate the optimal operating parameters of the proposed equipment.

1.4 DEVELOPMENT OF EQUIPMENT FOR THE REGENERATION OF ION-EXCHANGE RESIN

Figure 1.5 shows a diagram of a twin-screw agitator, in which the electric motor (1), through the planetary gearbox (2) and the open gear transmission (3), rotates the screws (4) and (5) located in the hopper (6). Resin and salt solution is also loaded into the bunker for regeneration. Due to the opposite direction of the turns of the screws 4 and 5, rotating in one and the same direction, a circular movement of the processed material along the walls of the bunker is ensured, not only in the longitudinal direction but also in the transverse direction. Thereby, conditions are created for maximum mobility of the resin particles, penetration of the regenerating solution between them, and their intensive recovery (Zhu et al., 2016).

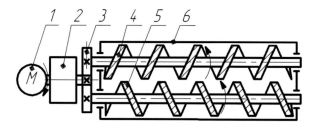

Figure 1.5 Diagram of a twin-screw mixer: (1) electric motor; (2) gearbox; (3) open gear; (4, 5) screws; (6) bunker.

The required drive power of the screws (4) and (5) of the mixer is determined by formula (Obertyukh et al., 2019):

$$N_n = 2 \cdot g \cdot Q_{max} L_{sc} \omega \cdot k_z \cdot 10^{-3} \tag{1.3}$$

where Q_{max} is the maximum performance of the mixer; L_{sc} is the length of the mixing screw, m ($L_{sc} = 2.2 \div 2.5$ m); ω is the coefficient of resistance to movement ($\omega = 4 \div 5$); k_z is the power safety factor ($k_z = 1.2 \div 1.25$).

Performance can be determined by taking into account the required minimum of movement speed for the processed material in the axial direction $v = 0.01$ m/s, cross-sectional area S_m of the flow of the material in one direction, and its density ρ_m shown in the formula (equation 1.4):

$$Q_{max} = \frac{S_m \cdot v}{2 \cdot \rho_m} \tag{1.4}$$

Substituting formula (equation 1.4) into formula (equation 1.3), we obtain:

$$N_n = \frac{g \cdot S_m \cdot v \cdot L_{sc} \cdot \omega \cdot k_z \cdot 10^{-3}}{\rho_m} = \frac{g \cdot \pi \cdot D_m^2 \cdot v \cdot L_{sc} \cdot \omega \cdot k_z \cdot 10^{-3}}{4 \cdot \rho_m}$$
$$= \frac{9.81 \cdot 3.14 \cdot 0.4^2 \cdot 0.01 \cdot 2.5 \cdot 5 \cdot 1.25 \cdot 10^{-3}}{4 \cdot 1.01 \cdot 10^{-3}} = 0.185 \, \text{kW} \tag{1.5}$$

On the basis of N_n, the power of the auger drive motor can be determined by the formula (equation 1.6):

$$N_{dm} = \frac{N_n}{\eta_{dsc}} = \frac{N_n}{\eta_c \eta_g \eta_{og} \eta_{gr}} = \frac{0.185}{0.98 \cdot 0.75 \cdot 0.94 \cdot 0.99 \cdot 0.99} = 0.273 \, \text{kW} \tag{1.6}$$

where η_{dsc}, η_c, η_g, η_{og}, η_{gr} are the efficiency of the augers drive, the coupling between the electric motor (1) and the gearbox (2) (not shown in Figure 1.1), the gearbox (2), the open gear train (3), and the bearings in which the screws are mounted, respectively.

As can be seen, the obtained value is significantly less than most types of known equipment of similar purpose, discussed above. At the same time, only the homogenizer (Figure 1.2) has a proportional power. However, this does not provide the necessary processing time for a portion of the resin.

Figure 1.6 shows a diagram of a centrifugal agitator. A portion of the processed material is loaded into it through the hatch (5), after which the rotation of the impeller (1) with blades (2) is activated. They ensure the material is pumped in the direction shown by arrows from the center of the wheel to the channel (3) and further along pipe (6) back to the center. As a result, there is intensive mixing and regeneration of the material.

In accordance with (Iskovich-Lototsky et al., 2019), the flow rate of a centrifugal pump, the analogue, which is this mixer, is determined by the formula (equation 1.7):

$$Q = 2 \cdot \pi \cdot R \cdot b \cdot \psi \cdot \eta_0 \cdot v_m, \tag{1.7}$$

8 Mechatronic Systems 2

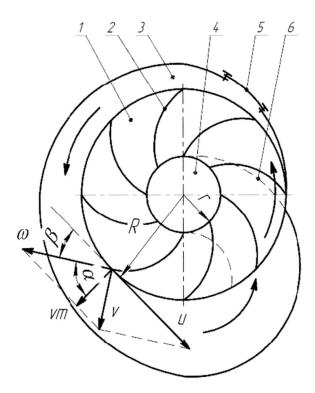

Figure 1.6 Centrifugal agitator: 1 impeller; (2) scapula; (3) channel; (4) inlet; (5) exit hatch; (6) return pipe.

where R is the radius of the impeller 1; b is the width of the impeller at the output; ψ is the compression ratio at the exit of the impeller; η_o is the volumetric efficiency; and v_m is the meridian component of the absolute velocity of the material v at the output.

The coefficient ψ was found according to the empirical formula (equation 1.8) (Iskovich-Lototsky et al., 2019):

$$\psi = 1 - \frac{z \cdot \delta}{2 \cdot \pi \cdot R \cdot \sin \beta}, \qquad (1.8)$$

where z is the number of blades; δ is the blade thickness.

Actual head H is provided by the agitator formula (equation 1.9) (Franceschinis et al., 2014):

$$H = \frac{u \cdot v \cdot \cos \alpha}{g} \eta_{mx} k_z, \qquad (1.9)$$

where u is the speed of the portable movement of the material; η_{mx} is the hydraulic efficiency of the mixer; k_z is the dimensionless coefficient of influence of a finite number of stirrer blades, determined by the following formula (equation 1.10) (Iskovich-Lototsky et al., 2019):

$$k_z = \frac{1}{1 + \frac{2\cdot\varphi}{z\left[1-(r/R)^2\right]}},\qquad(1.10)$$

where φ is the coefficient taking into account the influence of the guide vane ($\varphi = 0.8$–1.0- in the presence of the guide vane, $\varphi = 1.0$–1.3- in its absence); r is the radius of the inner edges of the blades.

The shaft power of the agitator formula (equation 1.11) is defined as:

$$N_n = \frac{Q\cdot\rho\cdot g\cdot H}{\eta},\qquad(1.11)$$

where ρ is the density of the material being processed; η is the overall efficiency of the mixer.

To calculate the estimated value of N_n taken, $v_m = v_{\min} = 0.01$ m/s. Then with such a low speed of the material, the impeller head H can be calculated based on atmospheric pressure as shown in equation (X):

$$H = \frac{p_a}{\rho_m\cdot g} = \frac{101{,}300}{1.04\cdot 10^{-3}\cdot 9.8} = 9.9\cdot 10^6\,\text{m},$$

and the compression of the material at the exit of the impeller is neglected ($\psi = 1$). Then by the formulas (equation 1.7) and (equation 1.11), we get:

$$N_n = \frac{2\cdot\pi\cdot R\cdot b\cdot \psi\cdot \eta_o\cdot v_{\min}\cdot \rho_m\cdot g\cdot H}{\eta}$$
$$= \frac{2\cdot 3.14\cdot 0.25\cdot 0.05\cdot 1\cdot 0.9\cdot 0.01\cdot 1.04\cdot 10^{-3}\cdot 9.81\cdot 9.9\cdot 10^6}{0.6}$$
$$= 1{,}189 \approx 1.2\,\text{kW}.$$

The resulting power value is also quite small. Figure 1.7 shows a drum mixer with work surfaces tilted at different angles. The principle of operation of the concrete mixer and the drum screen was described by Morgano et al. (Morgano et al., 2015). After loading a portion of the processed material (4) into the drum (3), the electric motor (1) is turned on, which, through the gearbox (2), causes the drum to rotate. As a result, the material in the drum interacts with work surfaces tilted at different angles. This contributes to its better mixing and recovery (Muschiolik & Dickinson, 2017).

The maximum torque of the drum formula (equation 1.12) is defined as:

$$M = m\cdot g\cdot r\qquad(1.12)$$

where m is the portion's mass of the processed material, and r is the radius of the center of the portion mass (see Figure 1.7).

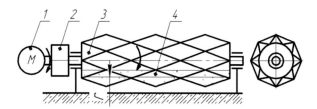

Figure 1.7 Diagram of a drum mixer with working surfaces tilted at different angles: (1) electric motor; (2) gearbox; (3) drum; (4) processed material.

The necessary power for mixing is calculated by the following formula (equation 1.13),

$$N_n = M \cdot \omega = M \frac{v}{r} \tag{1.13}$$

where ω and v are the minimum angular frequency and speed necessary for mixing and effective recovery of the material being processed, respectively (Obertyukh et al., 2018).

Then, after substituting formula equation (1.12) into formula equation (1.13), we obtain the following formula (equation 1.14):

$$N_n = m \cdot g \cdot v \tag{1.14}$$

The optimum drum speed (equation 1.15) can be calculated as:

$$n = \frac{2 \cdot v}{\pi \cdot r} \tag{1.15}$$

The mass m is determined based on the approximate dimensions of the drum (3) and the density ρ_m using the formula below:

$$m = \frac{\pi \cdot D_b^2}{8} L_b \cdot \rho_m = \frac{3.14 \cdot 0.7^2}{8} \cdot 2 \cdot 1.04 \cdot 10^{-3} 0.01 = 4 \cdot 10^{-6} \, \text{kg}$$

Then, in accordance with formula equation (1.14), we obtain $N_n = 4 \cdot 10^{-6} \cdot 9.81 \cdot 0.01 = 3.92 \cdot 10^{-10}$ kW, which is negligible power.

Figure 1.8 shows a structurally simpler scheme compared with the previous mixer. After loading a portion of the processed material (5) into the drum (3), its rotation is activated, which is provided by the electric motor (1) by means of the reducer (2). When the drum rotates, the blades (4) attached to its inner surface ensure that the material is mixed. To determine the required mixing power, formulas (equations 1.12–1.15) can be used.

The mixer with unbalance vibrators (Figure 1.9) provides more intensive and dynamic mixing of the portion of the processed material (2) in the container (1). Vibrators (3) are unbalanced with drives from the electric motors, during the rotation of which vertical power pulses are transmitted to the container (2) alternately up and then down (Polishchuk et al., 2016). As a result, the container, mounted on the springs (4), makes periodic vertical reciprocating movements (Kozlov et al., 2019, Polishchuk & Kozlov, 2018; Polishchuk et al., 2019).

Regeneration of industrial filters 11

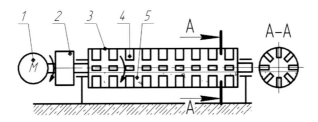

Figure 1.8 Diagram of a drum mixer with straight blades: (1) electric motor; (2) gearbox; (3) drum; (4) blades; (5) processed material.

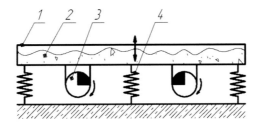

Each container movement cycle can be divided into four stages (Iskovich-Lototsky et al., 2019):

Stage I: Moving the container from its lowest position up to the middle position with fully compressed springs. This position is characterized by the absence of deformation of the springs.
Stage II: Moving the container further up from the middle to the highest position before stopping and stretching the springs.
Stage III: Moving the container from the top to the middle position.
Stage IV: Moving the container from the middle to the lowest position.

The motion equations of the container at each of these stages are:

For Stage I (equation 1.16):

$$M \cdot \ddot{z} = \frac{m \cdot v^2}{r} k + c \cdot l \cdot \left(\frac{A}{2} - z\right) - M \cdot g; \quad 0 \leq z \leq \frac{A}{2}; \quad 0 \leq t \leq \frac{15}{n} \tag{1.16}$$

For Stage II (equation 1.17):

$$-M \cdot \ddot{z} = \frac{m \cdot v^2}{r} k - c \cdot l \cdot z - M \cdot g; \quad 0 \leq z \leq \frac{A}{2}; \quad \frac{15}{n} < t \leq \frac{30}{n} \tag{1.17}$$

For Stage III (equation 1.18):

$$-M \cdot \ddot{z} = -\frac{m \cdot v^2}{r} k - c \cdot l \cdot \left(\frac{A}{2} - z\right) - M \cdot g; \quad 0 \leq z \leq \frac{A}{2}; \quad \frac{30}{n} < t \leq \frac{45}{n} \tag{1.18}$$

For Stage IV (equation 1.19):

$$M \cdot \ddot{z} = -\frac{m \cdot v^2}{r} k + c \cdot l \cdot z - M \cdot g; \quad 0 \leq z \leq \frac{A}{2}; \quad \frac{45}{n} < t \leq \frac{60}{n} \quad (1.19)$$

Where M is the mass of the container with a portion of material (2) and vibrators (3); z is movement of the container; m is the unbalance mass of the vibrator (3); v is the angular velocity of rotation of the unbalance, which is selected in accordance with the minimum speed required for the material regeneration process; r is the radius of the center of the unbalanced mass of the vibrator relative to its axis of rotation; k is the number of vibrators; c is the stiffness coefficient of the spring (4); l is the number of springs; A is the amplitude of oscillation of the container; n is the rotational speed of the unbalance, and m/min; t is a time.

From the analysis of equations (1.13) through (1.15), it is obvious that the greatest value of the driving force is necessary at the end of Stage I at $z = A/2$; $t = 15/n$ (see equation (1.15). Then the corresponding mixing power can be found from equation (1.20):

$$N_n = F_{\max} v = \frac{m \cdot v^3}{r} k = \frac{1 \cdot 0.01^3}{0.1} 2 = 2 \cdot 10^{-7} \, \text{kW} \quad (1.20)$$

This power is also insignificant compared with the power of the known equipment.

1.5 CONCLUSION

The available equipment that can be used to mix the ion-exchange resin in saline solution is, in most cases, complex and low-tech to manufacture. It often does not ensure uniform mixing of the entire mass of the resin being restored and its circulation in the working chamber, which is of great importance for the quality of the recovery of the resin.

This chapter proposed ideas for special equipment for the regeneration of ion-exchange resin. In accordance with the calculations given, it provides, in comparison with the known equipment, a reduction in the required power of the driving motor by 1.5–48 times and the speed of mixing of the processed material, which is 90–400 times. This provides a corresponding reduction in energy consumption for the implementation of the workflow and an increase in its efficiency.

The dependencies for determining the basic operating parameters of the proposed equipment are given. These dependencies allow one to choose the most rational option and the methodology whereby the design calculation is created.

REFERENCES

Bartholomai, A. 1987. *Food Factories: Processes, Equipment, Costs.* Weinheim: VCH.
Concidine, D.M. 2012. *Foods and Food Production Encyclopedia.* Springer Science & Business Media, USA, 2301 p.
Dragilev, A.I., & Drozdov, V.S. 1999. *Technological Machines of Food Production.* Moscow: Kolos.
Franceschinis, E., Santomaso, A.C., Trotter, A., & Realdon, N. 2014. High shear mixer granulation using food grade binders with different thickening power. *Food Research International* 64: 711–717.

Geissler, M., Li, K., Zhang, X.F., Clime, L., Robideau, G.P., Bilodeau, G.J. & Veres, T. 2014. Integrated air stream micromixer for performing bioanalytical assays on a plastic chip. *Lab on a Chip* 14(19): 3750–3761.

Hakansson, A, Askaner, M. & Innings, F. 2016a. Extent and mechanism of coalescence in rotor-stator mixer food-emulsion emulsification. *Journal of Food Engineering* 175: 127–135.

Hakansson, A., Chaudhry, Z., & Innings, F. 2016b. Model emulsions to study the mechanism of industrial mayonnaise emulsification. *Food and Bioproducts Processing* 98: 189–195.

Iskovich-Lototsky, R., Kots, I., Ivanchuk, Y., Ivashko, Y., Gromaszek, K., Mussabekova, A. & Kalimoldayev, M. 2019. Terms of the stability for the control valve of the hydraulic impulse drive of vibrating and vibro-impact machines. *Przeglad Elektrotechniczny* 4(19): 19–23.

Kozlov, L.G., Polishchuk, L.K., Piontkevych, O.V., Korinenko, M.P., Horbatiuk, R.M., Komada, P., Orazalieva, S. & Ussatova, O. 2019. Experimental research characteristics of counter balance valve for hydraulic drive control system of mobile machine. *Przeglad Elektrotechniczny* 95(4): 104–109.

Morgano, M.A., Milani, R.F. & Perrone, A.A.M. 2015. Determination of total mercury in sushi samples employing direct mercury analyzer. *Food Analytical Methods* 8(9): 2301–2307.

Muschiolik, G. & Dickinson, E. 2017. Double emulsions relevant to food systems: Preparation, stability, and applications. *Comprehensive Reviews in Food Science and Food Safety* 16(3): 532–555.

Obertyukh, R.R., Slabkyi, A.V., Marushchak, M.V., Koval, L.G., Baitussupov, D. & Klimek, J. 2018. Dynamic and mathematical models of the hydraulic-pulse device for deformation strengthening of materials. *Proceedings on SPIE: Photonics Applications in Astronomy, Communications, Industry, and High-Energy Physics Experiments* 10808: 108084Y.

Obertyukh, R.R., Slabkyi, A.V., Marushchak, M.V., Kobylianskyi, O.V., Wójcik, W., Yerkeldessova, G. & Oralbekov, A. 2019. Method of design calculation of a hydropulse device for strain hardening of materials. *Przegląd Elektrotechniczny R* 95(4): 65–75.

Polishchuk, L., Bilyy, O. & Kharchenko, Y. 2016. Prediction of the propagation of crack-like defects in profile elements of the boom of stack discharge conveyor. *Eastern-European Journal of Enterprise Technologies* 6(1): 44–52.

Polishchuk, L.K. & Kozlov L.G. 2018. Study of the dynamic stability of the conveyor belt adaptive drive. *Proc. SPIE, Photonics Applications in Astronomy, Communications, Industry, and High-Energy Physics Experiments* vol. 10808: 1080862.

Polishchuk, L.K., Kozlov, L.G. & Piontkevych, O.V. 2019. Study of the dynamic stability of the belt conveyor adaptive drive. *Przeglad Elektrotechniczny* 95(4): Pages 98–103.

Saravacos, G.D. 2002. *Handbook of Food Processing Equipment*. New York: Kluwer/Plenum.

Sevostyanov, I. 2013. The analysis of methods and the equipment for clearing of the damp disperse waste of food productions. *Tehnomus – New technologies and products in machine manufacturing technologies; Proc. intern. conference*, Suceava, 17–18 May 2013: 44–49.

Sevostyanov, I., Polischuk, O. & Slabkiy. 2015. A. Development and research of installations for two-component vibro-impact dewatering of food industry wastes. *Eastern-European Journal of Enterprise Technologies* 5/7(77): 40–46.

Zhu, B., Wang, X., Tan, L. & Ma Z. 2016. Investigation on flow characteristics of pump-turbine runners with large blade lean. *Journal of Fluids Engineering – Transactions of the ASME* 140(3): FE-16-1113.

Chapter 2

Intelligent implants in dentistry

Realities and prospects

S. Zlepko, S. Tymchyk, O. Hrushko, I. Vishtak, Z. Omiotek, S. Amirgaliyeva, and A. Tuleshov

CONTENTS

2.1 Introduction 15
2.2 Analysis and classification 16
2.3 Method of choosing informative indicators 20
2.4 Experiments and results 21
2.5 Conclusions 24
References 24

2.1 INTRODUCTION

In recent years, there has been an active trend in the development of materials aimed at creating artificial tissues that replace damaged skin, muscle tissue, blood vessels, nerve fibers, bone tissue, etc. These materials are called biomaterials. "Biomaterials these are materials designed to serve as a distribution boundary with biological systems in order to evaluate, treat, build up or replace any tissue, organ or function of the body" (Popkov, 2014). Such work is most successfully carried out in the treatment of the pathology of the musculoskeletal system, aided by the development of the industry of endoprosthetic replacement of large joints.

Biomaterials, which are used as implants, that replace a section of bone (endoprostheses) or as temporary fixatives for a broken bone (periosteal plates, intramedullary rods) are evaluated in terms of their impact on the reparative ability of the latter. These can be (Popkov & Popkov, 2012):

- biotolerant materials (stainless steel and cobalt–chromium alloys) – the surface of such implants is separated from the adjacent bone by a layer of fibrous tissue; reparative regeneration of the damaged bone occurs at a normal rate and a certain distance from the implant (distant osteogenesis);
- bioinert materials (oxides of titanium and aluminum) – do not cause the formation of fibrous tissue; reparative osteogenesis takes place in direct contact with the surface of the implant, but consolidation occurs at the usual rate;
- bioactive materials (calcium-phosphate ceramics and silicon-based bioglass) – characterized by the formation of a very close chemical bond with the bone (combined osteogenesis), they enhance bone tissue formation reactions starting from the surface of the implant and induce the creation of a continuous bond from the tissue to its surface (Popkov & Popkov, 2012).

DOI: 10.1201/9781003225447-2

In modern traumatology, there are two fundamentally different approaches to solving the problem of treating injuries and diseases: (1) simple replacement of a damaged section of bone with an implant, up to the creation of a large bio-engineering structure that replaces the bone and its adjacent joints or (2) creating conditions for regeneration (recovery) of bones in the damaged areas with the help of an implant.

2.2 ANALYSIS AND CLASSIFICATION

An analysis of modern literature shows that both areas are increasingly associated with bioceramics, whose utilization in medicine is expanding with the development of chemistry and improvements in the production technology of materials that are similar in properties to bone tissue.

From the practical point of view of a surgeon, the management of the process of bone formation is viewed from two sides (Popkov & Popkov, 2012):

1. active impact on the rate of reparative regeneration of damaged bone tissue;
2. the possibility of impact on skeletal development in the postnatal period, which for unknown reasons has slowed down in the embryonic period.

Simultaneously, such materials should have several necessary properties of the bone (Popkov & Popkov, 2012):

- to perform and maintain (scaffold) the amount of defect;
- exhibit osteogenic activity, that is, actively facilitate osteoid regions in bone matrix formation and aid mesenchymal cells in transforming into osteogenic cells;
- exhibit good indicators of biocompatibility, namely be biodegradable and not cause inflammatory reactions in the recipient.

The combination of these properties allows such materials, in parallel with their supporting function, to ensure the biointegration of the ingrowth of cells and vessels into the implant structures (osteoconductive) and to stimulate the formation of bone tissue around the implant for a considerable length (osteoinductive) (Popkov & Popkov, 2012).

At the present stage of development of dental implantology, the main common methods for assessing the condition of intraosseous implants, apart from the clinical method, are radiological – torque-testing using a torque wrench when installing implants, gnathodinometry, periotestometry using a Periotest instrument (Gulden Medizintechnik, Germany); frequency resonance analysis of the stability of implants using an Osstell-mentor device (Integration Diagnostics, Sweden) (Ramazanov, 2009).

The frequency resonance analysis method (FRA), developed by N. Meredith in 1997, is of particular interest as a criterion for predicting the effectiveness of implants loaded with temporary prostheses immediately after placement. The method is based on registering the electromagnetic oscillation resonance of an implant and the surrounding bone when exposed to an electromagnetic field using a magnetized pin. The results of the study are expressed in units of the implant's stability coefficient – Implant Stability Quotient (ISQ) on a scale from 1 to 100 (Ramazanov, 2009).

Recently, a new trend has emerged in implantology – direct load of implants with a temporary prosthesis immediately after placement of implants. With this method, the requirements for identifying the degree of implant stability have increased, which became the basis for developing a new method for determining implant stability – frequency-resonance analysis (RFA) (Ramazanov, 2009).

No less important is the assessment of implant stability – when performing two-stage implantation after the development of full-fledged osseointegration, which is a fundamental condition for the long-term success of prosthetics with a focus on dental implants. The process of osseointegration depends on a significant number of external and internal factors: the building and structure of the jaw bone tissue; bone volume at the implantation location; the characteristics of tools used to form the space for an implant; surgical methods; the term of functional implant load; the implant's material and surface features; properties and methods of using osteopathic materials; the patient's osteoreparative potential. All of these conditions may affect the durability of implants. Additionally, the assessment of the intraosseous implants' osseointegration degree is a crucial factor in the choice of prosthesis designs, tactics of implant functional load, and predicting the effectiveness of orthopedic treatment (Ramazanov, 2009, Pavlov et al., 2017).

In implantology, there are several possibilities for indirectly assessing the degree of osseointegration and stability of implants: clinical (manual control, implant stability); radiological (including through the determination of bone density around the implant); echoosometry; periotestometry; gnathodinometry, torque test using a torque wrench; FRA.

The implanted biotelemetric complex is designed to study the physiological processes occurring in the internal organs, in which the parameters of these processes are measured using the radiotelemetric method. The complex consists of an implanted radio transmitter, a radio receiver, information from which is fed into a computer, and software necessary to visualize and further process the received biophysical data. These include: temperature, pressure, ECG, EEG, EMG (Ramazanov, 2009).

The operation of implantable systems has a number of features, primary among which is the ability to function near (inductive) radiation zones in specific environmental conditions, which determine the basic requirements for transmission methods and the design of the implanted system parts. When a radio signal passes through living tissue, it is weakened more significantly the shorter its wavelength; therefore, low frequencies are used in order to transmit information from the internal parts of the body. However, the use of a low-frequency range of radio waves is associated with significant technical difficulties, in particular, with the difficulties of creating highly efficient emitters with severe restrictions on the size and weight of the transmitter. The latter circumstance is particularly important for systems that are implanted. Therefore, systems located close to the receiving antenna use a frequency range of 0.05–100 MHz. The use of high frequencies in the range of hundreds of MHz allows creating a sufficiently effective radiating system. Limiting factors in increasing frequencies here are the biomedical requirements, according to which frequencies above 10 MHz have a harmful effect on the body.

Figure 2.1 shows the block diagram of the transmitting module of the biotelemetry complex.

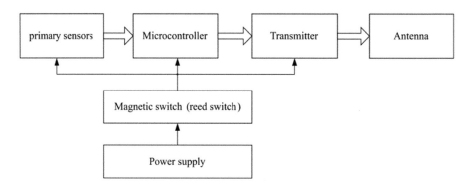

Figure 2.1 Block diagram of the implantable transmitting module of the biotelemetric complex.

Before the developer of the transmitting module of the biotelemetry complex had a goal: to synthesize the optimal structure of this module. To achieve this goal, the following main tasks were solved:

- selection of optimality criteria;
- synthesis of a model for the transmission module;
- analysis of the synthesized model.

Each of these tasks was broken down into several components. The joint solution of all components allowed to assess the optimal solution to the main tasks. The primary criteria for assessing the optimality for an implantable transmission module were: weight and size parameters; power usage; compatibility of the body material of the device with a biological organism; information criterion; communication channel compatibility; possibility of remote control of the implanted module (Grushko et al., 2012; Mashkov et al., 2014).

One of the promising areas in the development of implantology is the replacement of lost teeth with implants, which are artificially made multicomponent structures used for insertion as prosthetics into the bone tissue of the human jaw with its subsequent splicing (osseointegration) (Ramazanov, 2009).

In order to restore the function and aesthetics of the dentition system, in general, the implant must meet the following requirements (Popkov, 2014): be comfortable at the establishment stage, provide ample opportunities for prosthetics, successfully integrate into the bone tissue, not hinder osseointegration or promote its potentiation, exhibit a high level of sterility and cleaning from mechanical particles, provide long-term preservation of consumer properties, exhibit mechanical strength.

In modern dentistry, intraosseous implants are the most common and reliable because they are implanted directly into the jaw bone. After the bone grows together with the implant installed in it and the gum tissue heals, repeated surgical procedures are performed in order to attach the pin to the installed implant. After the installation of crowns or artificial teeth, the patient has the opportunity, subject to oral hygiene and regular visits to the dentist, to live with such implants for 10–15 years (Sorochan, 2018).

However, the large loads placed on implants, not always high-quality work of doctors, the impact of unpredictable internal and external factors sometimes leads to friction between the implants, gum, and bone, and as a result – to inflammation.

A proposed solution to this problem is to develop a new direction in dental implantology – the creation of intelligent implants, with the functions of continuous or periodic monitoring of their stability and the ability to predict undesired complications and to inform the physician about them in a timely manner.

Failures in treatment using dental implants are due to two factors: impairments in the phase of osseointegration, and errors at the stage of manufacturing the superstructure (early and late treatment failures). Impairments of the processes of osseointegration in the pre-functional and functional periods occur due to insufficient biocompatibility and corrosive properties of the material being implanted; errors in the design selection and evaluation of the ratio of mechanical loads, differences in deformation characteristics, and the structure of the surfaces of the implant and bone tissue. In the event of complications during or after orthopedic treatment, there is talk of systemic bad luck, which is a consequence of the erroneous actions of the orthopedic surgeon (Ramazanov, 2009).

Preventing the above is possible, at a minimum, by periodically controlling the speed of reparative regeneration of damaged bone tissue, as well as the biocompatibility level of the implant during its life cycle, which will make it impossible to untimely determine the moment of occurrence of the inflammatory process in the patient.

All of the above is potentially attainable, provided that a new functional class of implantable biotelemetry modules is created and used, which together with the implant forms a single whole. The development of a class of intelligent implants, as already noted, is first aimed at solving the problem of monitoring their stability, predicting their life cycle, and preventing the occurrence of undesirable complications and inflammatory processes and, second, at solving the issues of interaction between biological and technological elements, obtaining a single biotechnical system, which involves interfacing biosensors with dental implants.

The basic components of intelligent implants are intelligent sensors, which we propose to consider as a new functional class in the existing classification of biosensors and sensors of biomedical information. To determine their place in the classification, formulate and justify the definitions of new terms, an analysis was conducted of the existing literary sources and the issues of compliance of intelligent implants with the requirements for biomaterials in general, and above all, of biocompatibility with bone tissue (Popkov & Popkov, 2012), which revealed the following:

a. chemical properties – no corrosion and undesirable chemical reactions with tissues;
b. mechanical characteristics of intelligent sensors must be such that they do not cause deterioration of elasticity and heat resistance, and even more so – bone resorption;
c. the bioactivity of the material – the lack of response from the immune system, osteoconductivity, and osteoinductivity, which contribute to "splicing" with bone tissue (Popkov, 2014).

Modern methods of informatization cover increasingly broad spheres of human activity. At the same time, the development rate of digital information is significantly ahead of the development of information on paper. Medical information has its own specifics, and the increase in its volume is accompanied by certain problems that justify the need to create medical information systems (MIS). They are different from economic or technical information systems, which creates additional difficulties in their design and implementation. The difference is manifested in the specifics of informatization objects, for example, diagnostic coding systems are becoming more universal now, but a detailed nomenclature of signs and symptoms, data recording formats, and the organization of records are determined individually; there is no standardization in terminology, format, scales of measurement of medical data, etc.

Very often, medical decisions are characterized by insufficient knowledge, limited time resources, the inability to attract competent experts, and incomplete information concerning the patient's condition. These factors are causes of medical errors, which may lead to further loss of a patient's health. Therefore, along with the development of an MIS, it is important to create medical decision support subsystems, which are information subsystems operating independently of or as part of an MIS.

2.3 METHOD OF CHOOSING INFORMATIVE INDICATORS

An important element for any measurement system is the optimization of the quantity and quality of informative indicators (Bezsmertnyi et al., 2018).

They are not known to measure human health since such a process requires objective estimates of health parameters, and it is practically impossible to achieve them using only the primary indicators. For that reason, WHO recommends integral indicators for determining the state of health of a person, such as physical development, functional status, physical capacity, physiological reserve, adaptive potential. At the same time, the procedure selecting informative indicators is nothing but a classification task, the solution to which is to obtain spatial informative indicators that characterize the physiological costs to a person in the performance of their respective activities.

Most informative indicators are integral indicators and are not directly measurable. In fact, they are latent variables that cannot be measured and may be determined only by a certain set of informative features. For the study of such latent variables, the apparatus of the latent variables measurement theory is based on the models of G. Rasha. It is possible to increase the informativeness level in the area of the obtained indicators by combining the latent variables measurement theory and the G. Rasha model. The method of choosing informative indicators is represented by a sequence of logically linked stages.

Stage I – expert assessment – expertly determined through the composition of informative indicators (indicator variables), which is formed from the set of possible primary and secondary health indicators.

Stage II – the conversion of measurement scales – the transfer of quantitative scales of the X_i indicator variables into the K_i qualitative scale and thus forms the only range of qualitative changes in the indicator variables.

Stage III – selection of informative features (indicators).

1. The theoretical dependences of G. Rasha's models for each indicator variable are calculated.
2. The mean values of the results of the experimental measurements according to the Xi-square criterion for each indicator variable are determined.

$$X^2 = \frac{(f_0 - f_e)^2}{f_e} \tag{2.1}$$

where f_0 – the actual frequency, or the frequency of "observation"; f_e – the "waiting" frequency.
3. Comparison $P_j\{x_{ij} = 1, \beta i\}$ with X^2 for each indicator variable.
  Indicative variables for which the condition X^2 exceeds $P < 0.05$ are excluded from the list of informative features (indicators).
4. The procedure in clause 3 is repeated as long as the list of informative attributes for each latent variable includes only those informative signs for which the condition X^2 is critical $P < 0.05$ is not fulfilled.

Stage VI – determining the measure of informativeness of the indicator variable (control).

The fulfillment of the above condition indicates that the system of informative features corresponds to the measured latent variable, and the set of certain informative features is effective for the purpose of measuring the generalized latent variable.

2.4 EXPERIMENTS AND RESULTS

The specificity of the constructive implementation of sensors, which provides comfort for the patient, has led to its transition to: the choice of construction principles and features of the technical implementation and practical application of sensors in a new subject area; calculation of their electrical and mechanical characteristics; a graphic representation of the placement and mounting options of sensors directly in the implant; solving the issues of interfacing the output parameters of sensors with the input blocks of implantable radio modules; transfer of data to an external PC; reduction of measurement errors and increase in the reliability of results.

Further development of intelligent implants has led to the creation of microcontroller systems (MCS), focused on solving problems of biosignal registration and preprocessing; research of the construction principles of computational functional information converters for biotechnical systems (BTS); improvement in synthesis methods of neural network biosignal converters with a learning structure; development of abilities and skills in the field of designing adapters and individual autonomous microcontroller modules for communicating implantable sensors with digital diagnostic systems.

The main theoretical and practical aspects of building and applying intelligent implants include general methodological issues of organizing implantable BTS, focused on solving the issues of obtaining, converting, preprocessing, and transmitting biomedical signals from sensors embedded in the implants to an external PC; providing a rationale for the introduction and definition of the concept of an intelligent implant as a primary transducer of biomedical information; structural and functional organization of the biosignal registration, transformation, and transmission subsystems, functional and structural approach to intelligent implant design; issues concerning the organization of interfaces, adapters, and data exchange protocols.

Future plans include the adaptation of neural network technologies for the development of intelligent implants, as well as justification of utilizing a neural network

ADC as a basis for creating neuro implants of human sensory systems (visual, auditory, tactile, etc.).

The basis for the project can be the study (Sorochan, 2018), which was carried out with the participation of the authors and dedicated to the development of an osteosynthesis biotechnical system. The feature of the proposed BTS is the presence in its structure of an implanted osteosynthesis biotelemetric module, structurally designed as an autonomous, functionally complete structure with an independent power source, a load cell, a temperature sensor, and an implantable radio frequency module in the Microsemi Med-Net standard, which provided timely detection of inflammation, displacement plates, and bones – information that was promptly communicated to the physician in real time.

The implanted ZL70321 module provides the necessary radio frequency functions of telemetry systems operating in the MICS radio band. The built-in antenna oscillating circuit allows the component to be used in conjunction with a wide range of antennas (the nominal antenna impedance is $100 + j150\,\Omega$) (Suchanek & Yoshimura, 1998).

This module contains the following main blocks:

- MICS-band radio frequency receiver based on a ZL70102 component, with integrated matching chains;
- SAW filters for suppressing unwanted obstacle and tuning the antenna;
- 2.45 GHz matching scheme for setting the receiver into work mode;
- integrated 24 MHz crystal oscillator;
- block capacitors (Suchanek & Yoshimura, 1998).

The radio frequency module of the base ZL70120 component also features all the radio frequency functions necessary for devices operating in the MICS frequency range. The module is designed to meet all the requirements of regulatory organizations – FCC, ETSI, and IEC.

Additional features of the module include (Suchanek & Yoshimura, 1998):

- integrated matching circuit, band-pass filters to suppress unwanted obstacles, and an additional IBE receiver to increase its sensitivity;
- 2.45 GHz transmitter, which switches the circuit out of low-power mode;
- matching scheme with antenna;
- received power indicator filter and a logarithmic amplifier, simplifying the channel estimation procedure in accordance with the MICS standard;
- fully shielded enclosure.

The semiconductor of the Microsemi ZL70102 receiver, which is used both in the implantable module and in the base component, makes it possible to transfer patient health data and data from the device at high speed and with minimal impact on the device's battery charge level. The component operates in the MICS radio frequency band (402–405 MHz). Various low-power options are supported, including the availability of a receiver for the ISM radio frequency band (2.45 GHz). The ZL70102 current is less than 6 mA when transmitting or receiving data, 290 nA in listening mode before transmission, and 10 nA in standby mode (Suchanek & Yoshimura, 1998).

The use of wireless information transmission between the registering unit and the server was also tested in the development of a telemedicine diagnostic system in the vital activity monitoring module, which provides round-the-clock online monitoring of the patient's functional state. Special attention was paid to the construction of an information database, which is based on medical indicators obtained using sensors (Moscovko et al., 2016; Tymchyk et al., 2016; Virozub, 2017). A diagnostic telemedicine system was developed to fully meet all the requirements for this class of equipment, the information support of which is built on a special medical application system. The presence of a switched input module block for registering biomedical data concerning the patient's condition allows for a full range of measures aimed at diagnosing, determining treatment tactics, predicting the development of diseases and possible complications.

In solving the problem of building the information system, considerable attention was paid to the organization of an information container for storing an array of data obtained in the process of filling the database, as well as the data contained in the knowledge base, and forming an array of expert "facts" and "rules."

In our case, a database is defined as a named, structured collection of interrelated data. These data characterize the subject area in the field of functioning of the aforementioned system and are controlled by the DBMS. Thus, it was allocated three main thematic components of the database – the service, diagnostic, and reference part. Each component, in turn, is divided into separate areas by content and information, access to which is only available to the relevant categories of employees working with the system. In terms of organizing connections in the database, the technology utilizes a key-based connection – minimum sets of attributes, whose meaning can uniquely identify each instance of the entity. This ensures adherence to database integrity, the properties that determine the completeness and correctness of the information contained in the database.

Because a physician needs to have an "instant" and up-to-date picture, which is achieved by increasing the frequency of condition assessment, reducing the time interval during which a rapid assessment of the patient's functional state is performed. The most significant requirements for the system include:

- minimizing the number of recorded psycho-physiological processes;
- minimizing the number of psychophysiological controlled parameters that adequately reflect the functional state during the examination;
- choice of informative methods for registration and primary processing of biomedical information;
- standardization of dimensions of psycho-physiological vital signs.

Implementation of certain requirements for the system's design process and structure will allow its use in the following modes: (1) information and measuring monitor complex; (2) information complex of making decisions and functional state predictions; (3) online telemedicine counseling.

The multivariance of types, modifications, and types of medical equipment results in a need to create a universal intelligent control interface for any existing medical equipment, which is based not on short binding to circuit solutions, but on building it at the software level, the flexibility of which determines the versatility and multifunctionality of the intelligent interface itself.

At the transition from one type of input medical equipment to another, the structure of the intelligent interface and its management as a whole does not change, because the variable part, in this case, will be the number of functional elements included in the interface structure, their speed, and software – the establishment of functional element interaction patterns. (Tymchyk et al., 2016).

This approach allows the design of medical equipment with automatic configuration when changing sensors and detectors, automated control objects, build up a system or complex, and change the circuitry according to new medical and technical concepts and ideas.

The use of intelligent implants should also take into consideration the nuances associated with elderly patients, which are characterized by: reduced immunity, which leads to an increased osseointegration time; increased looseness of the jawbones, which requires additional measures and means to stabilize the implant; selection or manufacture of special veneers and crowns for a specific patient.

Another feature of the BTS is that, in making a final decision on the tactics of treatment and prognosis, the so-called reverse procedure is proposed, when a decision is made based on the condition's development forecast, and not vice versa (Kovalenko et al., 2017). This approach, on the one hand, contributes to further improvement and development of implantology based on clinical and biomechanical research, and on the other hand, it increases the reliability, veracity, and adequacy of surgical interventions by providing the physician with information support in the form of a biotechnical system based on a biotelemetric radio module.

2.5 CONCLUSIONS

Solving the problem of development, implementation, and use of intelligent sensors to assess the stability and biocompatibility of implants in dentistry will provide the physician with new opportunities and will provide them:

a. the means of controlling the load on the implants immediately after their installation with the subsequent identification and assessment of the implant's degree of stability;
b. the means to monitor and control (continuously or periodically) the stability and condition of implants during two-stage implantation, after full osseointegration, which is the basic condition for early successful prediction based on dental implants.

REFERENCES

Bezsmertnyi, Y.O, Shevchuk, V.I., Grushko, O.V. et al. 2018. Information model for the evaluation of the efficiency of osteoplasty performing in case of amputations on below knee. *Proceedings of SPIE: Photonics Applications in Astronomy, Communications, Industry, and High-Energy Physics Experiments 2018* vol. 10808: 108083H.

Grushko, A.V., Sheykin, S.E. & Rostotskiy, I.Y. 2012. Contact pressure in hip endoprosthetic swivel joints. *Journal of Friction and Wear.* 33(2): 124–129.

Kovalenko, A.S., Tymchyk, S.V., Kostyshyn, S.V. et al. 2017. Concept of information technology of monitoring and decision-making support. *Proceedings of SPIE: Photonics Applications in Astronomy, Communications, Industry, and High-Energy Physics Experiments 2017* vol. 10445: 130337.

Mashkov, V., Smolarz, A., Lytvynenko, V. & Gromaszek, K. 2014. The problem of system fault-tolerance. *Informatyka, Automatyka, Pomiary w Gospodarce i Ochronie Srodowiska* 4: 41–44.

Moscovko, M.V., Mirzub, R.M. & Lepokhina, G.S. 2016. Medical equipment for a family doctor: problems and ways of development. *Proceedings of Conference: Topical Issues of Modern Medicine: A Collection of Theses of the XIII International Scientific Conference Students and Young Scientists*, Kharkiv, 14–15 April 2016: 159–163.

Pavlov, S.V., Kozhukhar, A.T., Titkov, S.V. et al., 2017. Electro-optical system for the automated selection of dental implants according to their colour matching. *Przeglad Elektrotechniczny* 93(3): 121–124.

Popkov, A.V. 2014. Biocompatible implants in traumatology and orthopedics (literature review). *The genius of orthopedics* 3: 94–99.

Popkov, A.V. & Popkov, D.A. 2012. *Bioactive Implants in Traumatology and Orthopedics*. Irkutsk: SC RVH SB RAMS.

Ramazanov, S.R. *Determination of implant stability as an objective method for predicting the effectiveness of treatment in dental implantology.* Ph.D. Thesis: Moscow.

Sorochan, O.M. 2018. *Device for plate osteosynthesis of the human musculoskeletal system.* Ph.D. Thesis: Vinnytsia.

Suchanek, W. & Yoshimura, M. 1998. Processing and properties of hydroxyapatite-based biomaterials for use as hard tissue replacement implants. *Journal of Materials Research and Technology.* 13: 94–117.

Tymchyk, S.V., Kostyshyn, S.V. Zlepko, S.M. & Virozub, R.M. 2016. Algorithmic software for IT monitoring and decision support for student health. *Bulletin of the Khmelnitsky National University* 1: 43–48.

Virozub, R.M. 2017. *Method and telemedicine diagnostic system for the family doctor.* Ph.D. Thesis: Vinnitsa.

Chapter 3

Modeling of the exhaustion and regeneration of the resource regularities of objects with different natures

V. Mykhalevych, V. Kraievskyi, O. Mykhalevych,
O. Hrushko, A. Kotyra, P. Droździel,
O. Mamyrbaev, and S. Orazalieva

CONTENTS

3.1 Introduction 27
3.2 Actuality 27
3.3 Formal problem statement 28
3.4 Literature review 29
3.5 Resource regeneration in the pause 32
3.6 Exhaustion and regeneration of the athlete's resource when running at a variable speed 35
3.7 Conclusions 37
References 37

3.1 INTRODUCTION

The functioning of objects of different natures is accompanied by their physical wear. If, as a result of physical wear, the state of an object reaches critical a value that is incompatible with the further use for its designated purpose, it is considered that the full exhaustion of the object's resources has occurred. The intensity of the physical wear greatly depends on the conditions of the object's functioning. A similar situation takes place in the technological processes of the parts used in manufacturing when the methods of plastic deformation are applied. It is natural that the problems of current and limit states object assessment correspond to the complete exhaustion of the resource, and this is of great importance. A solution of similar problems is realized by applying a damage summation theory approach.

3.2 ACTUALITY

The basis of the damage summation theory is a general methodological approach; however, depending on the specific conditions of the object functioning, particular determining relations may have large distinctions. Today, the theories most well known and in the highest demand are those of high-cycle and low-cycle fatigue (Troshchenko et al., 1993): the creep rupture strength theory (Kachanov, 1999); the fracture theory at cold plastic deformation; and the ultimate strains theory under hot deformation (Mykhalevych, 1998). Fundamental results on the creep rupture strength theory at uniaxial creep were obtained using Kachanov's (1999) suggestions. The monograph provided by Lemaitre's (1992) research is a fundamental work on the theory of damage accumulation with an emphasis on the physical foundations of phenomenological models. Theoretical and experimental results of the nonlinear theory of damage summation at creep were presented in Golub's work (2000). A systematic analysis of theoretical and experimental results on creep rupture strength theory in conditions of a complex stress state and stationary loading was founded in Lokoshchenko's research (2012). In Mykhalevych (1998), a unified mathematical apparatus was developed for constructing models of the creep rupture strength theory, limiting plastic strain to fracture during cold and hot deformation. In Bao and Wierzbicki (2004a), a solution was proposed to the important problem of the mathematical description of a non-monotonic change in limiting deformation in the region of positive values of stress triaxiality. Another, perhaps even more interesting approach to solving this problem was proposed by Hooputra et al.'s model (2004). The use of these models for solving applied problems was proposed in Ogorodnikov et al. (2018). A comparative analysis of fracture models under the conditions of a triaxial stress state is given in Bao and Wierzbicki (2004b).

In the last two decades, a drastic increase of the volume of scientific research devoted to the experimental and theoretical studies of the equivalent plastic strain to fracture at cold plastic deformation was observed. This can be explained by the increases of the role and practical value of mathematical modeling in the investigation of plastic deformation processes. The latter is connected, in our opinion, with the number of circumstances. First of all, the emergence of the powerful finite-element analysis programming complexes of Ansys LS-DYNA, DEFORM-3D, SIMULIA™ Abaqus, and QForm type software, intended for calculation of the stress–strain state of the materials and forecasting of their properties, should be noted, as well as the creation of models and algorithms enabling the replacement, at least partially, of expensive crash tests of road cars.

A great number of publications show a high dependence of the equivalent plastic strain to fracture on the strain rate at high-temperature straining. However, the regularities of the fracture, based on the given dependence and peculiarities of isothermal high-temperature straining, developing at different models, are understudied.

3.3 FORMAL PROBLEM STATEMENT

The aim of this research is to out from the totality of damage summation models, relationships, taking into account the dependence of the equivalent plastic strain at fracture on the history of the strain rate change against the background of competing processes of accumulation and micro-crack healing. The description of the basic features of the processes, which could be described by similar models, and the analysis of main regularities described by these models, in comparison with the experimental data and observations, are presented.

3.4 LITERATURE REVIEW

Among the most popular damage summation models at cold plastic deformation, one should note the models suggested in Bao and Wierzbicki (2004a). In general form, all of these models can be presented as:

$$\psi(\bar{\varepsilon}) = \int_0^{\bar{\varepsilon}} \frac{dx}{\bar{\varepsilon}_{fs}[\eta(x)]} \tag{3.1}$$

where ψ is a level of damage accumulation in material particle,

$$0 < \psi(\bar{\varepsilon}) < 1, \quad \bar{\varepsilon} \in (0, \bar{\varepsilon}_f), \quad \psi(0) = 0, \quad \psi(\bar{\varepsilon}_f) = 1 \tag{3.2}$$

$\bar{\varepsilon}$ – an equivalent plastic strain; $\bar{\varepsilon}_{fs}$ – limit equivalent plastic strain to fracture at $\eta = \text{const}$; $\bar{\varepsilon}_{fs}$ – limit equivalent plastic strain to fracture at $\eta \neq \text{const}$; the stress triaxiality η,

$$\eta = \frac{3 \cdot \sigma_m}{\bar{\sigma}} \tag{3.3}$$

$\sigma_m, \bar{\sigma}$ – means stress and equivalent stress (von Mises), respectively.

$$\sigma_m = \frac{\sigma_{ii}}{3} \tag{3.4}$$

$$\bar{\sigma} = \sqrt{\frac{3}{2} \cdot s_{ij} \cdot s_{ij}} \tag{3.5}$$

σ_{ij} is a stress tensor, s_{ij} is a stress deviator,

$$s_{ij} = \sigma_{ij} - \delta_{ij} \cdot \sigma_m \tag{3.6}$$

$$\delta_{ij} = \begin{cases} 1, & i = j \\ 0, & i \neq j \end{cases} \tag{3.7}$$

In the Rice–Tracey model,

$$\bar{\varepsilon}_{fs}(\eta) = \bar{\varepsilon}_{f1} \cdot \exp\left(\frac{\eta_1 - \eta}{2}\right) \tag{3.8}$$

in the Cockcroft and Latham–Oh model,

$$\bar{\varepsilon}_{fs}(\eta) = \bar{\varepsilon}_{f1} \cdot \frac{f(\eta_1)}{f(\eta)} \tag{3.9}$$

$$f(\eta) = \eta + 2 \cdot \cos\left[\frac{1}{3} \cdot \arccos\left(0.5 \cdot \eta \cdot (3 - \eta^2)\right)\right] \tag{3.10}$$

in the authors' model (Bao and Wierzbicki, 2004a) for 2024-T351 aluminum alloy,

$$\bar{\varepsilon}_{fs}(k) = \begin{cases} \dfrac{0.125}{\left(k+\dfrac{1}{3}\right)^{0.46}} & -\dfrac{1}{3} < k \leq 0 \\ 1.9 \cdot k^2 - 0.18k + 0.21 & 0 < k \leq 0.4 \\ \dfrac{0.15}{k} & 0.4 < k \leq 0.95 \end{cases} \quad (3.11)$$

in the Oyane–Sato model,

$$\bar{\varepsilon}_{fs}(k) = \dfrac{C}{1 + \dfrac{k}{k_1} \cdot \left(\dfrac{C}{\bar{\varepsilon}_{f1}} - 1\right)} \quad (3.12)$$

$$k = \dfrac{\eta}{3} \quad (3.13)$$

where $\bar{\varepsilon}_{f1}$ and C are determined from the experiments to be at =1 or $k = k_1$.

The form of the relations (3.8–3.10) and (3.12) presentations is the variant of the original relations, adapted by the authors to provide a more convenient application and investigation. In the literature, one can find a great number of models similar to the models, shown above. The analysis of certain models is given in the research of Bao and Wierzbicki (2015), Hooputra et al. (2004), and Mohr et al. (2015). The model suggested by the authors (Hooputra, 2004), deserve a separate study.

It should be noted that model (3.11) is presented by piecewise continuous function on $k \in (-1/3; 0.95)$. In fact,

$$\bar{\varepsilon}_{fs}(0.4) = 0.442 \neq \lim_{k \to 0.4+} \bar{\varepsilon}_{fs}(k) = 0.375 \quad (3.14)$$

is seen in the graph in Figure 3.1a.

However, it is not difficult to correct the given relationship to obtain a piecewise continuous function on $k \in (-1/3; 0.95)$. In the relations, determining the function in the ranges of the argument change (0.4; 0.95), the value of the coefficient should be the changed 0.15–0.177 (Figure 3.1b). In this case, the calculations will be matched with the graphs, as shown in Bao and Wierzbicki (2014a).

The authors (He et al., 2013) suggested the generalization Oyane–Sato model for the account of the strain rate $\dot{\varepsilon}_i$ and temperature T,

$$\int_0^{\bar{\varepsilon}_f} \left(1 + \dfrac{k(x)}{B}\right) \cdot dx = C_f(\dot{\varepsilon}_i, T) \quad (3.15)$$

where the constant B and function C_f are determined from the results of the experiments. The given model, following the approach accepted in this research, can be presented in the following form:

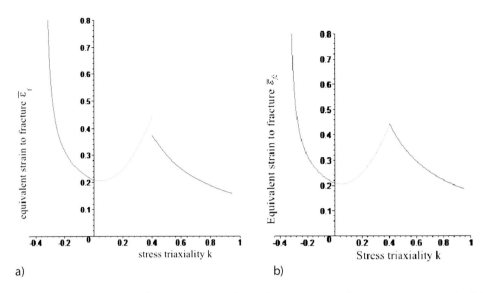

Figure 3.1 Dependence of the limit equivalent plastic strain to fracture on stress triaxiality k: a – Eq. (11), b – Equation (3.11) (corrected by change 0.15–0.177).

$$\bar{\varepsilon}_f + \frac{1}{k_1}\left(\frac{C_f(\dot{\varepsilon}_i,T)}{\bar{\varepsilon}_f(\dot{\varepsilon}_i,T)} - 1\right)\int_0^{\bar{\varepsilon}_f} k(x)\cdot dx = C_f(\dot{\varepsilon}_i,T) \qquad (3.16)$$

The given model reflects the impact on the equivalent plastic strain to fracture $\bar{\varepsilon}_f$ of the history of stress triaxiality k (or η) change and also takes into account the averaged value of strain rate $\dot{\varepsilon}_i$ and temperature T, but does not reflect the impact of the history of their change.

The model, suggested in (Novella et al., 2015)

$$\int_0^{\bar{\varepsilon}_f} C_f(\dot{\varepsilon}_i(x),T(x))\cdot(1+B\cdot k(x))\cdot dx = 1 \qquad (3.17)$$

enables to repel the impact of the history change of the strain rate $\dot{\varepsilon}_i$ and temperature T but the corresponding studies were not carried out; moreover, this model does not reflect the processes of partial or full regeneration of the equivalent plastic strain to fracture, which characterize high-temperature straining (Kraievskyi et al., 2018a; Kraievskyi et al., 2018b; Perig et al, 2018).

In Mikhalevich (1991) and Mykhalevych (1998), the tensor variant of the model is suggested:

$$0 < \psi(t) = \sqrt{\psi_{ij}(t)\cdot\psi_{ij}(t)} < 1, \quad t\in(0,t_f), \quad \psi(0)=0, \quad \psi(t_f)=1 \qquad (3.18)$$

$$\psi_{ij}(t) = \int_0^t \varphi(t-\tau)\cdot\beta_{ij}(\tau)\cdot g(\eta(\tau),\dot{\varepsilon}_i(\tau))\cdot d\tau \qquad (3.19)$$

where t, τ are time; ψ_{ij} is a damage tensor; φ is a hereditary function; g is a some function; and β_{ij} is a direction tensor for strain rate,

$$\beta_{ij} = \sqrt{\frac{2}{3}} \cdot \frac{\dot{\varepsilon}_{ij}}{\dot{\varepsilon}_i} \tag{3.20}$$

$\dot{\varepsilon}_{ij}$ is a strain rate deviator; $\dot{\varepsilon}_i$ is a strain rate intensity,

$$\dot{\varepsilon}_i = \sqrt{\frac{2}{3} \cdot \dot{\varepsilon}_{ij} \cdot \dot{\varepsilon}_{ij}} \tag{3.21}$$

These relationships (3.18 and 3.19) show that the damages increment at the moment of time τ decrease with the distance from this moment. In the presented variant model (3.18), model (3.19) takes into account the history of the strain rate at isothermal high-temperature straining change. For the account of the history of strain rate and temperature change, the relation in model (3.19) should be presented in the following form (Bao & Wierzbicki, 2004b; Bao & Wierzbicki, 2015; Ogorodnikov et al., 2018):

$$\psi_{ij}(t) = \int_0^t \varphi(t-\tau) \cdot \beta_{ij}(\tau) \cdot g(\dot{\varepsilon}_i(\tau), T(\tau)) \cdot d\tau \tag{3.22}$$

Fragmental studies of the models' adequacy, describing various by nature physical processes, determining the relation of which are particular cases of the relations in models (3.18) and (3.19) are known. It is necessary to carry out the analysis of these results and identify the main regularities, which can be described on the phenomenological level for different processes (Kukharchuk et al., 2017).

3.5 RESOURCE REGENERATION IN THE PAUSE

For the process, meeting the requirement:

$$\beta_{ij} = \beta_{ij}^{(0)} = \text{const} \tag{3.23}$$

Considering the relations (3.18), (3.19), we can write:

$$\psi_{ij}(t) = \int_0^t \varphi(t-\tau) \cdot g(\dot{\varepsilon}_i(\tau)) \cdot d\tau \tag{3.24}$$

Choosing functions φ and g by means of one of the possible variants, we obtain:

$$\int_0^{t_f} \frac{n \cdot \exp\left(-n \dfrac{t-\tau}{t_{fs}\dot{\varepsilon}_i(\tau)}\right)}{t_{fs}\dot{\varepsilon}_i(\tau) \cdot \left(1 - e^{-n}\right)} \cdot d\tau = 1, \quad (n > 0) \tag{3.25}$$

where n is a material constant.

Usage of the given model in general case of the nonstationary straining $\dot{\varepsilon}_i(t) \neq \text{const}$ assumes the presence of the dependence of the time prior to fracture on strain rate at stationary straining ($\dot{\varepsilon}_i(t) = \text{const}$),

$$t_{fs} = t_{fs}(\dot{\varepsilon}_i) \tag{3.26}$$

Taking into account that in case of simple deformation $\bar{\varepsilon}_{fs} = t_{fs}(\dot{\varepsilon}_i) \cdot \dot{\varepsilon}_i$, it is easy to construct the dependence (3.26) on the base of the dependence,

$$\bar{\varepsilon}_{fs} = \bar{\varepsilon}_{fs}(\dot{\varepsilon}_i) \tag{3.27}$$

Dependences (3.27) can be found in the literature. In (He et al., 2013), on the base of the experimental data for 30Cr2Ni4MoV, ultra-super-critical rotor steel in some range of strain rate and temperature measurement, the dependence can be obtained,

$$\bar{\varepsilon}_{fs} = k(T) \cdot \dot{\varepsilon}_i^{\alpha} \tag{3.28}$$

where $k(T)$, α are the material constants.

We will consider the process where the strain rate changes in accordance with the following law:

$$\bar{\varepsilon}_i(t) = \begin{cases} \dot{\varepsilon}_i = \dot{\varepsilon}_0 = \text{const}, & 0 \leq t \leq t_1 < t_{fs}(\dot{\varepsilon}_0) \\ \dot{\varepsilon}_i = 0 & t_1 < t \leq t_1 + t_p \\ \dot{\varepsilon}_i = \dot{\varepsilon}_0 = \text{const}, & t_1 + t_p < t \leq t_2 \end{cases} \tag{3.29}$$

For the given case on the base of the model (3.25), we obtained:

$$\Delta\psi = \frac{1}{n} \cdot \ln\left(1 + e^{n\psi_1}\left(e^{n\Delta_p - 1}\right)\right) - \Delta_p \tag{3.30}$$

where ψ_1 is the resource, used at the first step,

$$\psi_1 = \frac{\bar{\varepsilon}_1}{\bar{\varepsilon}_{fs}(\dot{\varepsilon}_0)} = \frac{\dot{\varepsilon}_0 \cdot t_1}{\dot{\varepsilon}_0 \cdot t_{fs}(\dot{\varepsilon}_0)} = \frac{t_1}{t_{fs}(\dot{\varepsilon}_0)} \tag{3.31}$$

$\Delta\psi$ is a regenerated resource in pause, equal to the relation of the difference of the limit strain at nonstationary straining according to (3.28) $\bar{\varepsilon}_{fp}(\dot{\varepsilon}_0, \psi_1, \Delta_p)$ and stationary straining $t_{fs}(\dot{\varepsilon}_0)$,

$$\Delta\psi = \frac{\bar{\varepsilon}_{fp}(\dot{\varepsilon}_0, \psi_1, \Delta_p) - \bar{\varepsilon}_{fs}(\dot{\varepsilon}_0)}{\bar{\varepsilon}_{fs}(\dot{\varepsilon}_0)} \tag{3.32}$$

On the base of the experimental data on the two-stage torsion of the solid cylindrical specimen with the intermediate pause between, the stages of the empirical dependence can be obtained:

$$\Delta\psi = \exp\left(0.241 \cdot \psi_1^{0.588} \cdot (T/1000)^{-1.361} \cdot t_p^{0.251}\right) - 1 \tag{3.33}$$

The analysis of the given relationship in comparison with the values involved in the relation (30) is performed; as a result, a more compact and universal relationship has been obtained:

$$\Delta\psi = \exp\left(0.495 \cdot \psi_1^{0.588} \cdot \Delta_p^{0.251}\right) - 1 \tag{3.34}$$

Figure 3.2 presents the comparison of the results of the calculation on empirical relation (3.34) and dependence (3.30), obtained on the base of the model (3.25). Compliance results are quite satisfactory.

One more important peculiarity of the modeling should be noted. Even in the cases when the constructed model does not allow acceptable qualitative assessments to be obtained, the model can be useful from the point of view of the methodology for the construction of more universal empirical relations (Novella et al., 2015; Mikhalevich, 1991).

Another important advantage of the theory construction, compared with the empirical relation is that relation (3.30), unlike (3.33), meets the requirements of a number of boundary conditions (3.34), following the physical essence of the described process. For instance,

$$\lim_{\Delta_p \to \infty} (\Delta\psi) = \lim_{\Delta_p \to \infty} \left(\frac{1}{n} \cdot \ln\left(1 + e^{n\psi_1}\left(e^{n\Delta_p - 1}\right)\right) - \Delta_p\right) = \psi_1, \quad \psi_1 \geq 0 \tag{3.35}$$

$$\lim_{\Delta_p \to \infty} (\Delta\psi) = \lim_{\Delta_p \to \infty} \left(\exp\left(0.495 \cdot \psi_1^{0.588} \cdot \Delta_p^{0.251}\right) - 1\right) = \infty, \quad \psi_1 > 0 \tag{3.36}$$

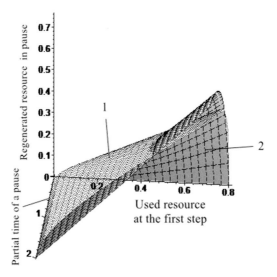

Figure 3.2 Dependence of the regenerated resource in pause $\Delta\psi$ on the used resource at the first step ψ_1 and pause time Δ_p: 1 – Equation (3.34), 2 – Equation (3.30).

3.6 EXHAUSTION AND REGENERATION OF THE ATHLETE'S RESOURCE WHEN RUNNING AT A VARIABLE SPEED

The problems of an athlete's forces distribution when running at medium and long distances remained actual until now. On the Internet, you can find numerous materials dealing with the promotion of interval running, characterized by alternations in higher and lower running speeds. For a more reasonable selection of interval running modes, it is natural to resort to mathematical modeling.

Model (3.24) can be applied for the description of the athlete's resource change when running a certain distance:

$$\psi(t) = \int_0^t \varphi(t-\tau) \cdot g(v(\tau)) \cdot d\tau \qquad (3.37)$$

where v is a travel speed. Performing the corresponding selection of the functions φ and g, we can obtain:

$$\int_0^t \frac{n \cdot (t-\tau)^{n-1}}{t_{fs}^n(v(\tau))} \cdot d\tau = 1, \qquad n > 0 \qquad (3.38)$$

For the linear model $g(v) = v$, from (3.38), there follows a power-law dependence of the limit time on the running speed:

$$t_{fs} = a \cdot v^{-1/n} \qquad (3.39)$$

In Figure 3.3 the similar dependence, constructed on the estimated normative data available on open access, is given. We will consider interval running, constructed according to the following scheme:

$$v(t) = \begin{cases} v_1 = \text{const}, & 0 \leq t \leq t_1 < t_{fs}(v_1) \\ v_2 = \text{const}, & t_1 < t \leq t_1 + t_f \\ v_1, & t_2 < t \leq t_f \end{cases} \qquad (3.40)$$

On the base of the model (3.38), taking into account (3.40), we obtain:

$$(1 + \psi_2 \alpha_{21} + \psi_{*3})^n - (\psi_2 \alpha_{21} + \psi_{*3})^n + \left(\psi_2 + \frac{\psi_{*3}}{\alpha_{21}}\right)^n - \left(\frac{\psi_{*3}}{\alpha_{21}}\right)^n + (\psi_{*3})^n = 1 \qquad (3.41)$$

where ψ_2 is a used resource at the second step, ψ_{*3} is the remaining lifetime at the third step:

$$\psi_2 = \frac{t_2 - t_1}{t_{fs}(v_2)}, \quad \psi_{*3} = \frac{t_f - t_2}{t_{fs}(v_1)}, \quad \psi_{*3} = \frac{t_{fs}(v_2)}{t_{fs}(v_1)} \qquad (3.42)$$

It can be seen that the solution $\psi_{*3} > 0$ of equation (3.41) exists only for $\psi_1 > 0$, $\alpha_{21} > 0$. According to (3.38), the change of the resource at the second stage occurs according to the law:

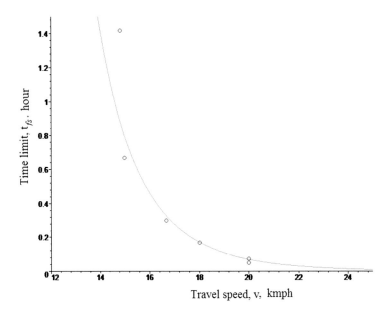

Figure 3.3 Dependence of the time limit t_{fs} on travel speed v: – Equation (3.39), $\alpha = 6 \cdot 10^9$, $n = 0.119$.

$$\psi = (1 + \psi_2 \alpha_{21})^n - (\psi_2 \alpha_{21})^n + (\psi_2)^n \tag{3.43}$$

We will consider the regularities shown by the graphs in Figure 3.4. To improve the visualization, the values measured on the vertical axis were replaced $\psi_{*3} \Rightarrow 30 \cdot \psi_{*3}$, $\psi \Rightarrow \psi \cdot 10^{7(\psi-1)}$. Such replacement changes the numerical values, but does not change the qualitative behavior of the corresponding curves.

Let at the first stage, the athlete, running at a speed of v_1 (e. q., $v_1 = 16{,}27$ km/h, $t_{fs}(16.27) = 0.4$ h), reached his limit state, where he cannot maintain his present running speed $\psi_1 = 1$. However, if the running speed is reduced to the value v_2 (e. q., $v_2 = 14.02$ km/h, $t_{fs}(14.02) = 1.4$ h), the athlete would be able to move further. The curve 2 in Figure 3.4 shows that in the initial period of the second stage, a sharp decrease of ψ value occurs, characterizing the athlete's fatigue degree, and then the fatigue is constantly accumulating, reaching the limit value if $\psi_2 = 0.66$.

The potential ability of the athlete to return to a greater running speed is characterized by the value ψ_{*3}. At the initial period of the second stage, ψ_{*3} increases, reaching at $\psi_2 = \psi_2^{opt}$ its maximum possible value, and then it starts to further decrease. At a value of $\psi_2 \approx 0.3$, the athlete essentially losses the ability to return to the initial higher running speed. At the same time, the athlete is able to run, maintaining the current running speed v_2 until the value of $\psi_2 = 0.66$ is achieved.

It should be noted that the model describes a fact, which initially seems to be a paradox: at the start of the second stage, the simultaneous increase and decrease of the athlete's resource occur. In fact, the resource is regenerated relative to the ability to travel at a higher speed and decreases relative to the ability to travel at the current speed.

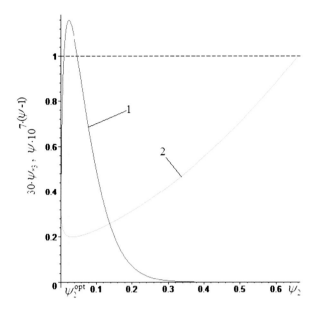

Figure 3.4 Dependences of the remaining lifetime at the third step ψ_{*3}, (1) and the value ψ (2) from the used resource at the second step: $n = 0.119$; $a_{21} = 3.5$.

These regularities can be observed by anyone who runs for medium or long distances. This indicates the adequacy of the model.

3.7 CONCLUSIONS

Using the theories of summation of damages, the authors were able to describe some key features of the behaviors of completely different objects of different physical natures: the accumulation of damages in metals during hot plastic deformation, and the accumulation of fatigue by an athlete running at medium or long distances. At the same time, the findings resulted in quite unexpected recommendations that followed from the models, and they require experimental studies. This concerns the superplasticity regime found in Kraievskyi et al. (2018a), the optimal distribution of the travel speed found in Kraievskyi et al. (2018b), and, possibly, the laws of learning forgetting offered by Perig (2018).

REFERENCES

Bao, Y. & Wierzbicki T. 2004a. On fracture locus in the equivalent strain and stress triaxiality space. *International Journal of Mechanical Sciences* 46(1): 81–98.
Bao, Y. & Wierzbicki, T. 2004b. A comparative study on various ductile crack formation criteria. *Journal of Engineering Materials and Technology* 126(3): 314–324.
Bao, Y. & Wierzbicki, T. 2015. A comparative study of three groups of ductile fracture loci in the 3D space. *Engineering Fracture Mechanics* 135: 147–167.

Golub, V.P. 2000. The nonlinear mechanics of continual damage and its application to problems of creep and fatigue. *International Applied Mechanics* 36(3): 303–342.

He, J., Cui, Z., Chen, F., Xiao, Y. & Ruan, L. 2013. The new ductile fracture criterion for 30Cr2Ni4MoV ultra-super-critical rotor steel at elevated temperatures. *Materials & Design* 52: 547–555.

Hooputra, H., Gese, H., Dell, H. & Werner, H. 2004. A comprehensive failure model for crashworthiness simulation of aluminium extrusions. *International Journal of Crashworthiness* 9(5): 449–464.

Kachanov, L.M. 1999. Rupture time under creep conditions. *International Journal of Fracture* 97(1–4): xi–xviii.

Kraievskyi, V. Mykhalevych, V., Dobranyuk, Y., Sawicki, D. & Mussabekov K. 2018b. Selection of optimal path of strain rate change in the process of multistage hot deformation under the condition of the equal duration of stages. *Proc. SPIE Photonics Applications in Astronomy, Communications, Industry, and High-Energy Physics Experiments* 10808: 108084T.

Kraievskyi, V., Mykhalevych, V., Sawicki D. & Ostapenko O. 2018a. Modeling of the materials superplasticity based on damage summation theory. *Proc. SPIE, Photonics Applications in Astronomy, Communications, Industry, and High-Energy Physics Experiments* 10808: 108084S.

Kukharchuk, V.V, Bogachuk, V.V, Hraniak, V.F. et al. 2017. Method of magneto-elastic control of mechanic rigidity in assemblies of hydropower units. *Proc. SPIE, Photonics Applications in Astronomy, Communications, Industry, and High Energy Physics Experiments* 10445: 104456A.

Lemaitre, J. 1992. *A Course on Damage Mechanics*. Berlin: Springer-Verlag.

Lokoshchenko, A.M. 2012. Dlitelnaya prochnost metallov pri slozhnom napryazhennom sostoyanii (obzor). *Mekhanika tverdogo tela* 3: 116–156.

Mikhalevich V. 1991. The model of ultimate strains during hot deformation. *Izvestia Akademii nauk SSSR, Metally* 5: 89–95.

Mohr, D. & Marcadet S. 2015. Micromechanically-motivated phenomenological hosford-coulomb model for predicting ductile fracture initiation at low stress triaxialities, *International Journal of Solids and Structures* 67(68): 40–55.

Mykhalevych, V.M. 1998. *Tenzorni modeli nakopychennia poshkodzhen*. Vinnytsia: Universum.

Novella, M.F., Ghiotti, A., Bruschi, S. & Bariani, P.F. 2015. Ductile damage modeling at elevated temperature applied to the cross wedge rolling of AA6082-T6 bars. *Journal of Materials Processing Technology*. 222: 259–267.

Ogorodnikov, V.A., Dereven'ko, I.A. & Sivak, R.I. 2018. On the Influence of Curvature of the Trajectories of Deformation of a Volume of the Material by Pressing on Its Plasticity Under the Conditions of Complex Loading: *Materials Science* 54(3): 326–332.

Perig, A. V., Golodenko, N. N., Skyrtach, V. M., & Kaikatsishvili, A.G. 2018. Hydraulic analogy method for phenomenological description of the learning processes of technical university students. *European Journal of Contemporary Education* 7(4): 764–789.

Troshchenko, L., Khamaza, V. et al. 1993. *High-Cycle Fatigue. Cyclic Deformation and Fatigue of Metals 78*. Elsevier Science. P.O. Box 211, 1000 AE Amsterdam, The Netherlands. https://www.elsevier.com/books/cyclic-deformation-and-fatigue-of-metals/bily/978-0-444-98790-7

Chapter 4

Increase in durability and reliability of drill column casing pipes by the surface strengthening

I. Aftanaziv, L. Shevchuk, L. Strutynska, I. Koval,
I. Svidrak, P. Komada, G. Yerkeldessova, and K. Nurseitova

CONTENTS

4.1 Introduction	39
4.2 Analysis of the recent research and publications	40
4.2.1 Objectives of the research	41
4.3 Presentation of the main research	41
4.4 Results and discussion	48
4.5 Conclusions	48
References	49

4.1 INTRODUCTION

In the course of drilling of a well, drill column casing pipes perform the function of protecting against depreciation of the rotating drill pipes as a result of their ground friction and also protecting the drill pipes against catching as a result of horizontal rock bed movements. Additionally, casing pipes increase the vertical rigidity of a drill column, preventing its deformations and vertical deviations from the well. They also serve as a rising channel from an oil or natural gas field. Quite often the durability and reliability of these significant pieces regulate parameters important to the process of well drilling such as the speed of drilling and the ability to pass the layers of increased hardness, the ability of a well column to resist horizontal shifts of the soil layers, and so on.

Unlike drill pipes, which are pieces of repeated use, casing pipes from the drilled well are not taken out from it and are left forever in the thickness of the drilled ground. In addition, there is a quite widespread engineering problem – for reasons of the considerable operational loads, the drill pipe has to be a detail of increased durability, on the other hand, taking into consideration the fact that it is a single-use element, it need not be material-intensive (Lototskaya 2008). In this problem, considering the high cost of drilling works and the equipment used in their performance, the preference is given to the increased reliability of casing pipes, neglecting the loss of the expensive metal casing pipe lowered into the well forever.

For these reasons, quite expensive type 40X and 30ХГС structural steels are used to manufacture casing pipes with a wall thickness of 10–15 mm. The transition to cheaper materials for the manufacturing of these pipes is not possible due to the complexity of ensuring their required durability. That is why the only available option of the cost reduction of casing pipes is decreasing their wall thickness as well as compensating for the weakening caused by it with the aid of specific technological resources.

4.2 ANALYSIS OF THE RECENT RESEARCH AND PUBLICATIONS

Casing pipes are made on rolling mills, by exposing them after shaping to natural cooling and aging for redistribution of foundry and technological residual stresses in the thickness of their material. Casing pipes belong to the category of long-length tubular parts with a ring transverse section.

An analysis of operational loads of casing pipes shows that the surface layers of metal on external and internal cylindrical surfaces are the most loaded on the thickness of the transverse section. At the same time, the metal of external surfaces of the pipe layers is exposed to the compression–tensile stresses caused by the ground shifting and the deflections of a drill column. The metal of the internal surface is mainly affected by abrasive wear caused by contact with the rotating drill pipes.

From numerous research in the field of materials science and engineering as well as from the experience of engine parts exploitation, it is known that in general, the external metal layers of parts resist compression–tensile stress as well as cyclic rotating loadings. The load on the inner layers of metal of such parts precipitantly decreases in the process of their removal from the outer surface (Kusyy & Kuk 2015). It gives grounds for a conclusion that from the perspective of material engineering theory, the decrease in material consumption of casing pipes without any loss in their strength parameters is exceptionally made possible thanks to the decrease of the pipes wall thickness at the increase of their inside diameter. At the same time, at the decrease of wall thickness, such an important parameter of casing pipes as the resistance of metal of their internal surface to abrasive wear doesn't have to necessarily decrease.

It is possible to solve this problem with the help of technological tools, for example, increasing the rigidity of metal on the internal surface of pipes. It is known that in the process of increasing the hardness of steel parts, the resistance of their material to abrasive wear increases (Kusyy & Topilnitsky 2009). It is possible to achieve this by methods of thermal hardening (Kusyy & Topilnitsky 2013) or by strengthening the superficial plastic deformation (Kusyy & Topilnitsky 2009, 2013). Taking into consideration the remarkable dimensions of casing pipes whose length can reach 13 meters, and weight in the thousands of kilograms, the extensional thermal hardening is inapplicable here, and face hardening using high-frequency currents is too energy-consuming (Kusyy & Topilnitsky 2013). There are particular problems connected with the choice of methods of the superficial plastic deformation, which are able to strengthen such specific parts. Due to the considerable length of casing pipes, such widespread methods of dynamic strengthening as shot blasting (Myronov & Redreev 2014, Kravchuk et al. 2016) become unsuitable for their internal surface processing. Because of the insignificant diameter of the processing of the internal surface of pipes using shock-vibrating strengthening (Kusyy & Topilnitsky 2009, Myronov & Redreev 2014) is also unsuitable.

An exception is the method of vibratory-centrifugal treatment (Athanasios et al. 2018), which was recently developed for channels of artillery gun barrels. Much like casing pipes, the gun barrels have a considerable length with an insignificant diameter of the internal surface, which is the object of the strengthening process. Some of the constructive diagrams of strengthening by the vibratory-centrifugal treatment are suitable for the strengthening processing of the internal surface of casing pipes, described in reference literature (Kusyy & Kuk 2015, Aftanaziv et al. 2018). However, they are difficult in regard to their constructive composition, and respectively, are insufficiently reliable for such mass parts as casing pipes.

At the same time, the positive experience of successful application of the vibratory-centrifugal strengthening treatment on an increase in the longevity of extended internal cylindrical surfaces operating in conditions of abrasive wear is already known (Kusyy & Topilnitsky 2013). Strengthening of an internal working area of drilling mud transport pump sleeves using this method allowed to double their longevity. Like casing pipes, the pump sleeves are exposed to abrasive wear when the piston that pushes the drilling mud (Kusyy & Kuk 2015) moves along them.

4.2.1 Objectives of the research

The objective of the research is to expand the technological capabilities of the progressive vibratory and centrifuging strengthening treatment to be suitable for strengthening the internal surface of extended long-length parts including drill column casing pipes as well as the development of an assessment modality of a possible decrease in casing pipe mass thanks to the strengthening of their internal surface.

The research included the following items and stages:

- an analysis of the operational loads affecting the material of drill column casing pipes;
- an analysis of technological capabilities and effectiveness of the known methods of the superficial plastic deformation in the context of their suitability for the strengthening of drill column casing pipes;
- the development of a constructive diagram of the new strengthening treatment equipment suitable for the effective strengthening processing of the internal surfaces of long-length parts, including drill column casing pipes;
- an analysis of the opportunities to decrease the wall thickness and the mass of casing pipes thanks to the strengthening of their internal surface.

4.3 PRESENTATION OF THE MAIN RESEARCH

For the high-quality strengthening of the internal surface of drill column casing pipes, the use of vibratory-centrifugal strengthening treatment is offered. The differential characteristic of the vibratory-centrifugal strengthening treatment is that thanks to the impact of the large surfaces of the processed detail and the tool at their contact through a small number of deformable bodies in the material of the processed detail in the places of shock contact, the considerable surface stresses, causing a high level of strengthening and formation of residual stresses of compression of the considerable gradient (Aftanaziv et al. 2018, Athanasios et al. 2018), are formed. The pressure loads generated as a result of such strengthening treatment of the internal surface of pipes in the thickness of their material actively counteract the layer separation and material continuity ruptures. Along with ensuring the creation of residual compressive stresses in the metal surface layer of the processed parts, such strengthening processing forms in them a surface layer with increased hardness and improves the structure of the metal in its surface layers (Aftanaziv et al. 2018). All this provides an active resistance of the strengthened material to wear including abrasive wear. This results in high-quality strengthening of the internal surface of casing pipes, thanks to an increase in resistance of the strengthened material to abrasive wear, which allows to proportionally reduce the thickness of the pipe walls without any loss in their strength characteristics (Aftanaziv & Shevchuk 2018a, 2018b, Vedmitskyi et al. 2017).

This newly formed method of the strengthening treatment passed successful trials in the industrial implementation for strengthening of aircraft wheel hubs and flanges, having shown a double increase in the motor potential of these important parts in comparison with unstrengthened ones (Aftanaziv et al. 2018). The resistance of steel cases of the drill pumps to abrasive wear due to the strengthening of their internal surface by the method of the vibratory-centrifugal treatment has increased by half or even doubled (Kusyy & Kuk 2015). The specific features and the essence of the vibratory-centrifugal strengthening processing method are more clearly exposed in the description of the equipment, which is used in the process of its realization.

Figure 4.1a–c schematically shows the placement of the vibratory-centrifugal strengthening processing device inside the processed casing pipe (Figure 4.1a), Figure 4.1b depicts section A–A perpendicular to the center line of a pipe (Figure 4.1a) at a low point of the strengthening device position, Figure 4.1b depicts a view in the direction shown by the arrow B (Figure 4.1a), with the reflection of forces having an effect on the strengthening device in the process of the strengthening treatment (Kukharchuk et al. 2017a, 2017b).

The strengthening of the internal surface of casing pipes by superficial plastic deformation in compliance with the vibratory-centrifugal strengthening treatment is carried out as follows. The strengthening device, which consists of an electric motor drive 1, cylindrical strengtheners 2 with deformable bodies placed with a possibility of rotation on its deformed outer surface 3 in the form of the steel tempered balls of high

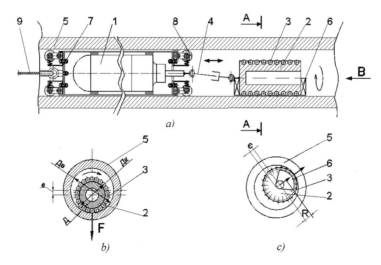

Figure 4.1 The device for the vibratory-centrifugal strengthening processing of the internal surface of casing pipes.

hardness and torque, and rotary actuator 4 connecting the electric drive shaft with the strengthener, is placed inside of the processed casing pipe 5.

The mechanism of rotary movement transfer 4 on the strengthener 2 also provides the radial movement of the strengthener in reference to the centerline of pipe 5. For this reason, as the mechanism of torque transmission and rotary movement, it is appropriate to use either a gimbal or flexible shaft (Figure 4.1a). On the strengthener 2, two eccentric weights are fixed 6 (Figure 4.1b). The electric drive 1 is placed in a cage 7, which, using rollers 8, is centered coaxially to the processed surface of casing pipe 5. The strengthening device is placed on the inside of the processed surface of the casing pipe 5, and for its movements along the strengthened surface is a cable 9 attached to the cage 7, which is spooled on the winch drum (not shown in Figure 4.1).

The maximum diameter of the circle D_{str}, which comprises deformed balls 3 placed on the strengthener 2, is equal to (Vedmitskyi et al. 2017):

$$D_{str} = (0.75 - 0.85) D_{int} \qquad (4.1)$$

where D_{int} is the diameter of the internal strengthened surface of casing pipe 5.

Eccentricity ε of the strengthener 2 with deformed balls 3 is equal to:

$$\varepsilon = \frac{D_{int} - (D_c + D)}{2} \qquad (4.2)$$

where D is the diameter of the spherical steel tempered balls 3.

D_c is the diameter of the circle of the spherical steel tempered balls 3 geometric center position.

From the condition of ensuring the given quality of the strengthening processing, which is regulated by the thickness of the strengthened metal layer, inseparably associated with the diameter of prints on the processed surface after its shock interaction with the deformable balls 3, and mechanical characteristics of the strengthened material, the quantity of the eccentricity is calculated as follows:

$$\varepsilon = \frac{D_{int} - (D_c + D)}{2} \qquad (4.3)$$

where:
- ε is the eccentricity of the strengthener mass decentration 2 relative to the geometrical axis of the processed surface of casing pipe 5;
- n is the rolling frequency of the strengthener 2 over the internal surface of casing pipe 5;
- σ_m is the limit of tensile stress of the material of casing pipe 5;
- m is the total weight to the strengthener 2 with the deformable balls 3 and the eccentric mass 6;
- d is the diameter of the print on the processed surface caused by the shock contact of a deformable ball 3 with the pipe 5;
- N is the number of the deformable balls 3, placed along the strengthener element 2;
- D is the diameter of the spherical deformable balls 3;
- l is the length of the element of the external cylindrical surface of the strengthener 2, where the deformable balls 3 are located.

The main operating movement of the strengthener 2 that provides its shock interaction with the processed surface of the casing pipe 5 is the mode of planetary rolling movement by the deformable bodies, which are placed on the strengthener's external cylindrical surface, on the internal processed surface of casing pipe 5. To ensure this rolling movement on the strengthener 2, eccentric weights 6 are fixed, whose weight and distance of the center mass from the strengthener's axis of rotation 2 are calculated using the following formula:

$$m_d = \frac{\varepsilon \cdot m}{2R - \varepsilon} \qquad (4.4)$$

where:
m is the mass of the strengthener 2 with the deformable balls 3;
ε is the eccentricity of the strengthener 2;
R is the distance from the center of the eccentric mass 6 to the axis of its rotation.

As a result, the danger of the strengthener rotation 2 under its own weight is eliminated in the treatment process being performed in the bottom of the processed casing pipe 5 transverse sections instead of the regular planetary rolling movement of the strengthener along all the length of the ring of the cross section of the processed surface. This means that the danger of the irregular strengthening along the entire length of the ring of the processed surface cross section of the casing pipe 5 is eliminated (Polishchuk et al. 2015).

In the figure, the directions of movement and rotation of the strengthener 2 are specified using arrows.

The process of strengthening the internal surface of casing pipes by superficial plastic deformation using this strengthening device is carried out as follows. The strengthening device is inserted into the casing pipe 5 fixed by the fasteners on a special table (not shown in the figure). The power supply is delivered on the electric motor of the strengthener 1 transmission (Figure 4.1a). When providing power to the electric driving motor 1, the operating torque from its shaft via the rotary movement transfer mechanism 4 (a drive shaft or a flexible shaft) is transferred to the strengthener 2, which at the same time engages in a rotary movement with a frequency equal to the rotation frequency of the motor drive shaft (Figure 4.1b). In the course of strengthener 2 rotation, the rotation of the eccentric weights 6 attached to it the strengthener is affected by the centrifugal force P (Figure 4.1c), which is equal to the product of the mass of the eccentric weights and the distance from their center mass to the axis of rotation. In the wind-powered units, this force is called "the first torque." The rotating vector of the action of the force P passes through the axis of rotation of the strengthener and the center of the mass of the eccentric weights 6. On the condition of observing of sizes of the eccentric weights 6 and the distance from the axis of rotation to the center mass associated with the dependence (1), the strengthener 2, under the influence of the operating torque transferred to it equal to the force P involves itself in the mode of the working planetary rolling movement by the deformable balls 3 located on its external cylindrical surface on the internal processed surface of the casing pipe 5.

At the rolling movement of the strengthener 2 on the internal processing surface of the pipe 5, the strengthener is affected by the centrifugal force F, whose rotating vector is directed from the center mass of the strengthener and passes perpendicular to the center axes of the strengthener and the processed surface (Figure 4.1c). In Figure 4.1b, the direction of action of the centrifugal force is shown with the aid of the arrow with the alphabetical reference F. The extent of this centrifugal force F is proportional to the mass m and the eccentricity ε of the strengthener 2 and the squared circular frequency n of its rolling movement and is defined by the following equation (Polishchuk, Kharchenko et al. 2016, Polishchuk, Bilyy et al. 2016, Polishchuk et al. 2018).

$$F = m \cdot \varepsilon \cdot \omega^2 \qquad (4.5)$$

where $\omega = 2\pi n$ – circular frequency of the strengthener 2 rolling movement.

At any time the strengthener 2 contacts with the processed surface of the casing pipe 5 through the deformable balls 3 placed along the forming cylindrical external surface of the strengthener. Contact with the next group of the deformable balls 3 placed along the generatrix of the strengthener occurs with the shock interaction. At the same time, the force of the blow, falling at each of the deformable bodies 3, is proportional to the centrifugal force F operating on the rotating strengthener and is inversely proportional to the quantity $N = l/D$ located along the generatrix of the deformable balls 3 of the strengthener, namely:

$$F_y = \frac{F}{N} = \frac{m \cdot \varepsilon \cdot D \cdot \omega^2}{l} \qquad (4.6)$$

where l – is the length of the generatrix cylindrical surface of the strengthener 2,

N – is the number of deformable balls 3, located along the generatrix of the strengthener 2.

The extent of the contact stresses, which have to be applied onto places of contact with the deformable balls 3 in the near-surface layers of the material of the processed surface in consequence of the shock interaction, is considered equal to the liquid limit σm of the processed material and defined as follows:

$$\sigma_{con} = \frac{F_y}{S} = \frac{10 \cdot m \cdot \varepsilon \cdot D \cdot n^2}{l \cdot d^2} \qquad (4.7)$$

where $S = \pi d^2/4$ is the area of a high-quality strengthening, which leaves on the processed surface of casing pipe 5 of the residual print after the shock contact with the deformable spherical body (ball).

The rolling movement of the strengthener 2 on the internal processed surface of casing pipe 5 occurs at the same time as the regular axial movement of the strengthening device along the generatrix of the strengthening surface. This axial movement is exercised by means of a cable 9 that is reeled up on the winch drum (not shown in Figure 4.1) and runs the entire device along the internal strengthening surfaces of the casing pipe 5. That way the regularity of processed surface strengthening on the circumference of its section's circle and the circumference of its generatrix is provided. In case of a need to increase the thickness of the strengthened material layer bedding on the processed surface, the repeated movements of the strengthening device along the axis of the casing pipe 5 are used or the mass of the strengthener 2 is increased.

After completion of the strengthening processing of casing pipe 5, which, depending on the need, can include one or several repeated movements of the strengthening device along the forming processed surface, the strengthening device is removed from within strengthened casing pipe 5 with the aid of the cable 9 and installed in the next pipe, which is subject to strengthening. Its strengthening processing is realized similarly to the processing of the previous pipe.

The suitability of the obtained mathematical relations for the connection of technological and constructive parameters of the offered device for strengthening the internal surfaces of steel tubular parts is realized on the working model of the strengthener. The casing pipe was imitated by a fragment of the thick-walled pipe with an internal surface diameter of $D_{int} = 125$ mm, made of type 30ХГС structural steel with a yield strength $\sigma_m = 750$ MP, which, according to its physical and mechanical properties, is close to a certain extent to the material that is used in the manufacture of drill column casing pipes.

Experimental conditions set the task of providing by the vibratory-centrifugal strengthening processing of a strengthened metal layer thickness of the processed internal cylindrical surface of a pipe within $h = 0.15-0.20$ mm, and the determination of the key design data of the strengthener capable to provide the given thickness of the strengthening.

At the first stage, the quality indicators of the strengthening treatment are specified, which provide the given thickness of the strengthened layer. From the primary literary sources (Kusyy & Topilnitsky 2013, Myronov & Redreev 2014, Kravchuk et al. 2016), it is known that at the superficial plastic deformation of steel parts by the spherical deformable bodies, the thickness of the strengthened layer of metal bedding is adjusted by the diameter of prints on the processed surface according to the correspondence $d = 2h$. Therefore, the diameter of the given ensured strengthening of the print thickness on the processed internal surface of the pipe has to be equal to $d = 2 \cdot 0.2 = 0.4$ mm.

In the following stage, the particular parameters of the strengthener are set and its main constructive size is determined, namely the eccentricity of the strengthener ε, which, under the dynamics of the strengthener, will provide an appropriate force of detail deformation and the extent of contact stresses.

The following geometrical sizes of the strengthener and the parameters of processing are determined:

$D = 10$ mm – is the diameter of the deformable bodies (steel tempered balls);
$l = 0.5$ m – is the length of the strengthener;
$m = 30$ kg – is the mass of the strengthener with the deformable bodies;
$N = 50$ – is the number of the deformable bodies along the generatrix of the strengthener;
$N = 940$ rotations/min = 16 1/s – is the electromotor shaft speed.

To determine the eccentricity of the strengthener, the numeric values are used in the following correspondence (2)

$$\varepsilon = \frac{\sigma_m \cdot d^2 \cdot l}{50 \cdot m \cdot D \cdot n^2} = \frac{750 \cdot 10^6 \cdot (0.4 \cdot 10^{-3})^2 \cdot 0.5}{50 \cdot 30 \cdot 10 \cdot 10^{-3} \cdot 16^2} \qquad (4.8)$$

Thus, the key geometrical parameter of the strengthener is determined, and its maintenance will provide high-quality strengthening of the casing pipe's internal surface using the offered method.

At the same time, the superficial microhardness of the reinforced internal surface material of the thick-walled pipe made of type 30ХГС structural steel that imitated the casing pipe increased to Hs.(str.) = 5.15 hPa. This exceeds the superficial microhardness of the unstrengthened pipe material by 10%–15%.

The researchers of the Odessa State Academy of Technical Regulation, Kravchuk V.S., Dashchenko A.F., Limarenko A.M. in their work (Kravchuk et al. 2016) based on research on the fatigue of strengthened steel samples developed a theory of the so-called "static similarities of the fatigue breakdown." They proved that thanks to the increase in the resistance to fatigue damage of the material parts, which underwent the strengthening treatment, it is possible to significantly reduce the cross sections of hard-loaded steel parts as well as their weight. At the same time, the leading material engineers, Kudryavtsev I.V., Kravchuk V.S., Kaledin B.A., Chepa P.A., and others proved in their research studies that the surface strengthening of the steel parts is able to raise not only their fatigue limit but also their resistance to wear and their durability.

These research studies suggest that there is a particular correlation between the increase in the superficial microhardness and the resistance to wear of material of the parts strengthened by the superficial plastic deformation and their weight.

Having taken known dependence (Aftanaziv et al. 2018, Kravchuk et al. 2016) as a basis that connects the change of the weight of the strengthened internal surface of a tubular detail using the strengthening processing parameters, it is possible to recommend the similar interpreted mathematical dependence for the strengthening of casing pipes. This mathematical dependence connects the possible decrease of the casing pipe wall thickness and, respectively, its mass thanks to the increase in resistance to wear of the internal pipe surface material, which underwent the strengthening treatment as a result of an increase in its superficial microhardness.

According to this dependence, the casing pipe wall thickness, and respectively its mass, can be reduced in proportion to the relation of the geometrical size coefficient of the strengthened and unstrengthened surfaces K and the efficiency coefficient of the strengthening processing β, that is:

$$m_2 = \frac{K \cdot m_1}{\beta} \qquad (4.9)$$

where т2 – is the mass of the casing pipe strengthened by the vibratory-centrifugal treatment;

$K = (d_1^2 + d_{2H}^2)(d_1^2 + d_{2y}^2)^{-1}$ – is the ratio of the geometrical sizes of the strengthened and unstrengthened surfaces;

d_1 – is the diameter of the outer surface of the casing pipe;

d_{2H} – is the diameter of the inner unstrengthened surface of the casing pipe;

d_{2y} – is the diameter of the inner strengthened surface of the casing pipe;

$\beta = Hs.(str.) / Hs.(un.)$ – is the efficiency coefficient of the strengthening processing;

$Hs.(str.)$ – is the superficial microhardness of the strengthened material on the inner strengthened surface of the casing pipe;

$Hs.(un.)$ – is the superficial microhardness of the original unstrengthened material on the inner surface of the casing pipe.

4.4 RESULTS AND DISCUSSION

The correspondence (1) reflects a possible mass decrease of the casing pipe as a result of the thickness decrease of its wall due to its high-quality strengthening by way of superficial plastic deformation in the vibratory-centrifugal strengthening processing. At the same time, the casing pipe does not lose its durability and ability to realize its functional tasks. Our experimental research shows that the value of the coefficient To is within the limits of 0.95–0.99, and the value of the strengthening coefficient is $\beta = Hs.(str.)/Hs.(un.)$, depending on the processing conditions, in the limits $\beta = 1.05$–1.055. Therefore, respectively, the mass of casing pipes due to their surface strengthening by vibratory-centrifugal processing can be reduced by 10%–15%, which is approximately 100–120 kg per one casing pipe, depending on its length. In a well with a depth of 2–2.5 km, this will provide savings ranging between 10 and 15 tons of expensive steel.

In summary, it is worth mentioning the following. First of all, attention should be paid to developing a new method of surface strengthening of the inner surfaces of the long-length parts of the circular cross section, which the authors called the vibratory-centrifugal strengthening treatment. The resulting positive aspect is the creation of a completely new constructive scheme of surface strengthening equipment by the vibratory-centrifugal strengthening treatment of internal surfaces of the long-length parts. Neither the new method of the long-length parts strengthening treatment nor the equipment used in its realization have any analogs in the global practices, which is confirmed by the Ukrainian patents obtained by the authors of the invention. A powerful advantage of the strengthening treatment method is its versatility. It can be successfully used together with casing pipes, for the strengthening of artillery and tank gun barrels, as well as for the strengthening of high-pressure pipes, for the landing gear of planes, the tubes of the drill pumps, and so forth.

4.5 CONCLUSIONS

The offered original construction of the long-length parts internal surface strengthener, due to shock interactions between the tool with the material of the processed part, is capable of providing a thickness of steel detail strengthening in the range of 0.15–0.20 mm at a 10%–15% increase in the superficial microhardness of the strengthened material. The strengthener has a simple construction, is energy-efficient, and does not require any highly skilled maintenance staff.

The construction of the internal surfaces strengthener for long-length parts is not only structurally simple but also universal. Apart from being used for the strengthening of casing pipes, it can be successfully used to increase the durability of such important parts as the tubes of the drill pumps, the landing gear of planes, artillery and tank gun barrels, and so on.

Along with the increase in the longevity of the strengthened parts, the high-quality strengthening of their internal surfaces allows reducing their materials consumption by 10%–15%. Due to the strengthening of the casing pipe internal surface with the offered strengtheners and reducing wall thickness, it is possible to provide savings of 10–15 tons of expensive steel on a 2–2.5 km well.

REFERENCES

Aftenaziv, I.S. & Shevchuk, L.I. 2018a. Device for strengthening superficial plastic deformation of internal cylindrical surfaces of long-length parts. *Patent no. 116268 Ukraine, МПК B24B 39/02 (2006.01), B23P 9/04 (2006.01)* Bul. no. 4.

Aftenaziv, I.S. & Shevchuk, L.I. 2018b. A method of strengthening the surface plastic deformation of internal surfaces of artillery gun barrels. *Patent no 116266 Ukraine, МПК B24B 39/04 (2006.01), B23P 9/04 (2006.01)* Bul. no. 4.

Athanasios, I.S., Shevchuk, L.I., Strogan, O.I. & Strot'yanska, L.R. 2018. Increase of durability and durability of the drums of the wheels of aircraft by strengthening processing. *National Aerospace University named after. N.U. Zhukovsky "Kharkiv Aviation University"* 5/149: 47–57.

Aftanaziv, I.S., Snevchuk, L.I., Strutynska, L.R. & Strogan, O.I. 2018. Vibrational-centrifugal surface strengthening of drill and casing pipes. *Naukovyi Visnyk Natsionalnoho Hirnychoho Universytetu* 5(167): 88–97

Kravchuk, V.S., Dashchenko, A.F. & Limarenko, A.M. 2016. Graphoanalytic method for determining the effect of hardening of surface hardened parts of machines. *Collection of scientific works of the Odessa State Academy of Technical Regulation*: 79–82.

Kukharchuk, V.V., Bogachuk, V.V., Hraniak, V.F., et al. 2017a. Method of magneto-elastic control of mechanic rigidity in assemblies of hydropower units. *Proceedings of SPIE, Photonics Applications in Astronomy, Communications, Industry, and High Energy Physics Experiments* 10445: 104456A.

Kukharchuk, V.V., Kazyv, S.S. & Bykovsky, S.A. 2017b. Discrete wavelet transformation in spectral analysis of vibration processes at hydropower units. *Przeglad Elektrotechniczny R* 93(5): 65–68.

Kusyy, Y. & Kuk, A.M. 2015. Development of the method of the vibration and information center for the technological zabolepechnya without child machine building. *East European Journal of Advanced Technology* 1/7(73): 41–51.

Kusyy, Y.A. & Topilnitsky, V.G. 2009. Calculations of vibratory-centrifugal strengthening treatment's dynamics by means of application software. *Conference Proc of XVII Polish-Ukrainian Conference "CAD in Machinery Design-Implementation and Educational Problems"*: 25–26.

Kusyy, Y.A. & Topilnitsky, V.G. 2013. Surveying the quality of the surface of vibrated parts of machines. *Bulletin of the National University "Lviv Polytechnic"* 772: 196–201.

Lototskaya, O.I. 2008. Improvement of operational properties of parts of printing machines. *Technology and Technique of Printing* (3–4): 16–20.

Myronov, A.V. & Redreev, H.V. 2014. On the question of strengthening the surface of the parts by plastic deformation. *Bulletin of the Omsk State Agrarian University* 3(15): 35–38.

Polishchuk, L., Bilyy, O. & Kharchenko, Y. 2015. Life time assessment of clamp-forming machine boom durability, *Diagnostyka* 4(16): 71–76.

Polishchuk, L., Bilyy, O. & Kharchenko, Y. 2016. Prediction of the propagation of crack-like defects in profile elements of the boom of stack discharge conveyor. *Eastern-European Journal of Enterprise Technologies* 6(1): 44–52.

Polishchuk, L., Kharchenko, Y., Piontkevych, O. & Koval, O. 2016. The research of the dynamic processes of control system of hydraulic drive of belt conveyors with variable cargo flows. *Eastern-European Journal of Enterprise Technologies* 2(8), pp. 22–29

Polishchuk, L.K., Kozlov, L.G., Piontkevych, V., et al. 2018. Study of the dynamic stability of the conveyor belt adaptive drive. *Proc. SPIE, Photonics Applications in Astronomy, Communications, Industry, and High-Energy Physics Experiments 2018* 10808: 1080862.

Vedmitskyi, Y.G., Kukharchuk, V.V. & Hraniak, V.F. 2017. New non-system physical quantities for vibration monitoring of transient processes at hydropower facilities, integral vibratory accelerations. *Przeglad Elektrotechniczny* 93(3): 69–72.

Chapter 5

Experimental research of forming machine with a spatial character of motion

I. Nazarenko, O. Dedov, M. Ruchynskyi, A. Sviderskyi, O. Diachenko, P. Komada, M. Junisbekov, and A. Oralbekova

CONTENTS

5.1 Introduction 51
5.2 Analysis of the dynamic processes of the model of the vibration unit 52
5.3 Materials and methods 53
    5.3.1 Studying the vibration machine with spatial movement 53
5.4 Results and discussion 55
5.5 Conclusion 58
References 59

5.1 INTRODUCTION

At the present stage of development of the construction industry, there is an urgent problem of introducing such technologies and machines that make it possible to ensure high quality of the finished product, achieve a significant reduction in energy costs, and increase productivity. The leading place among the equipment of the construction industry belongs to vibrating machines. They are widely used to perform a variety of technological processes: comminuting, sorting, mixing, and compaction. These machines work, as a rule, above resonance modes. Such modes are characterized by the considerable energy consumption of technological processes. As a rule, such machines are designed and created based on calculation models that take into account the characteristics of the processing medium and the machine with discrete parameters. The use of such models gives reliable results only within the framework of the conducted research. As a result, to date, there is a significant difference between the existing physical and mathematical models that describe the motion of vibration machines and sealing environments. This lack of generally accepted computational models adequately reflects the real picture of the movement of the machine, and the movement of the processing material makes it difficult to develop an effective vibration technique. The importance of creating an effective vibration technique is the estimation of the movement and influence of this movement of metal structures of machines. Therefore, the actual task is the research and assessment of the shaping design on the efficiency of the technological process. This is important because the trends in the construction industry are aimed at reducing the cost of energy resources with high-quality technological processes. In particular, the process of compacting concrete is one of the

DOI: 10.1201/9781003225447-5

5.2 ANALYSIS OF THE DYNAMIC PROCESSES OF THE MODEL OF THE VIBRATION UNIT

A study of the process of forming and sealing concrete and reinforced concrete mixtures, as well as types of equipment and methods for bringing it into operation, is devoted to a number of works. In Nazarenko et al. (2012), the design scheme of the machine is considered a system with discrete parameters, and the environment is modeled by a system with distributed parameters. The given method takes into account not only the elastic but also the dissipative properties of the medium processed during oscillation. In Nesterenko (2015), oscillations of a vibration platform of a frame structure in a horizontal plane are investigated. However, a simplified design model of the machine is used in the form of a discrete one, and the concrete mix is taken into account by the empirical coefficient. Effects of wave processes are not considered. Such a method is reliable only within the limits of the performed studies and identical in design and parameters of the vibration machines. Thus, in the considered studies, no studies exist that study the amplitude–frequency spectra of oscillations of vibration machines where simplified calculation models are applied. Application of continual models is more effective. It makes it possible to take into account the propagation of waves, both in the construction of the frame of the vibration machine and in the mixture compaction. This approach is the basis for determining the real distribution of the amplitudes and frequencies of oscillations and applying multimode effects. Under spatial load, complex oscillations arise, and division into torsion and bending oscillations is a kind of assumption. In this case, there is no evidence of rational design taking into account the stress state of the plate.

In Bui et al. (2016), it was noted that the analysis of natural oscillations of plate structures is an important area of research due to the wide application in equipment. The influence of various ratios of sides and boundaries of thick and thin plates was considered. The influence of other materials in contact with plates was not taken into account. No stress studies were given. This confirms the necessity and relevance of the study of stresses and strains in the shaping surface. In Dedov (2018) and Kozlov et al. (2019), an analytical method is proposed for determining the influence of the processed medium on the dynamics of the "machine-medium" system. The analytical dependences for estimating the influence of the medium resistance at poly-frequency oscillations are obtained. In Nesterenko et al. (2018), studies of a shock vibration machine for molding concrete products are given. Studies are based on the determination of the reduced mass and the equivalent coefficient of resistance of a concrete mix. As a result, the dependencies for the description of wave phenomena in the medium are obtained. However, the results of the experimental determination of the dynamic parameters of the investigated units in the above works are absent. The authors of Andò et al. (2015) consider a dynamic system capable of accumulating internal energy. Phenomena in complex nonlinear systems, as the authors note, are a promising direction and require additional research.

Experimental studies of the oscillatory system using measurements of accelerations are given in Kavyanpoor & Shokrollahi (2019). The study is based on a certain oscillation spectrum and the identification of natural oscillation frequencies. The method can be used to study more complex dynamic systems. Improvement of the calculation model based on the received dynamic characteristics is considered in the works of Bendjama et al. (2012) and Ghandchi et al. (2013). Such an approach can be used to verify the conformity of the mathematical and experimental models of the investigated complex dynamic systems. In Pawelczyk and Wrona (2016) and Yue-min et al. (2009), particular cases of rational constructions of planar elements are given, and optimization methods can be applied in the study of small and simple products. The stress–strain state of structures in nonlinear analysis is rather complicated and requires a lot of resources, as indicated in Yue-min et al. (2009). In Yamamoto et al. (2016), the method is applied to nonlinear active vibration control systems. The advantage of such an integral method is that there is no need to know the system parameters, such as mass, attenuation, and stiffness coefficients, which are usually obtained by finite element methods. According to the measurement of oscillations, acceleration measurement sensors are usually used (Jia & Seshia, 2014). Strain–stress state of construction at the linear analysis is quite difficult and needs many resources, which are noted in the works of Yue-min et al. (2009) and Wójcik & Kisała (2009). That is why for difficult computing models, the author recommends the use of finite element modeling (Nazarenko, Dedov, & Zalisko, 2017; Nazarenko, Gaidaichuk, Dedov, & Diachenko, 2017).

5.3 MATERIALS AND METHODS

5.3.1 Studying the vibration machine with spatial movement

The object of this research is the movement process of forming structures of a vibration unit with spatial oscillations. The main problem of such vibration systems is the lack of data on the mutual influence of the machines and the machining environments relating to the technological load on the machine. To implement the research of the vibration unit, the following sequence of research works by Iskovich-Lototsky et al. (2019), Nazarenko, Dedov, & Zalisko (2017), Nazarenko, Gaidaichuk, Dedov, & Diachenko (2017), and Polishchuk et al. (2018), is assumed:

- Analysis of calculations of the structural elements of the machine in terms of accounting for all types of loads that are carried out in the machine design
- Development of a computer model of the object of research (general or some of the most loaded nodes, structural elements)
- Carrying out additional modelling and calculations to determine the behavior of structural elements and the machine as a whole, while the various loads act simultaneously
- Development on a computer model of a matrix of control points of limiting values of integral characteristics of the structure state for further use in field tests
- Carrying out field tests by applying certain loads on its model
- Adjustments of the computer model until the comparison of the integral characteristics obtained by measuring at the control points during the experiment and in

the simulation differ among themselves within the limits of the permissible error. The computer model obtained in this way will be adequate to the real construction within the limits of adequacy—the points of the integral characteristics control.

The vibration unit simultaneously performs the function of a mold for a concrete mix and consists of a welded box-section frame that is installed on rubber elastic supports on a concrete foundation. The vibration unit is equipped with two installed centrifugal generators of high-frequency oscillations; the generators are not symmetrical. Two nonremovable sides and one movable side are fixed onto the frame. A geometric 3-D model is created to study the vibration unit on the basis of which the design finite element model is developed (Figure 5.1a). The mathematical model of the construction of a vibration system is built on the following assumptions:

- The machine frame is a shaping surface and modeled by distributed parameters.
- Metal structures perceive only elastic deformations.
- The concrete mixture, which is on the shape-forming surface, is modeled by a system with distributed parameters.

In the equations of motion of the general system (vibration machine–concrete mixture), the constructive mass and mass of the concrete mixture are taken into account. When creating a computer model of the investigated system, principles are applied that will ensure the simplicity and adequacy of the model, as well as the possibility of further research—the solution of other types of problems.

The study is carried out in three stages. At the first stage, a static analysis of the stress–strain state of the structure under the action of all external forces (static analysis) is carried out in the nonlinear theory. At the second stage, based on the method of modes analysis, the basic shapes and frequencies of oscillations are determined. The third stage of the research is performed using dynamic analysis in the implementation of one of the forms of vibration (transient analysis), which is defined in the model analysis. The dynamic analysis determines the oscillation amplitude of the structure in different regions (sections) and estimates the change in the amplitude–frequency spectrum of the vibration action. The experimental model of the vibration unit is developed on the basis of studies of the computational model (Iskovich-Lototsky et al., 2019; Nazarenko, Dedov, & Zalisko, 2017; Nazarenko, Gaidaichuk, Dedov, & Diachenko, 2017). The vibration experimental unit is made of metal. The design consists of a welded tubular frame with forming surfaces. The frame rests on rubber elastic supports.

A general view of the experimental unit with the measurement system is shown in Figure 5.1b (Kukharchuk et al., 2016; Kukharchuk, Bogachuk, Hraniak, & Wójcik et al., 2017; Kukharchuk, Kazyv, & Bykovsky, 2017; Vedmitskyi et al., 2017).

For the study of the vibration unit, records of the continuous fixation of the distribution of active oscillations of the forming surface were used. Data readings from sensors and their subsequent processing are carried out using the developed circuit based on a 32-bit controller with two independent analog–digital converters. Such a system provides the speed of sampling signals from 20 kHz sensors. Data processing is carried out using a computer. To control the excitation frequency and the unbalance position in space, the vibration exciters are equipped with unbalance position sensors (Figure 5.2a). To determine the amplitude of oscillations, inductive-type displacement

Experimental research of forming machine 55

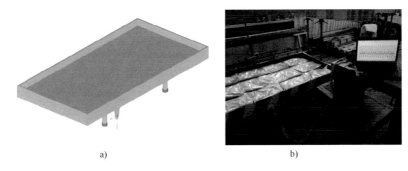

Figure 5.1 Vibration machine for the formation and compaction of concrete mixtures: (a) calculation of 3D model; (b) experimental vibration unit.

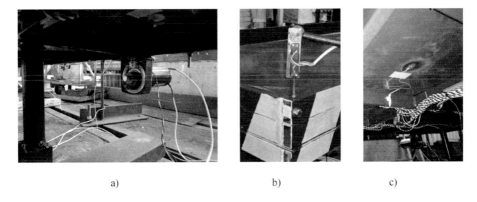

Figure 5.2 Sensor: (a) unbalance position; (b) displacements; (c) deformations.

sensors are used (Figure 5.2b). The condition of the deformations of the forming surfaces is monitored using strain sensors (Figure 5.2c).

Defined points for measuring the parameters of the vibration process of the experimental installation are shown in Figure 5.3. The distribution of vibration amplitudes over the surface of the plate is analyzed from the values of the displacement of the nodes of the finite element grid in these sections (Kukharchuk, Bogachuk, Hraniak, Wójcik et al., 2017; Kukharchuk, Kazyv, & Bykovsky, 2017; Vedmitskyi et al., 2017; Wójcik et al., 2008).

5.4 RESULTS AND DISCUSSION

Figure 5.4 shows the oscillogram of the block movement under the action of vibration in the transient process. The presence of a large number of harmonics with different degrees of influence on the process of movement is noted. This is clearly documented by sensors located on the vibration unit in the corresponding locations (see Figure 5.3).

An important factor in the behavior of the block in the mode from 12.5 to 18.6 Hz, where there are explicitly obtained results of the transition regime, is noted in Figure 5.5.

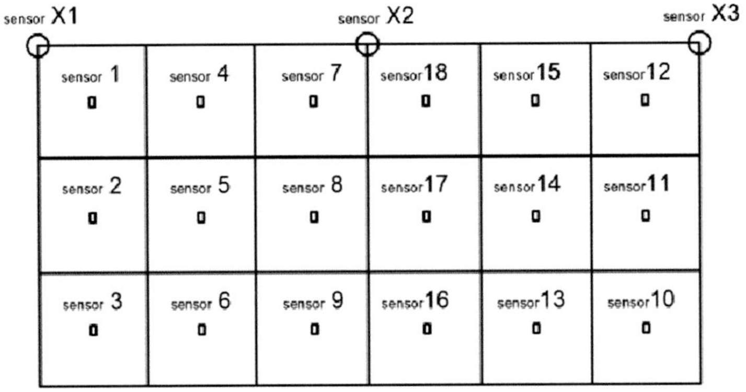

56 Mechatronic Systems 2

sensor X1		sensor X2		sensor X3	
sensor 1	sensor 4	sensor 7	sensor 18	sensor 15	sensor 12
sensor 2	sensor 5	sensor 8	sensor 17	sensor 14	sensor 11
sensor 3	sensor 6	sensor 9	sensor 16	sensor 13	sensor 10

Figure 5.3 Layout of measurement sensors.

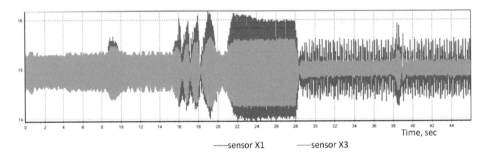

Figure 5.4 Oscillogram of vibration unit movement of the transition process.

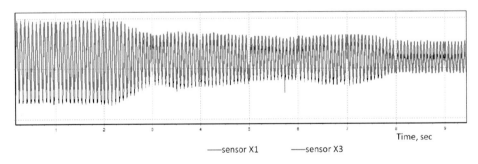

Figure 5.5 Oscillogram of vibration unit movement of the transition process mode from 12.5 to 18.6 Hz.

After processing the received oscillograms of vibration unit movements, the main oscillation frequencies are determined. Also, the waveforms that are carried out at these frequencies are analyzed. So, with an excitation frequency of 12.5 Hz, the forming surfaces perform vertical oscillations (Figure 5.6). The movement of the unit occurs in the common mode, which indicates the realization of the oscillation form when the whole structure moves progressively in the vertical direction (Figure 5.7).

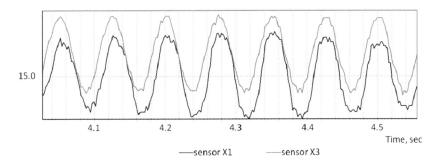

Figure 5.6 Oscillogram of vibration unit movement of the transition process at excitation frequency of 12.5 Hz.

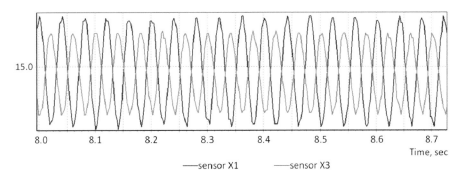

Figure 5.7 Oscillogram of the vibration unit movement at an oscillation frequency of 24.3 Hz.

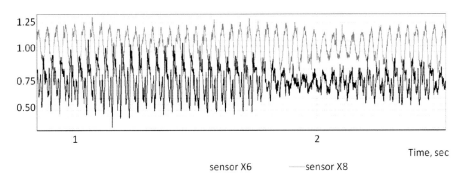

Figure 5.8 Deformations of the forming surfaces at an oscillation frequency of 24.3 Hz (sensors 6, 8).

To assess the stress–strain state of forming structures, the data of strain sensors are analyzed. Based on the obtained results, the following can be noted. When implementing the operating mode at a frequency of 24.3 Hz, a complex stress–strain state arises in the forming surface. This is evidenced by various forms and values of deformation in the corresponding parts of the surface. It should also be noted about the presence of wave phenomena (Figures 5.8–5.10), which occur in the forming surface.

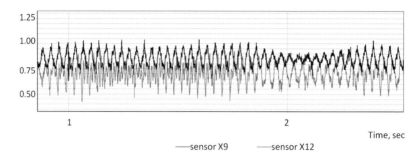

Figure 5.9 Deformations of the forming surfaces at an oscillation frequency of 24.3 Hz (sensors 9, 12).

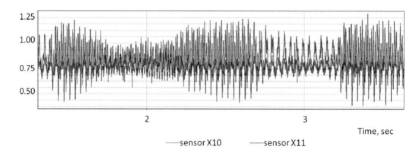

Figure 5.10 Deformations of the forming surfaces at an oscillation frequency of 24.3 Hz (sensors 10, 11).

5.5 CONCLUSION

Based on preliminary calculations and modeling of frame-bearing elements by beam finite elements, which were elastically deformed under the action of longitudinal force, with bending moments in two planes, and torque, an experimental model of a vibration unit with active forming surfaces was developed. In the investigation of the system, principles were applied that ensured the model adequacy, as well as the possibility of further research, solving other types of problems. The main oscillation frequencies were determined, which were realized at 12.50, 18.60, and 24.30 Hz. The study and determination of stresses and strains in time confirmed the hypothesis of a significant effect of structural features of the shaping structure on the compaction process. The numerical values and the nature of stress distribution in the shaping structure were obtained, depending on the angle of the instantaneous action of the external force of vibrators. Conditions for the implementation of phased and antiphase directions of stresses and acting external force were determined. At the same time, oscillation forms with the complex movement of forming surfaces were realized. The presence of wave phenomena in the forming surface was experimentally proven when implementing modes of operation at the main frequencies of oscillations. The amplitudes of oscillations of the unit in the range of 0.0006–0.0003 m were determined at the excitation frequencies of 18.60 and 24.30 Hz.

REFERENCES

Andò, B., Bagilo, S., Bulsara, A. R., Marletta, V., & Pistorio, A. 2015. Experimental and theoretical investigation of a nonlinear vibrational energy harvester. *Procedia Engineering 120*:1024–1027.

Bendjama, H., Bouhouche, S., & Boucherit, M. S. 2012. Application of wavelet transform for fault diagnosis in rotating machinery. *International Journal of Machine Learning and Computing 2*(1): 82–87.

Bui, T. Q., Doan, D. H., Van Do, T., Hirose, S., & Duc, N. D. 2016. High frequency modes meshfree analysis of Reissner-Mindlin plates, *Journal of Science: Advanced Materials and Devices 1*(3): 400–412.

Dedov, O. 2018. Determining the influence of the environment on the dynamics of the machine on the basis of spectral analysis, *Control, Navigation and Communication Systems 4*(50): 69–72.

Ghandchi, T. M., Wilmshurst, L., & Elliott, S. J. 2013. Receptance method for active vibration control of a nonlinear system, *Journal of Sound and Vibration 332*(19): 4440–4449.

Iskovich-Lototsky, R., Kots, I., Ivanchuk, Y., Ivashko Y., Gromaszek, K., Mussabekova, A., & Kalimoldayev, M. 2019. Terms of the stability for the control valve of the hydraulic impulse drive of vibrating and vibro-impact machines, *Przegląd Elektrotechniczny 4*(19): 19–23.

Jia, Y. & Seshia, A. A. 2014. An auto-parametrically excited vibration energy harvester. *Sensors and Actuators A: Physical 220*: 69–75.

Kavyanpoor, M. & Shokrollahi, S. 2019. Dynamic behaviors of a fractional order nonlinear oscillator, *Journal of King Saud University-Science 31*(1): 14–20.

Kozlov, L. G., Polishchuk, L. K., Piontkevych, O. V., Korinenko, M. P., Horbatiuk, R. M., Komada, P., & Ussatova, O. (2019). Experimental research characteristics of counter balance valve for hydraulic drive control system of mobile machine, *Przeglad Elektrotechniczny 95*(4): 104–109.

Kukharchuk, V. V., Bogachuk, V. V., Hraniak, V. F., Wójcik, W., Suleimenov, B., & Karnakova, G. 2017. Method of magneto-elastic control of mechanic rigidity in assemblies of hydropower units, *Proc. SPIE 10445*: 104456A.

Kukharchuk, V. V., Hraniak, V. F., Vedmitskyi, Y. G., Bogachuk, V. V., Zyska, T., Komada, P., & Sadikova, G. 2016. Noncontact method of temperature measurement based on the phenomenon of the luminophor temperature decreasing, *Proc. SPIE 10031*: 100312F.

Kukharchuk, V. V., Kazyv, S. S., & Bykovsky, S. A. 2017. Discrete wavelet transformation in spectral analysis of vibration processes at hydropower units. *Przegląd Elektrotechniczny 93*(5): 65–68.

Nazarenko, I. I., Dedov, O., & Zalisko, I. 2017. Research of stress-strain state of metal constructions for static and dynamic loads machinery. *The IX International Conference Heavy Machinery HM 2017*, Zlatibor: Serbia, B: 13–14.

Nazarenko, I. I., Gaidaichuk, V., Dedov, O., & Diachenko, O. 2017. Investigation of vibration machine movement with a multimode oscillation spectrum. *Eastern-European Journal of Enterprise Technologies* (Восточно-Европейский журнал передовых технологий) 6(1), 28–36.

Nazarenko, I. I., Sviderskiy, A. T., & Dedov, O. P. 2012. Design of new structures of vibroshocking building machines by internal characteristics of oscillating system. *Research & Development 18*(2): 22–36.

Nesterenko, M. P. 2015. A vibrant development of vibration plants with spacious spikes for the formation of high-grade concrete cutting down (Prohresyvnyi rozvytok vibratsiynykh ustanovok z prostorovymy kolyvanniamy dlia formuvannia zalizobetonnykh vyrobiv), *Zbirnyk naukovykh prats. Ser.: Haluzeve mashynobuduvannia, budivnytstvo 2*(44): 177–181.

Nesterenko, M., Nesterenko, T., & Skliarenko, T. 2018. Theoretical studies of stresses in a layer of a light-concrete mixture, which is compacted on the shock-vibration machine. *International Journal of Engineering & Technology 7*(3.2): 419–424.

Pawelczyk, M. & Wrona, S. 2016. Impact of boundary conditions on shaping frequency response of a vibrating plate - modeling, optimization, and simulation, *Procedia Computer Science 80*: 1170–1179.

Polishchuk, L. K., Kozlov, L. G., Piontkevych, V., Gromaszek, K., & Mussabekova, A. 2018. Study of the dynamic stability of the conveyor belt adaptive drive, *Proc. SPIE 10808*: 1080862.

Vedmitskyi, Y. G., Kukharchuk, V. V., & Hraniak, V. F. 2017. New non-system physical quantities for vibration monitoring of transient processes at hydropower facilities, integral vibratory accelerations. *Przeglad Elektrotechniczny 93*(3): 69–72.

Wójcik, W. & Kisała, P. 2009. The application of inverse analysis in strain distribution recovery using the fibre bragg grating sensors, *Metrology and Measurement Systems 16*(4), 649–660.

Wójcik, W., Lach, Z., Smolarz, A., & Zyska, T. 2008. Power supply via optical fibre in home telematic networks, *Przegląd Elektrotechniczny 84*(3): 277–279.

Yamamoto, G. K., da Costa, C., & da Silva Sousa, J. S. 2016. A smart experimental setup for vibration measurement and imbalance fault detection in rotating machinery. *Case Studies in Mechanical Systems and Signal Processing 4*(1): 8–18.

Yue-min, Z., Chu-sheng, L., Xiao-Mei, H., Cheng-Yong, Z., Yi-bin, W., & Zi-ting, R. 2009. Dynamic design theory and application of large vibrating screen. *Procedia Earth and Planetary Science, 1*, 776–784.

Chapter 6

Research of ANSYS Autodyn capabilities in evaluating the landmine blast resistance of specialized armored vehicles

S. Shlyk, A. Smolarz, S. Rakhmetullina, and A. Ormanbekova

CONTENTS

6.1 Introduction 61
6.2 Mathematical model of the vehicle body explosive load 62
6.3 Development of the finite element model: experimental and simulation results ... 68
6.4 Results and research of the vehicles body destruction mechanism 69
6.5 Conclusions 74
References 74

6.1 INTRODUCTION

According to the United Nations Office for the Coordination of Humanitarian Affairs (OCHA), the Convention on the Prohibition of the Use, Stockpiling, Production and Transfer of Anti-Personnel Mines and on their Destruction since 1997 did not reduce their global application, as it is estimated that more than 100 million mines are installed in 60 countries around the world. Landmines (including anti-vehicle mines) are also widely used in modern conflicts (Aslybek et al. 2017). To date, the actual standard for armored military vehicles is in compliance with the MRAP class (Mine Resistant Ambush Protected). The NATO STANAG 4569 standardization agreement defines crew protection necessary to the MRAP class vehicles in the event of a landmine blast (charge weight of 6 kg in TNT equivalent) under any of the wheels or tracks and under the hull center (NATO AEP-55).

Finite element analysis (FEA) is widely used in research related to defense industries, such as high-speed collision and penetration. Numerical simulation of processes allows obtaining additional information on complex physical phenomena, which is not available in experimental research methods. ANSYS Autodyn is an analytical tool for solving problems with explicit statements for simulation of complex nonlinear dynamics of solids, liquids, gases, and their interaction. It represents a powerful tool for interdisciplinary calculations in problems with explicit statements, which provides a wide range of possibilities for simulation, including high-speed strikes or explosions (Trotsko & Shlyk 2018).

The purpose of the work is a theoretical assessment of the structural strength of the armored body of specialized vehicles KrAZ "Shrek" and KrAZ "Fiona" against the explosive load caused by explosive blasts with mass 6, 8, 10, 14, and 20 kg in TNT equivalent.

DOI: 10.1201/9781003225447-6

6.2 MATHEMATICAL MODEL OF THE VEHICLE BODY EXPLOSIVE LOAD

KrAZ "Shrek" and KrAZ "Fiona" (Figure 6.1) are a family of Ukrainian armored vehicles with a V-hull developed by PJSC "AutoKrAZ" in cooperation with the STREIT Group based on the KrAZ-5233BE. The vehicles are developed following the MRAP standard and were introduced in 2014. The vehicles are intended for the prompt delivery of military personnel and their fire support. In addition, they can be used as carriers for various weapons and military equipment.

The assessment of resistance to anti-vehicle mine blasts was carried out in accordance with the NATO AEP-55 STANAG 4569 standardization agreement using the finite element method (FEM) simulation. The simulation system includes atmospheric air, a 3D model of the vehicle body (Figure 6.2), the explosive charge, and the soil array. The locations of charges relative to the vehicle body in the model are shown in Figure 6.3.

The process of detonation is numerically described by the general system of differential equations. The material models play an important role in relating the deformation stress and internal energy. Liquids and gases (in this study, the detonation products and air) are sufficiently modeled by the equation of state (EOS), which expresses the relationship between pressure p, specific volume V, and specific energy e. Additional equations are required for solids simulation (in the presented study represented by the body material) since solids possess a shear resistance.

The loess loam is used as the soil in the calculation model. It is the loam species, which is characterized by the high content of clay particles, the presence of coarse sand, and (less) pebble material. To determine the soil model, the data obtained in (Fiserova 2006, Laine & Sandvik 2001, Mahdi & Banadaki 2010) were used. The model used in this study was obtained for sand with a moisture content of 6.57% by the three-dimensional compression, which made it possible to measure the velocity of the waves in the sand sample.

The atmospheric air was modeled using the ideal gas EOS. The initial density was $1.3\,kg/m^3$ and the internal energy was 192.31 kJ/kg, which is equal to an atmospheric pressure of 100 kPa (one atmosphere) at 0°C. In addition, an alternating pressure was given, which allows achieving zero pressure in the air during the simulation. This makes it possible to avoid undesired starting velocities (Laine et al. 2001).

Figure 6.1 General view of the KrAZ "Shrek" and KrAZ "Fiona" multipurpose vehicles.

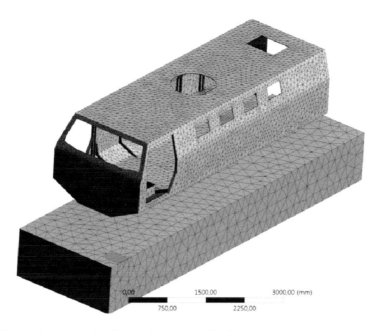

Figure 6.2 General view of the finite element model (the explosive charge is located under the vehicle's left front wheel; atmospheric air is not shown.

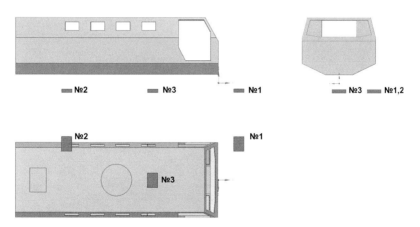

Figure 6.3 Location of the explosive charge during the simulation: 1 – under the left front wheel; 2 – under the left rear wheel; 3 – under the vehicle center.

Properties of the applied explosive model (trinitrotoluene) are specified according to (Mahdi & Banadaki 2010, Vorobyov et al. 2017). When considering the process of loading and interaction of flat material (plate) with a shockwave formed during the detonation of an explosive, the process is considered as two-stage: (1) the acquisition of the initial velocity at the passage of the shockwave on the surface and its output on the free surface; (2) further acceleration of the plate under the pressure of detonation products of the explosive.

The material of the plate is divided into elements with a mass concentrated in one point of reduction. Those elements are interconnected by elastic–plastic joints. The plate elements equation of motion under the action of detonation products pressure can be written in the equation:

$$Dm\frac{d^2 x_1(t)}{dt^2} + \rho c \frac{dx_1(t)}{dt} = P(t), \qquad (6.1)$$

where m = mass of the square plate unit element; dx_1/dt = velocity of the plate in the direction of displacement; ρ_1, c = the density and speed of sound in the environment behind the plate; $P(t)$ = pressure, which describes the detonation products action on the plate.

The detonation front pressure changes according to the equation:

$$Dm\frac{d^2 x_1(t)}{dt^2} + \rho c \frac{dx_1(t)}{dt} = P(t), \qquad (6.2)$$

where P_0 = the pressure of the detonation products on the plate at the moment the shockwave is released onto the free surface; θ = the time constant for the decline of detonation products; t = the time of the process.

The solution of (1) when neglecting the second term has the following form:

$$\dot{x}_1(t) = \frac{P_0 \theta}{m}\bigl(1 - \exp(-t/\theta)\bigr) + v_o, \qquad (6.3)$$

where v_0 = the initial velocity of the plate element at the moment the shockwave is released onto the free surface.

To determine P_0 and $\dot{x}_0$ the Chapman–Jouguet condition (Dragobetskii et al. 2015, 2017) for the pressure at the detonation front is used:

$$P_\mathcal{H} = \rho_0 D^2 / (k+1) \qquad (6.4)$$

where ρ_0 = the initial density of the explosive; D = detonation velocity of the explosive; k = indicator of the explosive adiabatic.

The pressure on the plate to the pressure of the detonation products ratio is determined by the following equation:

$$P_x / P_\mathcal{H} = \bigl[0.5(3k-1)k^{-1}\bigr]^{2k/(k-1)} \qquad (6.5)$$

At the moment the shockwave is released onto the free surface, the pressure of the detonation products on the plate is determined by the following equation:

$$P_x = \bigl[0.5(3k-1)k^{-1}\bigr]^{2k/(k-1)} P_\mathcal{H} \bigl[H \cdot D^{-1}(H/D + t_o)\bigr] \qquad (6.6)$$

where $t_0 = \delta_M/c_M$; t_0 = the time of a shockwave passing on a plate; c_M = the shockwave's velocity on the plate; δ_M = plate thickness.

$$C_M = a + \lambda u_x \qquad (6.7)$$

where u_x – the mass velocity of the plate; a – the speed of sound; λ – the shock compression coefficient of the plate material.

The shockwave velocity at the moment of release onto the plate's free surface is determined by the ratio:

$$c_M = P_o \cdot (\rho_M \cdot u_x)^{-1}. \qquad (6.8)$$

The system of equations for calculating Px, u_x, and c_M at the moment the shockwave is released onto the plate's free surface has the following form:

$$\begin{cases} P_x = \left[0.5(3k-1) \cdot k^{-1}\right]^{\frac{2k}{k-1}} P\left[H \cdot D^{-1}(H/D+t_0)^{-1}\right]^k \\ c_M = a + \lambda u_x \\ t_0 = \delta_M / (a + \lambda u_x) \\ P_x = \rho_M u_x \cdot c_M \end{cases} \qquad (6.9)$$

At the explosive charge blast, the maximum pressure at the front of the shockwave is determined by the following empirical formula (Dragobetskii et al. 2015, 2019):

$$P_m = \frac{1{,}08}{(r_o)^{1.08}} \cdot 10^4, \left(0.0773 \cdot 10^{-4} \le r \le 1.082 \cdot 10^{-4}\right) \qquad (6.10)$$

where $r_o = r/\sqrt{q}$ = reduced explosion distance; r = explosion distance, m; q = explosion energy per unit of length, J/m.

The change in pressure, depending on the location of the shockwave's front point, is approximated by the dependence:

$$P_m = \left(31.14 + 89.86\lambda + 380.69\lambda^2\right) \cdot 10^5 \qquad (6.11)$$

where λ is the length of the arc along the shockwave's front from a given front point assigned to the length of an arc from the axis of the charge toward the front point equidistant from both ends.

According to the foregoing, the field of the shockwave's peak pressures can be described with sufficient accuracy by the following formula:

$$P_m = 0.1241 \cdot 10^7 \left(\frac{q}{q_0}\right)^{0.572} \cdot r^{-1.144}$$
$$+ \left[0.1166 \cdot 10^8 \cdot \left(\frac{q}{q_0}\right)^{0,4} \cdot r^{-0,805} - 0.1241 \cdot 10^7 \left(\frac{q}{q_0}\right)^{0,572} \cdot r^{-1.144}\right] \cdot \lambda \qquad (6.12)$$

The body of the vehicles is the integral type supporting structure body, which is assembled from 8 mm Quardian 500 steel armored plates. Quardian 500 is a protective steel sheet that combines high ballistic resistance with high strength and an average

Brinell hardness of 500 HBW. It is used in the public sphere (protection of embassies, government and public buildings, banking institutions), special protection, and military applications (armoring of helicopters, boats, demolition vehicles, and armored personnel carriers). The main characteristics of Quardian 500 steel are given in Table 6.1.

Considering the vehicle's flat body surface, we cover it with the spatial Lagrangian mesh, which is associated with the median surface, and use the "node scheme," in which all the required quantities are determined at the nodal points of the calculated grid. It should be noted that with such a construction of the node scheme, the second derivatives with respect to the spatial coordinates are approximated satisfactorily. In conditions of complex loading, the formation process is associated with a transition from a stationary deformation to a nonstationary one. In a nonstationary deformation, the change in the Lagrange coordinates of the mesh element during the formation process is accompanied by a simultaneous change in the deformation cell. As a result, the magnitude of the stresses in the mesh element does not change in the deformation process. The material is considered isotropic elastic–plastic with hardening. The mesh acceleration in the median plane, its velocity, and displacement are determined based on the equilibrium equations for each node (Trotsko & Shylyk 2018, Vorobyov et al. 2017]:

$$\begin{cases} \nabla_\gamma M_{mn}^{\beta\alpha} - Q_{mn}^\beta R_{\gamma mn}^\beta + P_{mn}^\alpha + T_{mn}^\alpha + S_{mn}^\alpha = \bar{\rho}\ddot{X}_{mn}^\alpha - \rho\dot{X}_{mn}^\alpha c \\ M_{mn}^{\beta\alpha} R_{\beta\alpha}^{mn} + \nabla_\beta Q_\beta^{mn} + P_{mn}^3 + T_{mn}^3 + S_{mn}^3 = \rho\ddot{X}_{mn}^3 - \rho\dot{X}_{mn}^3 c , \\ \nabla_\beta L^{\alpha\beta} - Q_{mn}^\alpha = 0 \end{cases} \qquad (6.13)$$

where ∇_β – the sign of a covariant differentiation; M_{mn} – membrane forces; L – bending moments; Q_{mn}^β – cutting forces; $\bar{\rho}$ – reduced weight; $\ddot{X}_{mn}^j$ – acceleration; P_{mn}^j – force effect of loading; T_{mn} – friction forces in the peripheral zone of the element; S_{mn} – braking forces of the resistance elements; P_{mn} – forces acting in the mesh element from the elements of resistance; R_{mn} – curvature tensor; c – speed of sound in the environment (Polishchuk et al. 2018, Kozlov et al. 2019).

Taking into account the fact that the formation process is associated with force intensification, it is more appropriate to submit the equilibrium equation in the forces and moments as shown in Figure 6.4.

The values of forces and moments acting on each element are determined by the system of equations (Dragobetskii et al. 2017):

Table 6.1 Mechanical properties of the steel quardian 500

Hardness, brinell	Impact strength (cutting, $-40°C$ (min.)) (J/m^3)	Yield stress (MPa)	Ultimate tensile strength (MPa)	Elongation A50 (%)
480–540	24	1,200	1,450–1,800	8

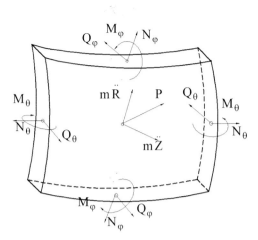

Figure 6.4 Forces in the element of the material.

$$\begin{cases} M_{mn}^{\alpha\beta} = \int_{-0,5\delta}^{+0.5\delta} \begin{bmatrix} \sigma_{\alpha 1}^{mn}\left(\delta_1^\beta - x^3 B_{1mn}^\beta\right) + \\ +\sigma_{\alpha 2}^{mn}\left(\delta_2^\beta - x^3 B_{2mn}^\beta\right) \end{bmatrix} \left(G_{mn} \cdot A_{mn}^1\right)^{0.5} \\ L_{mn}^{\alpha\beta} = \int \begin{bmatrix} \sigma_{\alpha 1}^{mn}\left(\delta_1^\beta - x^3 B_{1mn}^\beta\right) + \\ +\sigma_{\alpha 2}^{mn}\left(\delta_2^\beta - x^3 B_{2mn}^\beta\right) \end{bmatrix} \cdot \left(G_{mn} \cdot A^{-1}\right)^{0.5} x^3 dx^3 \\ Q_{mn}^\alpha = \frac{\partial L_{mn}^{\alpha\beta}}{\partial x^\beta} + \overline{A}_{mn}^\alpha \cdot \frac{\partial \overline{A}_1^{mn}}{\partial x^\beta} \cdot L_{mn}^{\gamma\beta} + \overline{A}_{mn}^\beta \frac{\partial \overline{A}_\gamma^{mn}}{\partial x^\beta} \cdot L_{mn}^{\alpha\beta} \end{cases} \quad (6.14)$$

where $A_{\alpha\beta}^{mn}$ – the metric tensor.

The acceleration of the grid nodes median surface in the next field of integration should be determined by the equation:

$$\ddot{x}_{mn}^j = \frac{A_{mn}^{0.5}}{\overline{\rho}_o} \cdot \left(P_{mn}^j + T_{mn}^j + S_{mn}^j + \Pi_{mn}^j\right) + \frac{A_{mn}^{0.5}}{\overline{\rho}_o} \cdot \left(\frac{\partial V_{mn}^{\beta j}}{\partial x^\beta} + \overline{A}_{mn}^\beta \frac{\partial \overline{A}_\gamma^{mn}}{\partial x^\beta} \cdot V^{\gamma j}\right), \quad (6.15)$$

where A_{mn} – the metric tensor determinant; $V_{mn}^{\alpha j}$ – the space-surface tensor.

The time interval Δt is chosen from the stability condition of the computational process (Dragobetskii et al. 2017):

$$\Delta t \le \Delta X_{5,j,0}\left[\rho_3^k(1-v_k)^2(E)^{-1}\right]^{0.5} \quad (6.16)$$

where ρ_3^k – the body material density; v_k – Poisson's ratio; E – Young's modulus of the body material.

To build the "stress–deformation" diagram that is specified as dependencies in describing the material in the ANSYS library, the following expressions were obtained:

$$\frac{\partial^2 \sigma_i}{\partial \varepsilon_i^2} = n(n-1)C\varepsilon_i \qquad (6.17)$$

$$\sigma_2 = \sqrt{\sigma_1 \cdot \sigma_3} \qquad (6.18)$$

$$\sigma_i = \frac{3}{2} H \varepsilon_i^n / \left[029 - 2D + \ln C_3 \left(n_c + 0.75 \varepsilon_i^n \right) \varepsilon_i^n \right] \qquad (6.19)$$

$$n = \frac{\lg \sigma_3 - \lg \sigma_1}{\lg(\varepsilon_o + \varepsilon_3) - \lg(\varepsilon_o + \varepsilon_1)} \qquad (6.20)$$

$$C = \frac{\sigma_1}{(\varepsilon_o + \varepsilon_i)^n} \qquad (6.21)$$

$$\begin{cases} \sigma_i = C(\varepsilon_o + \varepsilon_i)^n \\ \sigma_1 = C(\varepsilon_o + \varepsilon_1)^n \\ \sigma_2 = C(\varepsilon_o + \varepsilon_2)^n \\ \sigma_3 = C(\varepsilon_o + \varepsilon_3)^n \end{cases}, \qquad (6.22)$$

where σ_1, σ_2, σ_3, ε_1, ε_2, ε_3 are the current values of stress and deformations passing through the points of the actual stresses curve; ε_1, C, n – the constants that satisfy the system of equations (6.22); σ_3, ε_1 – the stress and the strain intensities.

6.3 DEVELOPMENT OF THE FINITE ELEMENT MODEL: EXPERIMENTAL AND SIMULATION RESULTS

An important criterion that determines the validity of the results obtained by the simulation and the adequacy of the developed mathematical model to the real process is

Figure 6.5 Experimental study of the Quardian 500 plate dynamic explosive load; (a) – the Quardian 500 plate with a mounted explosive charge, striker plate, and detonator; (b) – deformation of the Quardian 500 plate by the blast.

the correspondence with data obtained as a result of experimental tests. For experimental studies, a series of 8 mm thick Quardian 500 steel plates 500×500 mm in size was used. The plates were placed in sandy soil (Figure 6.5) and subjected to a dynamic load by an explosion of Amatol and Ammonium Nitrate mixture with an aggregate capacity of 10 kgs in a TNT equivalent. In order to better transfer the explosive load onto the Quardian 500 steel plate and ensure the possibility of more accurate deflections readings, a 5 mm thick striker plate made from Mangalloy (Hadfield steel) in the size of 125×375 mm was used. The striker plate was located on test plates under the explosive charges. During the experiment, the dynamic displacement values of the deformed plate were recorded (Dragobetskii et al. 2018, 2019, Polishchuk et al. 2018).

In addition, a numerical simulation of the Quardian 500 steel plate explosive load was performed in the ANSYS Autodyn system using the data, dependencies, and assumptions described in Section 1. The simulation system included the atmospheric air, explosive charge (9.9 kg of TNT), a striker plate, Quardian 500 plate, and soil. The relative error of the deflection differences values at the control points obtained by the field and numerical experiments is within the precision of the numerical method solution and makes up no more than 1.66% in the middle area of the plate. Thus, the developed mathematical model for the numerical solution of explosive loading allows to simulate the process of the Quardian 500 test plate explosive load with high precision and can be used for estimation of landmine blast resistance of the KrAZ "Shrek" and KrAZ "Fiona" armored MPVs body.

The shape of the finite elements for the explosive charge, soil, and the air was generated automatically; for the vehicle body, the shape of elements is tetrahedron generated by the Patch Conforming algorithm. Parameters of the developed finite element model are given in Table 6.2. The general view of the simulation model with the explosive charge located under the left front wheel is shown in Figure 6.2.

The explosive charge mass in the model was determined by changing the size and volume of the charge model. In the model, the Flow Out boundary was used as the boundary condition. An additional condition in the form of gravitation, the vector of which is directed along the Z-axis and opposite to it, is also applied to the whole system (accounting for gravity is important at the stage of calculation of soil emissions and movement of the body under the action of an explosive wave). A restriction to the movement in the direction opposite to the Z-axis is also applied to the lower edge of the V-shaped body's bottom, which imitates the chassis on which the body is mounted and prevents the body from falling down under the gravitation influence from the moment of the test's commencement.

The selective results of the explosive load simulation of the KrAZ "Shrek" and KrAZ "Fiona" body are presented in Figure 6.6 and Table 6.3.

6.4 RESULTS AND RESEARCH OF THE VEHICLES BODY DESTRUCTION MECHANISM

The theoretical assessment of the KrAZ "Shrek" and KrAZ "Fiona" vehicles' landmine blast resistance allows predicting the destruction of the body in the event of an explosion of a charge with a mass of 8, 10, and 20 kg under the left rear wheel. Figure 6.7 shows the locations of the body destruction areas relative to the initiation points of the explosive's detonation.

Table 6.2 Simulation model parameters

Object (material)	Solver type	Overall dimensions			Mass-dimensional properties		FE – mesh parameters	
		X-axis (mm)	Y-axis (mm)	Z-axis (mm)	Volume (mm)	Mass (kg)	Nodes	Elements
Vehicle body (Quardian 500)	Lagrangian	2,126	5,876	1,607	$5.1861 \cdot 10^8$	4,060.7	19,329	58,852
Soil	Lagrangian	2,304	7,876	1,000	$1.8134 \cdot 10^{10}$	44,247	1,881	7,848
Air, atm	Eulerian	2,304	7,876	3,011	$5.4639 \cdot 10^{10}$	66,932	6,840	5,698

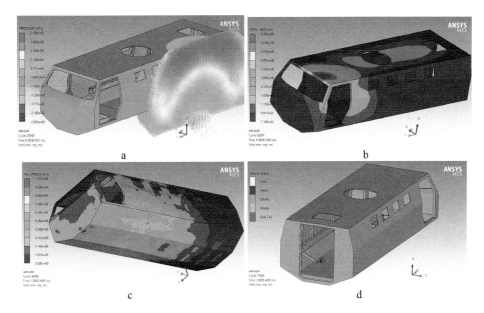

Figure 6.6 Selective simulation results (charge mass and its location; time after detonation): (a) diagram of the explosive pressure distribution on the vehicle body (the shockwave is shown) (14 kg under left rear wheel; 0.8 ms); (b) diagram of the body's total deformations (10 kg under left front wheel; 1.6 ms); (c) diagram of the equivalent (von Mises) stress on the vehicle's body (20 kg under the vehicle center; 1.2 ms); (d) destruction of the vehicle's body (10 kg under left rear wheel; 2 ms).

However, the destruction does not occur in the case of a charge mass of 14 kg. The formation of irreversible plastic deformations exceeding the body material yield stress requires constant or increasing load for a certain time. As it can be seen from the systems of equations (6.9), (6.14) and expressions (6.10) and (6.12), such loads are possible provided that the vector of the largest explosive wave's pressure is located perpendicular to the loaded surface.

In addition, it is important to note that the body moves under the influence of shockwave propagation. Given the scalar values of the body displacement velocity, the lateral surface of the body's bottom constantly moves relative to the detonation point. Constructing a perpendicular from the body's bottom to the detonation point, we can see that the perpendicular constantly passes through the stress formation area (Figure 6.8). The body material destruction is possible when its stress gradually begins to exceed the body material's yield stress limit. Thus, the condition for the body's irreversible deformations and its destruction is fulfilled.

From the diagrams in Figure 6.8, it can be seen that in case of detonation of an explosive with a mass of 14 kg at the time of 1.2 ms after detonation, the perpendicular passes almost past the lateral surface of the bottom. At a time of 1.6 ms, it passes through the lateral surface, in connection with which it creates an area of stress concentration; and further, it shifts down and back. The maximum achieved stress reaches 1,052,106 kPa (1,052 MPa), which does not exceed the specified material yield stress boundary of 1,200 MPa and is not sufficient for its destruction.

72 Mechatronic Systems 2

Table 6.3
The largest calculated deformations (mm) of the vehicle's body and its condition

Deformation zones	Level of threat according to NATO AEP-55 STANAG 4569 and its location															
	Left front wheel					Left rear wheel					Hull center					
	6 kg	8 kg	10 kg	14 kg	20 kg	6 kg	8 kg	10 kg	14 kg	20 kg	6 kg	8 kg	10 kg	14 kg	20 kg
L. board	1.58	2.23	3.11	6.45	7.80	2.37	13.5	22.9	9.70	12.2	9.27	11.1	13.6	14.3	19.9
R. board	0.20	3.75	3.07	2.46	5.28	2.02	2.58	4.01	7.08	7.27	9.27	11.1	13.6	14.3	19.9
Roof	4.16	7.26	9.32	16.7	19.4	10.2	28.1	30.1	27.5	34.7	2.29	5.01	4.49	5.12	29.9
Bottom	0.95	1.38	1.63	2.80	3.47	2.67	3.06	15.4	10.2	14.6	13.5	17.7	24.3	25.3	46.7
Destruction	No	No	No	No	No	No	Yes	Yes	No	Yes	No	No	No	No	No

Research of ANSYS Autodyn capabilities 73

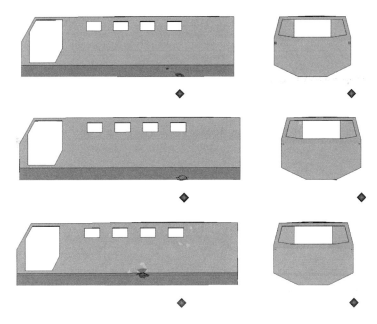

Figure 6.7 Location of the destruction areas of the body relative to the detonation initiation points of the explosive with a mass of: (a) 8 kg; (b) 10 kg; (c) 20 kg in TNT equivalent.

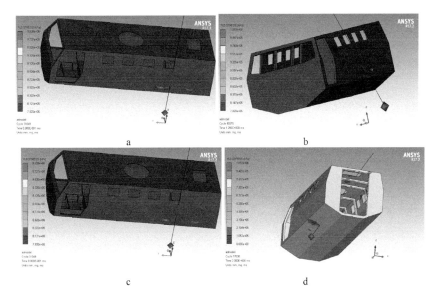

Figure 6.8 Location of the destruction areas of the body relative to the detonation initiation points of the explosive with mass in of: (a) 8 kg; (b) 10 kg; (c) 20 kg TNT equivalent.

6.5 CONCLUSIONS

As part of the study, the mathematical models of the simulation system's components for the KrAZ "Shrek" and KrAZ "Fiona" vehicle body's dynamic explosive load analysis were developed. Additionally, the mathematical apparatus was developed for calculating the shockwave parameters such as the detonation front pressure, its change in time, and the shockwave velocity at the time of its exit to the surface. The conditions and destruction mechanism of the vehicle's body exposed to explosive load were studied and theoretically substantiated.

REFERENCES

Aslybek Kyzy, G., Hofmann, U., Jung Y. & Rapillard, P. 2017. *Global Mapping and Analysis of Anti-Vehicle Mine Incidents in 2017*. Geneva: GICHD–SIPRI.

Dragobetskii, V., Shapoval, A., Mos'pan, D., Trotsko, O. & Lotous, V. 2015. Excavator bucket teeth strengthening using a plastic explosive deformation. *Metallurgical and Mining Industry* 4: 363–368.

Dragobetskii, V., Shapoval, A., Naumova, O., Shlyk S., Mospan, D. & Sikulskiy, V. 2017. The technology of production of a copper – aluminum – copper composite to produce current lead buses of the high–voltage plants. *IEEE International Conference on Modern Electrical and Energy Systems*: 400–403.

Dragobetskii, V., Zagirnyak, M., Naumova, O., Shlyk, S. & Shapoval, A. 2018. Method of determination of technological durability of plastically deformed sheet parts of vehicles. *International Journal of Engineering & Technology* 7(4.3): 92–99.

Dragobetskii, V., Zagirnyak, V., Shlyk, S., Shapoval, A. & Naumova, O. 2019. Application of explosion treatment methods for production items of powder materials. *Przegląd Elektrotechniczny* 95(5): 39–42.

Fiserova, D. 2006. Numerical analyses of buried mine explosions with emphasis on effect of soil properties on loading. *Cranfield University. Defence College of Management and Technology, Engineering Systems Department*.

Kozlov, L.G., Polishchuk, L.K., Piontkevych, O.V., Korinenko, M.P., Horbatiuk, R.M., Komada, P., Orazalieva, S. & Ussatova, O. 2019. Experimental research characteristics of counterbalance valve for hydraulic drive control system of mobile machine. *Przeglad Elektrotechniczny* 95(4): 104–109.

Laine, L., Ranestad, Ø., Sandvik, A. & Snekkevik, A. 2002. Numerical simulation of anti-tank mine detonations. Shock compression of condensed matter. *American Institute of Physics*: 431–434.

Laine, L. & Sandvik, A. 2001. Derivation of mechanical properties for sand. In *4th Asia-Pacific Conference on Shock and Impact Loads on Structures*: 361–368.

Mahdi, M. & Banadaki, D. 2010. Stress-wave induced fracture in rock due to explosive action. *Department of Civil Engineering, University of Toronto*.

NATO AEP-55 STANAG 4569. 2001. Protection levels for occupants of logistic and light armoured vehicles.

Polishchuk, L.K., Kozlov, L.G., Piontkevych, O.V., Gromaszek, K. & Mussabekova, A. 2018. Study of the dynamic stability of the conveyor belt adaptive drive. *Proc. SPIE 10808, Photonics Applications in Astronomy, Communications, Industry, and High-Energy Physics Experiments 2018* 1080862.

Polishchuk, L.K., Kozlov, L.G., Piontkevych, O.V., Horbatiuk, R.M., Pinaiev, B., Wójcik, W., Sagymbai, A. & Abdihanov A. 2019. Study of the dynamic stability of the belt conveyor adaptive drive. *Przegląd Elektrotechniczny* 95(4): 98–103.

Trotsko, O. & Shlyk, S. 2018. Development of the sheet blanks forming mathematical model for calculation using simulation in ANSYS software. *IEEE International Scientific and Technical Conference on Computer Sciences and Information Technologies (CSIT)*: 169–172.

Vorobyov, V., Pomazan, M., Vorobyova, L. & Shlyk, S. 2017. Simulation of dynamic fracture of the borehole bottom taking into consideration stress concentrator. *Eastern-European Journal of Enterprise Technologies* 3/1(87): 53–62.

Chapter 7

Phenomenological aspects in modern mechanics of deformable solids

V. Ogorodnikov, T. Arkhipova, M.O. Mokliuk, P. Komada, A. Tuleshov, U. Zhunissova, and M. Kozhamberdiyeva

CONTENTS

7.1 Introduction 77
7.2 Analysis of methods of dynamic and quasistatic loading conditions 78
7.3 Results and discussion 79
    7.3.1 Actual plasticity resource in the process of backward extrusion 79
    7.3.2 Experimental-computational component of the calculation model in the condition of road accidents 81
    7.3.3 Energy absorption of the reinforced concrete constructions 82
7.4 Conclusions 83
References 84

7.1 INTRODUCTION

Phenomenological theory of metals deformability without fracture, developed since the 1970s, enables, at the stage of the technological processes design, fracture forecasting and assessment to the used plasticity resource of the blank, worked mechanically in the conditions of the multiaxial load (Ogorodnikov et al., 2018).

The most important conditions are the phenomenological aspects of these problems. Among these conditions, the basic conditions are the additional parameters obtained experimentally and introduced in the calculation tool of the plasticity theory. Modern phenomenological criteria contain experimental data about the mechanical characteristics of the materials, thus a "sheet of the material" is formed (Ogorodnikov et al., 2005; Kukharchuk et al., 2016, 2017).

The aim of the research is to examine the phenomenological methods of the mechanics of deformable solids. The study reviews the development of methods of deformation energy assessment in several situations under the conditions of quasistatic and dynamic loading problems in plastic metal working, as well as in traffic accident emergency situations, and impact load problems of reinforced concrete structures, reinforced by mill bars of the crescent-shaped profile (Kukharchuk et al., 2016, 2017; Vedmitskyi et al., 2017).

DOI: 10.1201/9781003225447-7

7.2 ANALYSIS OF METHODS OF DYNAMIC AND QUASISTATIC LOADING CONDITIONS

It is known that in the conditions of static loading, the metals are strengthened, in the theory of plasticity, and the dependence of the stress intensity (by von Mises) σ_i depends on the strain intensity ε_i by power function (by Ludwig). The curve $\sigma_i = f(\varepsilon_i)$ is described in equation (7.1):

$$\sigma_i = A\varepsilon_i^n \tag{7.1}$$

where $A = \sigma_i$ if $\varepsilon_i = 1$, n – index $n = \varepsilon_i$ at normal stress of the conventional stress–strain curve.

In Ogorodnikov et al. (2005), the hardening curve in the conditions of the dynamic load was approximated by the equation, where the coefficient A_v can vary in the conditions of the dynamic load, following the equality:

$$\sigma_i = A_v \varepsilon_i^{n_v} \tag{7.2}$$

where the coefficient A_v can change in the conditions of the dynamic load, following the equality:

$$A_v = A\left[1.045 + \frac{\ln(0.0027 + \dot{\varepsilon}_i)}{135}\right] \tag{7.3}$$

and the coefficient n_v in the formula (7.2) – following the equality:

$$n_v = n \exp\left[-0.1273\ln(1+\dot{\varepsilon}_i)\right] \tag{7.4}$$

In formulas (7.2–7.4), A_v – coefficient of the hardening curve approximation, taking into account the impact of the deformation rate; the $\dot{\varepsilon}_i$ – strain intensity rate, s^{-1}, n_v – index, taking into account the impact of the deformation rate (Kukharchuk, Bogachuk et al., 2017, Kukharchuk et al., 2017, Vedmitskyi et al., 2017).

When forming the material sheet, in addition to the hardening curve, it is necessary to construct the dependences of the deformations accumulated by the moment of the fracture, by the deformations of the dimensionless indexes of the stress state. One of the indexes of the stress state is the ratio of the sum of the main stresses to their intensity (Polishchuk et al., 2019)

$$\eta = \frac{\sigma_1 + \sigma_2 + \sigma_3}{\sigma_u} \tag{7.5}$$

Index $\eta = -1$ in case of linear compression, $\eta = -0$ in case of a shift, and $\eta = +1$ in case of stretching. The dependence $e_p = f(\eta)$ is called a plasticity diagram (Ogorodnikov et al., 2018).

In Ogorodnikov et al. (2005), the approximation of the plasticity diagram is suggested in the form (Kozlov et al., 2019, Ogorodnikov, Dereven'ko, & Sivak, 2018.; Ogorodnikov, Zyska, & Sundetov, 2018; Polishchuk, Bilyy, & Kharchenko, 2016):

$$\varepsilon_p = \varepsilon_{p(\eta=0)} \exp(\lambda_i \cdot \eta) \tag{7.6}$$

where $\varepsilon_{p(\eta=0)}$ = ultimate strain under shift; λ_i = correspondingly:

$$\lambda_i = \ln \frac{\varepsilon_{p(\eta=0)}}{\varepsilon_{p(\eta=1)}};$$

λ_1 = the sensitivity coefficient of the plasticity to the change of the strain state scheme in the region of the index $0 \leq \eta \leq 1$;

$$\lambda_i = \ln \frac{\varepsilon_{p(\eta=-1)}}{\varepsilon_{p(\eta=0)}}$$

λ_2 = the sensitivity coefficient of the plasticity to the change of the strain state scheme in the region of the index $-1 \leq \eta \leq 0$.

Further metal treatment under pressure was classified according to the path curvature $e_p(\eta)$ – called "strain pass" of the particles of the material in dangerous sections of the deformed blanks. Calculations of the limiting deformations were formulated according to G. A. Smirnov-Aliaev (1970 and Polishchuk et al. (2018):

$$\psi = \frac{\varepsilon_i}{\varepsilon_p(\ )} \leq 1 \qquad (7.7)$$

According to Kolmogorov (1970),

$$\psi = \int_0^\varepsilon \frac{\varepsilon_i}{\varepsilon_p(\eta)} \leq 1 \qquad (7.8)$$

and (Ogorodnikov, 2018), see equation (7.9), which shows that in case of the stresses, close to simple, it is expedient to execute the criterion equation (7.7), for the conditions of the combined loading criteria equations (7.7) and (7.8), which provides more accurate results. For the case of the non-monotonic deformation, taking into account the Bauschinger effect, it is expedient to use the criterion (Aliev, 1988; Mikhalevich, 1996). The considered criteria are applied in the problems of the metal treatment under pressure on the example of the backward extrusion process (Polishchuk, Bilyy, & Kharchenko, 2016, Polishchuk et al., 2016; Polishchuk, 2018).

7.3 RESULTS AND DISCUSSION

In (Aliev, 1988), Alieva et al. (1988), and Renne and Sumarokova (1987), the technological processes of the metal treatment under pressure, based on combined loading, combination of the radial and axial extrusion, and radial extrusion with further support, were considered.

7.3.1 Actual plasticity resource in the process of backward extrusion

In order to assess the plasticity resource used in the process of cavity extrusion in the strain-hardening metal (Figure 7.1), it is necessary to form the material sheet, which, in addition to the flow curve, also includes the plasticity diagrams.

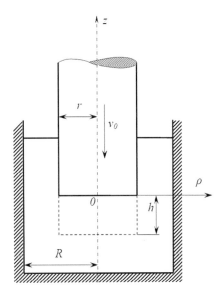

Figure 7.1 Design scheme of the backward extrusion.

Figure 7.2 shows the plasticity diagram of the steel 40×13. The experimental data for the construction of the plasticity diagram was obtained using the criterion of Ogorodnikov et al. (2018). The diagram was constructed from the results of the steel 40×13 compression test ($\varepsilon_{p(\eta = -1)} = 3.3$), shift ($\varepsilon_{p(\eta = 0)} = 0.84$), and is approximated by equation (7.6).

Calculation of the plasticity resource was used ψ according to the criterion of Grushko (2013) and Polishchuk et al. (2018),

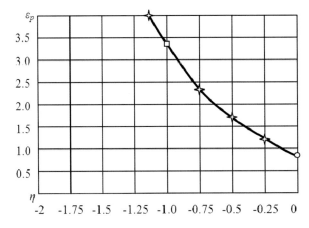

Figure 7.2 Plasticity diagram of the steel 40×13.

$$\psi = \int_0^{e_u'} \left(1 + 0.2\mathrm{arctg}\frac{d\eta}{de_u}\right) \frac{e_u^{0.2\mathrm{arctg}\frac{d\eta}{de_u}} de_u}{\left[e_{p(e_u)}\right]^{1+0.2\mathrm{arctg}\frac{d\eta}{de_u}}} \leq 1. \tag{7.9}$$

which enables the possibility of assessing the limiting deformation in the processes of the metal pressure treatment (Romanyuk et al., 2013).

It should be noted that the strain path of the material particles, as shown in Ogorodnikov et al. (2018), does not greatly depend on the properties of the material, which is why, if we know the material sheet, the limiting form changing can be assessed.

7.3.2 Experimental-computational component of the calculation model in the condition of road accidents

In the conditions of impact loading of a transport vehicle in different sections in the elements, the impact of the deformation rate on energy absorption depends on the proximity of the location of the constructions elements from the place of the impact – this impact decreases with distance from the place of the stroke (Ogorodnikov, Dereven'ko, & Sivak, 2018; Ogorodnikov, Zyska, & Sundetov, 2018).

Deformation rate distribution, for instance, in the process of the crash test of a transport vehicle () is if $V_0 = 64,4$ km/h near the place of the impact – $200\,\mathrm{s}^{-1}$, at a certain distance – $100\,\mathrm{s}^{-1}$, and at a distance where minor plastic deformations are observed – $70\,\mathrm{s}^{-1}$.

We will consider the example of the deformation energy calculation, absorbed by a Volvo motor vehicle at a speed at the moment of the stroke as $V_0 = 85$ km/h. The results of the calculation showed the increase of energy, taken into account the dynamic loading 1.3 times ($W_{\mathrm{def}}^{\mathrm{static}} = 74,308$ J; $W_{\mathrm{def}}^{\mathrm{dynam}} = 101,239$ J).

Figures 7.3 and 7.4 show the dependence of the approximation coefficient A_V and n_V on the deformation rate for the material of the Volvo motor vehicle hood (Rovira et al., 2013).

As it follows from Figure 7.3, the hardening coefficient of the flow curve increases with the increase of the deformation rate, according to the equation:

$$\overline{A_v} = B\dot{\varepsilon}_i^m \tag{7.10}$$

where $B = 825$ MPa, $m = 0,0163$ – are the approximation coefficients.

In Figure 7.4, the dependence $n_V = f(\dot{\varepsilon}_i)$ is approximated by the expression

$$n_V = n \cdot \exp(-\rho\dot{\varepsilon}_i) \tag{7.11}$$

where $n = 0.19$ – power exponent n of the hardening curve (1), ρ – approximation coefficient.

For isotropic substance n coefficient in (3), as was mentioned above, equals the intensity of the deformations at the maximum strain on the tensile diagram. As it follows from Figure 7.2, the value of the deformation's intensity decreases considerably

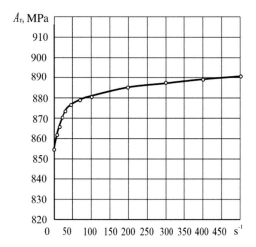

Figure 7.3 Dependence of the dynamic hardening coefficient A_V on the deformation rate.

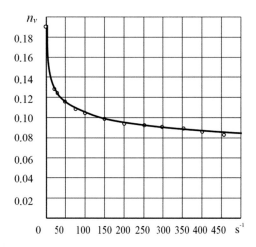

Figure 7.4 Dependence of n_V on the deformation rate.

with the increase of the deformation rate. However, the modulus of resilience for the considered steels increases:

$$W_{\text{def}} = \int \sigma_i d\varepsilon_i = A \int \varepsilon_i^n d\varepsilon_i = \frac{A\varepsilon_i^{n+1}}{n+1} \tag{7.12}$$

7.3.3 Energy absorption of the reinforced concrete constructions

The authors also considered the deformation energy calculation in the conditions of dynamic loading of reinforced concrete construction, reinforced with the crescent-shaped A500 class rolled products, made of 3Гпс steel by means of the compression test in

Table 7.1 Dependence of K_V coefficient and modulus of resilience W_{sp} on the deformation rate for the reinforced steel

Deformation intensity $\dot{\varepsilon}_i$	Coefficient K_v			Modulus of resilience W_{cp}			
	$\dot{\varepsilon}_i = 100$ (c^{-1})	$\dot{\varepsilon}_i = 150$ (c^{-1})	$\dot{\varepsilon}_i = 200$ (c^{-1})	Statics	$\dot{\varepsilon}_i = 100$ (c^{-1})	$\dot{\varepsilon}_i = 150$ (c^{-1})	$\dot{\varepsilon}_i = 200$ (c^{-1})
0.2	1,446	1,477	1,499	105,68	152,06	155,347	157,640
0.3	1,377	1,403	1,420	176,024	242,413	246,901	250,028
0.4	1,330	1,352	1,367	253,676	337,490	342,997	346,832
0.5	1,295	1,314	1,327	336,809	436,242	442,619	447,059
0.6	1,267	1,284	1,296	424,587	538,024	545,144	550,100
0.7	1,244	1,259	1,269	516,426	642,398	650,147	655,541
0.8	1,224	1,238	1,247	611,891	749,042	757,320	763,083

static conditions to different degrees of the deformation of the cylindrical samples, the hardening curve was plotted.

Coefficients A and n of formula (7.3) are correspondingly equal to: $A = 1,032$ MPa, $n = 0.27$.

Table 7.1 details the results of the calculations of the specific potential energy obtained in the dynamic conditions to the energy calculated in the conditions of the statics. Coefficient K_v in Table 7.1 varies from 1.25 to 1.5 for different degrees of deformations at a deformation rate of $200\,\text{s}^{-1}$.

By the results, presented in the table, the dependence of the modulus of resilience on the deformation intensity rate for different degrees of deformation is obtained. These dependences are approximated by the exponential functions:

$$W_{\text{cap}} = B\dot{\varepsilon}_i^m \tag{7.13}$$

where $B = 812$ MPa for static conditions, $B = 983$ MPa – for $\dot{\varepsilon}_i = 100\,\text{s}^{-1}$, $B = 977$ MPa for $\dot{\varepsilon}_i = 150\,\text{s}^{-1}$, and $B = 968$ MPa for $\dot{\varepsilon}_i = 200\,\text{s}^{-1}$.

Power exponent m of equation (7.9) turned out to be a value weakly dependent on the deformation rate ($m = 1.1$; $m = 1.15$; $m = 1.15$ for the corresponding deformation rate).

For static loading conditions, the authors used coefficient $m \approx 1.23$.

The example of the calculation of the energy absorbed by the reinforced steel in the dynamic conditions of loading shows that the deformation rate in the investigated limits increases the absorption energy at the dynamic loading by 25% as compared with quasistatic loading.

7.4 CONCLUSIONS

When considering problems relating to the mechanics of deformable solids, it is necessary to emphasize that the phenomenological methods of investigation, provided solutions to the problems of assessing the limiting form changing of the blanks, which, when worked mechanically, are the main factors. The findings of this study have shown that in emergency situation conditions such as those posed by traffic accidents, at the dynamic loading of the construction elements in the dynamic problems of metal forming, as well as at dynamic loading of reinforced concrete constructions, the deformation rate and its impact on the energy absorption effect of the construction elements should be taken into account. In the case of traffic accidents, the deformation rate causes an increase in the deformation energy by 20%–30% in the range of the speed change from 200 to 70 s^{-1}. At dynamic loading of reinforced concrete constructions in the range of the deformation rates from 100 to 200 s^{-1}, the energy absorption coefficient is $K_v \approx 1.5$.

REFERENCES

Aliev, I.S. 1988. Radial extrusion processes. Soviet forging and sheet metal stamping technology. *Allerton Press* 6: 1–4.

Alieva, L., Aliev, I. & Kartamyshev, D. 2017. Combined radial-forward extrusion of hollow parts like cups. *XVIII International Scientific Conference 'New Technologies and Achievements in Metallurgy, Material Engineering, Production Engineering and Physics'* (68): 108–113.

Grushko, A.V. 2013. Phenomenological aspects of the materials sheets creation for the processes of cold plastic strain. *Metal Working: Proceedings* 1(34): 85–95.

Kolmogorov, V.L. 1970. *Strain, Deformation, Fracture*. Moscow: Metallurgy.

Kozlov, L.G., Polishchuk, L.K., Piontkevych, O.V., Korinenko, M.P., Horbatiuk, R.M., Komada, P., Orazalieva, S. & Ussatova, O. 2019. Experimental research characteristics of counterbalance valve for hydraulic drive control system of mobile machine. *Przeglad Elektrotechniczny* 95(4): 104–109.

Kukharchuk, V.V., Bogachuk, V.V., Hraniak, V.F., Wójcik, W., Suleimenov, B. & Karnakova, G. 2017. Method of magneto-elastic control of mechanic rigidity in assemblies of hydropower units. *Proc. SPIE 10445, Photonics Applications in Astronomy, Communications, Industry, and High Energy Physics Experiments 2017* 104456A.

Kukharchuk, V.V., Hraniak, V.F., Vedmitskyi, Y.G., Bogachuk, V.V., Zyska T., Komada, P. & Sadikova, G. 2016. Noncontact method of temperature measurement based on the phenomenon of the luminophor temperature decreasing. *Proc. SPIE 10031, Photonics Applications in Astronomy, Communications, Industry, and High-Energy Physics Experiments 2016* 100312F.

Kukharchuk, V.V., Kazyv, S.S. &, Bykovsky, S.A. 2017., Discrete wavelet transformation in spectral analysis of vibration processes at hydropower units. *Przegląd Elektrotechniczny* 93(5): 65–68.

Mikhalevich, V.M. 1993. Models of defects accumulation for solids with original and strain-induced anisotropy. *Izvestia Akademii nauk SSSR. Metally* (5): 144–151.

Mikhalevich, V.M. 1996. Tensor models of rupture strength. report no. 3. Criterional relations for loading with a change in stress state and the directions of the principal stresses. *Strength of Materials* 28(3):238–246.

Ogorodnikov, V.A., Dereven'ko, I.A. & Sivak, R.I. 2018. On the influence of curvature of the trajectories of deformation of a volume of the material by pressing on its plasticity under the conditions of complex loading. *Materials Science* 54(3): 326–332

Ogorodnikov, V.A., Kiselyov V.B. & Sivak I.O. 2005. *Energy, Strain, Fracture (Problems of the Motoengineering Expert Evaluation)*. Vinnitstya: Universum.

Ogorodnikov, V.A., Zyska T. & Sundetov, S. 2018. The physical model of motor vehicle destruction under shock loading for analysis of road traffic accident. *Proc. SPIE 10808, Photonics Applications in Astronomy, Communications, Industry, and High-Energy Physics Experiments 2018* 108086C.

Polishchuk, L., Bilyy, O. & Kharchenko, Y. 2016. Prediction of the propagation of crack-like defects in profile elements of the boom of stack discharge conveyor. *Eastern-European Journal of Enterprise Technologies* 6(1): 44–52.

Polishchuk, L.K., Kozlov, L.G., Piontkevych, O.V., Horbatiuk, R.M., Pinaiev, B., Wójcik, W., Sagymbai, A. & Abdihanov A. 2019. Study of the dynamic stability of the belt conveyor adaptive drive. *Przegląd Elektrotechniczny* 95(4): 98–103.

Polishchuk, L.K., Kozlov, L.G., Piontkevych, V., Gromaszek, K. & Mussabekova, A. 2018. Study of the dynamic stability of the conveyor belt adaptive drive. *Proc. SPIE 10808, Photonics Applications in Astronomy, Communications, Industry, and High-Energy Physics Experiments* 1080862.

Renne, I. P. & Sumarokova, A.I. 1987. Technological capabilities of the process of free extrusion (without die) of hollow parts. *Forging and Stamping Production* (6): 25–26.

Romanyuk, N., Pavlov, S.V., Dovhaliuk, R.Y., Babyuk, N.P., Obidnyk, M.D., Kisała, P. & Suleimenov, B. 2012. Microfacet distribution function for physically based bidirectional reflectance distribution functions. *Proc. SPIE 8698, Optical Fibers and Their Applications 2012* 86980L.

Rovira, R.H., Pavlov, S.V., Kaminski, O.S. & Bayas, M.M. 2013. Methods of processing video polarimetry information based on least-squares and Fourier analysis. *Middle-East Journal of Scientific Research* 16(9): 1201–1204.

Smirnov-Aliaev, G.A. 1970. *Mechanic Fundamentals of Metals Plastic Working: Engineering Methods of Calculation*. Leningrad: Machine building.

Vedmitskyi, Y.G., Kukharchuk, V.V. & Hraniak, V.F. 2017. New non-system physical quantities for vibration monitoring of transient processes at hydropower facilities, integral vibratory accelerations. *Przegląd Elektrotechniczny* 93(3): 69–72.

Chapter 8

The determination of deformation velocity effect on cold backward extrusion processes with expansion in the movable die of axisymmetric hollow parts

I. Aliiev, V. Levchenko, L. Aliieva, V. Kaliuzhnyi, P. Kisała, B. Yeraliyeva, and Y. Kulakova

CONTENTS

8.1 Introduction 87
8.2 Analysis of methods of traditional cold backward extrusion of hollow parts 88
8.3 Materials and methods of studying backward extrusion with expansion in a movable die 91
8.4 Results and discussion 95
8.5 Conclusions 98
References 98

8.1 INTRODUCTION

When manufacturing axisymmetric hollow parts made of steel and nonferrous metals, including hollow semifinished ones to further obtain special-purpose parts, cold extrusion is used in mass production. Currently, the scheme of longitudinal backward extrusion has found widespread application, for which technology and die tooling are designed based on production experience (Semenov 1987).

The factor limiting the technological capabilities of the cold extrusion processes is the high values of forces and pressures of metal deformation in a cold state. To reduce the value of extrusion forces, various techniques have been proposed, among which we can highlight the use of extrusion schemes with active friction forces, combined schemes with increased degree of freedom of metal outflow, and extrusion with expansion (Ogorodnikov 2004; Ovchinnikov 1980; Ovchinnikov 1983).

The scheme of cold backward extrusion with expansion in a moving die provides the reduction of deformation forces and the increase in deforming tool resistance (Dmitriev 1984; Grechnikov 1985; Kaliuzhnyi 2013). The mobility of the die provides the extrusion of metal with the expansion in the deformation center and thereby the change of the stress–strain state of the workpiece. Combined deformation with expansion reduces the energy intensity of the deformation process by declining the stiffness of the stress state and replacing the stress state scheme with full compression to the contralateral stress state scheme with one (circumferential) stretching stress. Thus, by creating the radial flow in the center of intense deformation, it is possible to reduce the

DOI: 10.1201/9781003225447-8

values of pressure and deformation forces of the metal in the cold state significantly (Aliieva 2017; Dmitriev 1984; Kalyuzhnyi 2017).

These new methods of deformation with the combined flow (radial-forward or radial-backward extrusion) are attracting the increasing attention of researchers. It is shown that radial-forward extrusion can be accomplished without direction of the metal by the die, and such "die-free" extrusion contributes to reducing the power parameters of case extrusion up to 30% (Renne 1987). The combined radial-forward extrusion is effective to obtain deep cases because it proceeds at the optimal power mode (Aliieva 2017). It should be noted that the use of radial-forward extrusion methods for the production of hollow articles in manufacture is somewhat limited due to the difficulty of removing the finished part from the die (Alieva 2015; Aliieva 2017; Kaliuzhnyi 2013).

The comparison of the force parameters of deformation by means of combined flow and traditional methods of forward and backward extrusion of hollow parts demonstrates the effectiveness of such combined technological methods (Kalyuzhnyi 2017).

The analysis of changing the deformed state during the part formation using the upper estimation method makes it possible to evaluate the degree and unevenness of the accumulated deformation quickly as well as to estimate the thermal state of the workpiece approximately corresponding to its deformed state (Alieva 2019). The study of heat generation and temperature fields in the deformed workpiece, performed by the finite element method (FEM) and presented in (Kaliuzhnyi 2013; Kalyuzhnyi 2017), shows the significant increase in workpiece temperature during cold deformation. The regularities of such a change in workpiece temperature and the reduction in deformation resistance related to it are of practical interest, since they can serve as the additional reserve to optimize the force regime of cold deformation processes.

It is known that one of the parameters that have a significant impact on the force and deformation mode of plastic deformation processes is the deformation velocity V_0. The experimental studies of heat generation during cold extrusion of steel cups are performed in (Gusinskiy 1978). To estimate the effect of the strain rate on the power parameters, extrusion is carried out in the equipment with different strain velocities. It is found that a significant decrease in the deformation forces due to the heating of the metal is observed during the transition from very low deformation velocities, corresponding to test presses, to deformation velocities, corresponding to hydraulic forging presses. With a further velocity increase up to the velocity of deformation in crank presses, this reduction in power parameters becomes insignificant.

In this regard, conducting research to identify the effect of strain velocity on the formation of hollow parts according to the backward extrusion scheme with expansion in a moving die is an important task.

The purpose of the work is to study the effect of strain velocity on the parameters of cold backward extrusion with the expansion of axisymmetric hollow parts in a moving die by the FEM.

8.2 ANALYSIS OF METHODS OF TRADITIONAL COLD BACKWARD EXTRUSION OF HOLLOW PARTS

To verify the adequacy of the obtained mathematical models based on the FEM and to identify the influence of deformation velocity on cold forming hollow parts, an analysis of traditional cold backward extrusion of hollow parts made of steel 10 for the

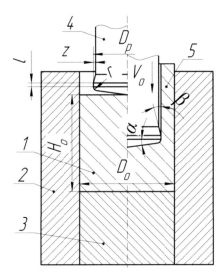

Figure 8.1 Scheme of traditional backward extrusion.

dimensions of the workpiece, and punch was performed in the DEFORM environment according to the reference data (Gusinskiy 1978). The diagram of true stresses for steel 10 is taken from (Kroha 1980). The scheme with designations of such extrusion is shown in Figure 8.1.

The position at the beginning of extrusion is given to the left of the symmetry axis and the one at the end of extrusion is presented to its right. The initial workpiece 1 is installed in the die 2 at the ejector 3. Extrusion is performed by lowering the punch 4, which moves at the velocity V_0, and the part 5 is obtained. The workpieces with dimensions $D_0 = 8$ mm and $H_0 = 20$ mm are subjected to deformation at the deformation rate $\varepsilon = 50\%$ and the deformation velocities $V_0 = 0.05$ and 5 mm/s. The velocity $V_0 = 0.05$ mm/s corresponds to the traverse velocity of a TsDMPu-200 testing machine, and $V_0 = 5$ mm/s is the working stroke velocity of the slider of a P459 hydraulic press. The punch dimensions are: $D_p = 19.8$ mm, $r = l = 2$ mm, $z = 0.15$ mm, $\alpha = \beta = 7°$. The workpiece material is considered to be a plastic with hardening and the tools are absolutely rigid. The value of the friction coefficient is $\mu = 0.1$.

The design schemes in the section for the traditional backward extrusion are shown in Figure 8.2.

The scheme for the beginning of extrusion is shown in Figure 8.2a. The initial workpiece 1 is installed in the die 2 at the ejector 3. The deformation force is applied by the punch 4. Figure 8.2b shows the scheme of the process for the end of extrusion. When lowering the punch 4, the part 5 is obtained (Kroha 1980; Polishchuk 2018; Polishchuk 2019). The force of extrusion as a function of the movement of the punch, the stress–strain state, and the temperature of the deformed metal is obtained by means of simulation. Figure 8.3 shows the force of extrusion as a function of the movement of the punch for extrusion at different deformation velocities.

When extruding at velocity of $V_0 = 0.05$ mm/s, the force reaches maximum value of 670 kN and remains almost constant until the end of the forming process (Figure 8.3a).

Figure 8.2 Design schemes in cross section: left – scheme for the beginning of extrusion, right – scheme for the end of extrusion.

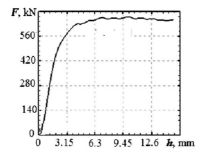 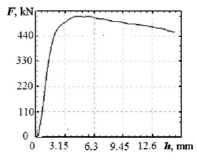

Figure 8.3 Extrusion force F vs. the movement of the punch h: (a) extrusion at $V_0 = 0.05$ mm/s, (b) extrusion at $V_0 = 5$ mm/s.

The increase in the deformation velocity to $V_0 = 5$ mm/s leads to another kind of function. At first, the force increases to a value of 530 kN and then somewhat decreases until the end of the extrusion process (Figure 8.3b). The decrease of the extrusion force is 21% in comparison with the previous case. Similar types of functions and the decrease in the extrusion force by 20% are obtained experimentally (Vedmitskyi 2017). Figure 8.4 shows temperature T, °C and mean stress $\sigma_m = (\sigma_r + \sigma_\theta + \sigma_z)/3$ distributions in deformed workpieces for the end of extrusion. Hereinafter, thin lines depict deforming tools, and the dimensions along the axes are given in millimeters. Figure 8.4a shows the temperature distribution for extrusion at the velocity $V_0 = 0.05$ mm/s. Due to the low velocity of deformation and heat transfer, the temperature of the deformed metal increases only by 2°C–3°C. The increase in velocity to $V_0 = 5$ mm/s leads to a significant increase in the temperature of the metal (Figure 8.4b).

The temperature at the place of wall transition to the bottom part and in the bottom part itself reaches values in the range of $T = 80°C–140°C$. The authors in (Vedmitskyi 2017) believe that the increase in the temperature of the deformed metal, which took

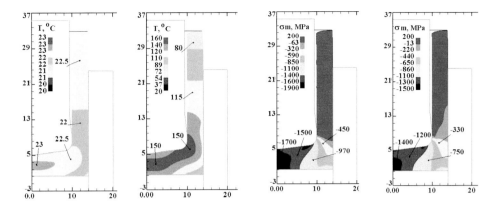

Figure 8.4 Temperature T and mean stress σ_m distributions in deformed workpieces: (a) distribution T at $V_0 = 0.05$ mm/s, (b) distribution T at $V_0 = 5$ mm/s, (c) distribution σ_m at $V_0 = 0.05$ mm/s, (d) – distribution σ_m at $V_0 = 5$ mm/s.

place in the experimental studies, is the reason for decreasing the extrusion force when increasing the deformation velocity. However, as simulation results show, increasing the velocity also leads to changing the stress state scheme of cold forming, which affects the hydrostatic pressure (σ_m) values in the deformation zone. When extruding at the velocity of $V_0 = 0.05$ mm/s, the values of this stress in the deformation zone are in the range of $\sigma_m = -450$ to $-1{,}700$ MPa (Figure 8.4c), and when extruding at the velocity of $V_0 = 5$ mm/s, $\sigma_m = -330$ to $-1{,}400$ MPa is obtained in the said place. The absolute decrease of hydrostatic pressure averages in the range of 18%–21%. Thus, the models created on the basis of the FEM allow the parameters of cold forming of hollow parts at different deformation velocities to be determined accurately enough.

8.3 MATERIALS AND METHODS OF STUDYING BACKWARD EXTRUSION WITH EXPANSION IN A MOVABLE DIE

The scheme of backward extrusion with expansion in a movable die with designations is shown in Figure 8.5. The scheme for the beginning of extrusion is to the left of the symmetry axis and the one for the end of extrusion is to the right. The initial workpiece 1 is installed in the movable die 2 at the ejector 3. When the punch 4 is moved at the deformation velocity V_0, the part 5 is formed. When extruding, the die 2 moves downward at the same velocity with the punch 4 (Kozlov 2019; Kukharchuk 2017; Vedmitskyi 2017).

When modeling the formation of axisymmetric hollow parts made of bronze BRASS 377 CDA, the metal of the workpiece is considered to be an elastically plastic one with hardening, and the deforming tools are absolutely rigid. The use of such a metal model makes it possible to determine the final shape and dimensions of the part as well as to analyze the processes of extracting the punch from the deformed workpiece and to push the workpiece out of the die after removing the punch. The influence of Coulomb friction at the friction coefficient of $\mu = 0.1$ was taken into account. The initial workpiece has dimensions $D_0 = 48$ mm and $H_0 = 52$ mm. The dimensions of

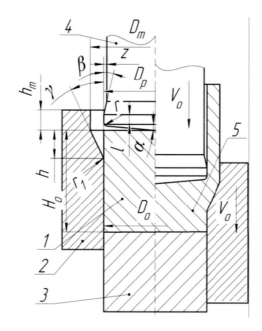

Figure 8.5 Scheme of backward extrusion with expansion in a movable die of hollow parts.

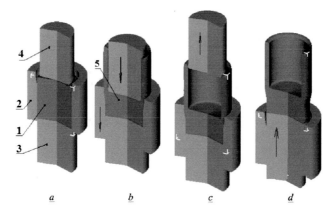

Figure 8.6 Design schemes in cross section: (a) beginning of extrusion, (b) end of extrusion, (c) after extracting the punch from the deformed workpiece, (d) after pushing the part out of the die.

the deforming tools are as follows: $D_p = 42$ mm, $\alpha = \beta = 7°$, $r = r_1 = 2$ mm, $z = 0.15$ mm, $l = 2$ mm, $D_m = 52$ mm, $h_m = 8$ mm, and $\lambda = 10°$. The deformation velocities are $V_0 = 0.14$, 2, 7, and 15 mm/s.

Figure 8.6 shows the design schemes in cross section obtained by simulation using the FEM.

The scheme of the process at the beginning of extrusion is shown in Figure 8.6a. The workpiece 1 is installed in the movable die 2 at the ejector 3. Deformation is performed by the punch 4. After extrusion, which is performed by lowering the punch 4 and die 2 simultaneously, the part 5 is obtained (Figure 8.6b). The scheme after removing the punch out of the deformed workpiece is shown in Figure 8.6c. In this case, the workpiece remains in the die. Figure 8.6d shows the scheme after pushing the part out of the die (Kukharchuk 2016a; Kukharchuk 2016b).

The extrusion force as a function of the movement of the punch for extrusion at different values of the deformation velocity is shown in Figure 8.7.

The increase in velocity V_0 leads to the decrease in the extrusion force. When extruding at velocity $V_0=0.14$ mm/s, the force constantly increases and reaches the value of 2,860 kN at the end of extrusion. The increase in velocity to $V_0=2$ mm/s provides the emergence of the steady-state stage when the force remains constant. When the displacement of the punch 7.5 mm occurs, the force reaches the value of 2,288 kN and then remains constant until the end of extrusion. The decrease in extrusion force is 20%. The further increase in deformation velocity to $V_0=7$ mm/s also leads to the decrease in force. The force reaches the value of 2,170 kN when the punch is moved 7.5 mm and then decreases somewhat until the end of extrusion. The decrease in force compared with the previous extrusion case is 4.8%. The deformation velocity $V_0=15$ mm/s also provides the decrease in the maximum force value to 2,115 kN. However, the intensity of decreasing the force decreases and accounts for only 2.7%. Thus, the rational values of the strain rates from the point of view of the formation force of hollow parts made of bronze are the values of the deformation velocity $V_0=2$–15 mm/s (Figure 8.8).

On the axis of the surface of the punch end face, $\sigma_n=1,750$ MPa was obtained with further decreasing to $\sigma_n=1,600$ MPa on the rounding radius of the punch end face. Stress $\sigma_n=600$ MPa appears on the surface of the punch calibrating belt. At the ejector, the normal stress reaches values in the range of $\sigma_n=1,080$–1,490 MPa. The highest stress value $\sigma_n=1,460$ MPa on the surface of the die occurs at the point of transition of the conical surface to the cylindrical surface; on the cylindrical surface itself, these stresses are in the range of $\sigma_n=290$–1,320 MPa. With such values of normal stress, the die should have one band.

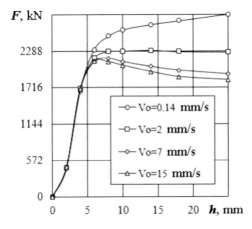

Figure 8.7 Extrusion force as a function of the movement of the punch for extrusions at different deformation velocity V_0.

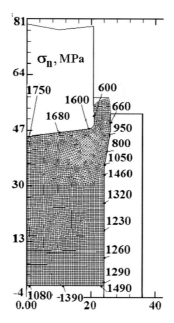

Figure 8.8 Distributions of normal stresses σ_n on the contact surfaces of the deformed workpiece with the tool.

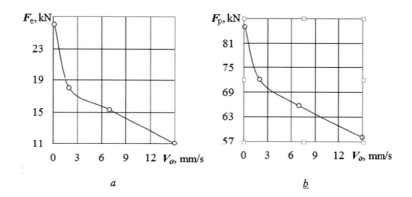

Figure 8.9 Maximum force values of extracting the punch out of the deformed workpiece and pushing out the part out of the die vs. the deformation velocity: *a* is the extracting force, *b* is the pushing out force.

The force of extracting the punch out of the die as a function of its movement and the force of pushing the part out of the die as a function of the ejector movement are determined by modeling as well. Figure 8.9 shows the maximum values of forces to extract the punch from the deformed workpiece and to push the part out of the die as a function of deformation velocity.

The increase in deformation velocity leads to the decrease in the extraction force (Figure 8.9a). The decrease is related to the decrease in extrusion force and stresses in the deformed workpiece during the reverse movement of the punch. The value of the extraction force of the punch F_e is approximately $F_e = (7.5-9.4) \cdot 10^{-3} \cdot F_{rev}$. Here, F_{rev} is the extrusion force. The extraction force must be considered when designing the punch and die holders. Figure 8.9b shows the maximum pushing out force as a function the deformation velocity. This force decreases with the increase in deformation velocity, which is also related to the decrease in the extrusion force. According to the data of the plot, the value of the ejection force F_p after backward extrusion in the movable die can be determined by the extrusion force F_{rev} : $F_p = (0.028-0.03) \cdot F_{rev}$.

8.4 RESULTS AND DISCUSSION

It has been determined by modeling that the value of the deformation velocity influences significantly the temperature of the deformed metal when extruding hollow parts made of bronze. Figure 8.10 shows the temperature distribution in deformed workpieces for the end of extrusion at different deformation velocities.

After extrusion at velocity $V_0 = 0.14$ mm/s (Figure 8.10a), the temperature of the deformed metal rises to $T = 29°C-35°C$ in the deformation zone under the punch and to $T = 33°C$ in the wall. Increasing the velocity up to $V_0 = 2$ mm/s results in the temperature within $T = 85°C-160°C$ in the deformation zone and $T = 123°C-160°C$ in the wall (Figure 8.10b).

The deformation velocity $V_0 = 7$ mm/s further increases the temperature of the deformed metal up to the values in the range of $T = 133°C-270°C$ in the deformation zone and up to $T = 170°C-270°C$ in the wall of the deformed metal. When extruding at the velocity $V_0 = 7$ mm/s in the deformed workpiece, the metal temperature reaches the values of $T = 325°C$ in the area of the wall transition to the bottom part of the workpiece and $T = 160°C-285°C$ in the wall itself. Such values of temperatures for cold extrusion are confirmed by the data of Ovchinnikov (1983). Thus, the use of backward extrusion in a movable die in the mass production of hollow parts made of bronze requires the forced cooling of punches and dies.

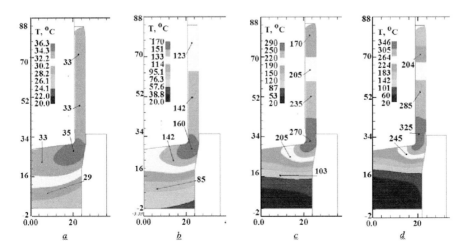

Figure 8.10 Temperature distributions in deformed workpieces after extrusion at different deformation velocities: (a) at $V_0 = 0.14$ mm/s, (b) at $V_0 = 2$ mm/s, (c) at $V_0 = 7$ mm/s, (d) at $V_0 = 15$ mm/s.

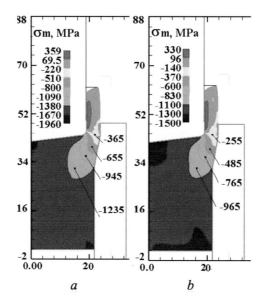

Figure 8.11 Average stress distributions in deformed workpieces: (a) at $V_0 = 0.14$ mm/s, (b) at $V_0 = 7$ mm/s.

When extruding with expansion in a moving die, as with traditional backward extrusion, the increase in the deformation velocity leads to the change in the stress state scheme and the value of hydrostatic pressure in the deformation zone. Figure 8.11 shows the stress distribution for extrusion at velocities $V_0 = 0.14$ mm/s and $V_0 = 7$ mm/s, which are obtained with the same movements of the punch.

The nature of the distributions is almost the same, but the stress values are different. When extruding at the velocity $V_0 = 0.14$ mm/s, these stresses in the deformation zone are within $\sigma_m = -365$ to $-1,235$ MPa (Figure 8.11a), and when extruding at $V_0 = 7$ mm/s, they are $\sigma_m = -255$ to -55 MPa (Figure 8.11b). The absolute decrease of hydrostatic pressure is up to 26%. Taking into account the results of steel extrusion (see Figure 8.4c and d), the reason for decreasing the force with increasing the deformation velocity is also the change of the stress state scheme in the deformation zone during formation.

The deformation velocity influences the metal structure after cold plastic deformation, which can be estimated by the distribution of effective strain ε_i. Figure 8.12 shows the distributions of the effective strain in the deformed workpiece, which are obtained at the deformation velocities $V_0 = 0.14$ mm/s and $V_0 = 7$ mm/s. The nature of the distributions is almost the same, but the effective strain values are different. When extruding at $V_0 = 0.14$ mm/s (Figure 8.12a), the value of the effective strain along the width of the wall varies from $\varepsilon_i = 3.3$ in the inner metal layers of the wall to $\varepsilon_i = 1.1$ in the outer metal layers of the wall. The increase in deformation velocity to $V_0 = 7$ mm/s decreases the unevenness of the effective strain along the wall width of the deformed workpiece. The value of effective strain is $\varepsilon_i = 3.1$ in the inner layers of the metal, and it is $\varepsilon_i = 1.5$ in the outer ones.

As an example, Figure 8.13 shows the shape and dimensions of the part after pushing it out of the die. The dimensions of the deforming tool are marked in brackets.

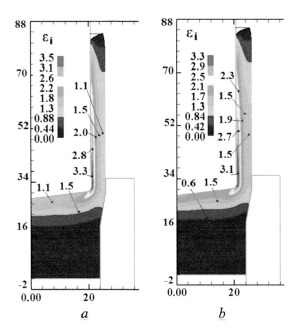

Figure 8.12 Effective strain distributions in deformed workpieces: (a) at $V_0 = 0.14$ mm/s, (b) at $V_0 = 7$ mm/s.

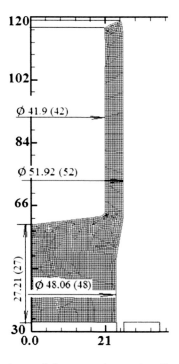

Figure 8.13 Shape and dimensions of the part obtained at $V_0 = 7$ mm/s.

The elastic deformation in height of the bottom part of the workpiece has the greatest value, which is 0.21 mm. Shrinkage depression and slight curvature of the wall occur at the end face of the wall.

8.5 CONCLUSIONS

The method of finite elements has been used to analyze cold extrusion with expansion in a movable die of axisymmetric hollow bronze parts at different deformation velocities. To verify the accuracy of the created models, a computational analysis of the traditional backward extrusion of such parts made of steel is performed, and the results of the deformation force and the temperature of the deformed metal are compared with known experimental data, which show good agreement. The force of extrusion, extraction of punches from the deformed workpiece, and the force of pushing the parts out of the die as a function of the displacement of the deforming tool for the process of extrusion with expansion have been found. The stress–strain state of the metal for the formation of parts and the temperature of the deformed metal are determined. The exact distribution of the specific force on the punch, ejector, and die has been revealed. Taking into account the elastic deformation, the final shape and dimensions of the parts are calculated. The increase in deformation velocity leads to the decrease in forces of the extrusion process, extraction of the punch, pushing out the part and to the increase in temperature of the deformed metal. Recommendations for the design of die tools are offered as well.

REFERENCES

Alieva, L., Zhbankov, Y. 2015. Radial-forward extrusion with a movable mandrel. *Metallurgical and Mining Industry* 11: 175–183.
Aliieva, L.I., Titov, A.V., Kordenko, M.Y. 2019. Simulation of cross side extrusion processes. *Materials Working by Pressure – Kramatorsk DSEA*. 1(48): 35–44.
Aliieva, L., Aliiev, I., Kartamyshev D. 2017. Combined radial-forward extrusion of hollow parts like cups. *Proceedings of XVIII International Scientific Conference 'New Technologies and Achievements in Metallurgy, Material Engineering, Production Engineering and Physics'* 68: 108–113.
Dmitriev, A.M. 1984. Studying of extrusion process with expansion. *News of higher education institutions. Machine Building* 4: 140–148.
Grechnikov, V.F., Dmitriev, A.M., Kukhar, V.D. 1985. Advanced cold forming processes. *Machine Building*: 1–184.
Gusinskiy, V.I., Mulin, V.P., Novikov, V.V. 1978. The effect of strain rate on the force of cold backward extrusion. *The improvement of Cold Forming Processes and Equipment. Voronezh. ENIKMASh*: 19–23.
Kaliuzhnyi, V.L., Aliieva, L.I., Kulikov, Y.P. 2013. Comparative analysis of the processes of backward extrusion and forward extrusion with expansion of parts with cavity of constant diameter. *Materials Working by Pressure. Kramatorsk: DSEA* 4(37): 87–92.
Kalyuzhnyi, V.L., Aliieva, L.I., Kartamyshev, D.A., Savchinskii, I.G. 2017. Simulation of cold extrusion of hollow parts. *Metallurgist*. 61(5–6): 359–365.
Kozlov, L.G., Polishchuk, L.K., Piontkevych, O.V., Korinenko, M.P., Horbatiuk, R.M., Komada, P., Orazalieva, S., Ussatova, O. 2019. Experimental research characteristics of counter balance valve for hydraulic drive control system of mobile machine. *Przeglad Elektrotechniczny* 95(4): 104–109.

Kroha, V.A. 1980. Hardening of metals during cold plastic deformation: The Handbook. *Machine Building*: 1–157.

Kukharchuk, V.V., Hraniak, V.F., Vedmitskyi, Y.G., Bogachuk, V.V. 2016. Noncontact method of temperature measurement based on the phenomenon of the luminophor temperature decreasing. *Proc. SPIE* 10031, 100312F: 968–974.

Kukharchuk, V.V., Kazyv, S.S., Bykovsky, S.A. 2017. Discrete wavelet transformation in spectral analysis of vibration processes at hydropower units. *Przeglad Elektrotechniczny* 93(5): 65–68.

Kukharchuk, V.V., Bogachuk, V.V., Hraniak, V.F., Wójcik, W, Suleimenov, B., Karnakova, G. 2017. Method of magneto-elastic control of mechanic rigidity in assemblies of hydropower units. *Proc. SPIE 10445*. 104456A: 1231–1238

Ogorodnikov, V.A., Savchinskij, I.G., Nakhajchuk, O.V. 2004. Stressed-strained state during the formation of the internal slot section by Mandrel reduction. *Heavy Machine Building* 12: 31–33.

Ovchinnikov, A.G., Khabarov, A.V. 1980. Forward extrusion of cylindrical cups. *Improving of Die Forging Processes. MDNTP*: 103–108.

Ovchinnikov, A.G. 1983. Fundamentals of the theory of stamping by extrusion on presses. *Machine Building*. Moscow: 1–200.

Polishchuk, L.K., Kozlov, L.G., Piontkevych, O.V., Gromaszek, K., Mussabekova, A. 2018. Study of the dynamic stability of the conveyor belt adaptive drive. *Proc. SPIE* 10808. 1080862: 1250–1262.

Polishchuk, L.K., Kozlov, L.G., Piontkevych, O.V., Horbatiuk, R.M., Pinaiev, B., Wójcik, W., Sagymbai, A., Abdihanov, A. 2019. Study of the dynamic stability of the belt conveyor adaptive drive. *Przeglad Elektrotechniczny* 95(4): 98–103.

Renne I.P., Sumarokova, A.I. 1987. Technological capabilities of the process of free extrusion (without die) of hollow parts. *Forging and Stamping Production* 6: 25–26.

Semenov, E.Y. (ed.) 1987. *Forging and die forging: The handbook. V. 3. Cold forging*. Machine Building. Moscow: 1–384.

Vedmitskyi, Y.G., Kukharchuk, V.V., Hraniak, V.F. 2017. New non-system physical quantities for vibration monitoring of transient processes at hydropower facilities, integral vibratory accelerations. *Przeglad Elektrotechniczny*. 93(3): 69–72.

Chapter 9

Stress state of a workpiece under double bending by pulse loading

V. Dragobetskii, V. Zagoryanskii, D. Moloshtan, S. Shlyk,
A. Shapoval, O. Naumova, A. Kotyra, M. Mussabekov,
G. Yusupova, and Y. Kulakova

CONTENTS

9.1 Introduction 101
9.2 Mathematical model of stress state of clad-bending areas during
      explosion welding 102
9.3 Development of the mathematical model: the shock wave peak
      pressures change during the explosion welding 104
9.4 Mathematical model of the clad stress state at the collision moment ... 105
9.5 Results and discussion 107
9.6 Conclusions 109
References 109

9.1 INTRODUCTION

Modern production equipment, technological and instrumental equipment, vehicles, military equipment, etc., work in conditions characterized by extreme operating parameters (high thermal and mechanical loads, aggressive environment, etc.). Therefore, the pieces of the equipment constantly experience necessities of their materials with a complex of incompatible mechanical properties: wear and abrasive resistance, high strength and low density, corrosion resistance, and high electrical and thermal conductivity. The required set of properties can be achieved by combining several materials into a single structural unit. Such materials are produced by combining dissimilar metals into monolithic compositions, in particular, into layered ones. Therefore, the demand for metal compositions continues to grow, and the requirements for their properties and quality are increasing. One of the effective methods for the production of layered compositions is explosion welding. Improvement of this process is associated with the development of methods for calculating its technological parameters. It is necessary to predict not only the mechanical properties of the compositions of the joint but also the operational characteristics of the cladding layer. For this, it is necessary to solve the problem of elastoplastic deformation of layers during the explosion welding. This will allow the calculation of the process of jet formation for elastoplastic units and the compressive stresses of the colliding workpieces time of existence. These data allow the determination of the quality indicators of the welded joint. In addition, this solution makes it possible to determine the deformations at

DOI: 10.1201/9781003225447-9

which the properties of the cladding layer (such as wear and corrosion resistance, fatigue strength, etc.) correspond to the limit of uniform deformations.

The aim of the study is to obtain the engineering dependencies for the calculation of the workpiece deformed state, which ensures the highest-quality indicators in terms of operating characteristics.

9.2 MATHEMATICAL MODEL OF STRESS STATE OF CLAD-BENDING AREAS DURING EXPLOSION WELDING

The solution of the problem of the collision of the flyer plate (clad) and the fixed target plate (substrate or base plate) is considered in three stages. The first is the movement of the element of the flyer plate before the collision with the target plate. At this stage, we will accept the assumption that it is possible to neglect the deformation energy of the flyer plate in comparison to the kinetic energy of its movement. The second stage is the inertial motion of the flyer plate and its collision with the surface of the fixed base plate. The dependencies of the second stage are caused by the duration of the explosion, which is tens of microseconds, and the workpiece deformation duration is a millisecond. The third stage is the deformation of the flyer workpiece after the collision with the fixed base plate (Dragobetskii 2007).

The explosive welding schematics in the case of obtaining two-layered and multilayered compositions are shown in Figure 9.1.

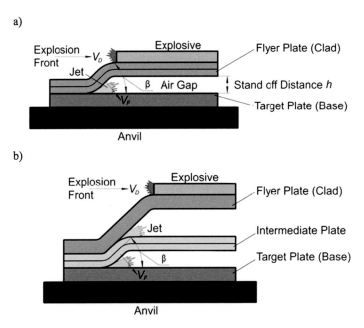

Figure 9.1 Explosive welding schematics: (a) – explosive welding of two-layered composition; (b) – explosive welding with an internal layer.

The flyer plate element equation of motion under the action of varying pressure has the form (Dragobetskii 2019):

$$\rho\delta\frac{d^2h}{dt^2} + \rho_1 c_1 \frac{dh}{dt} = P_m \exp(-t/\theta), \qquad (9.1)$$

where: δ – flyer plate element thickness; ρ – flyer plate density; ρ_1, c_1 – density and speed of sound in the environment behind the plate; h – gap between the welded plates; P_m – pressure of the detonation products on the plate at the moment of the release of the shock wave onto the free surface; θ – time constant for the decline of detonation products; and t – time of the process.

The solution of (equation 9.1) has the form:

$$h = r + \frac{\theta}{m} \cdot P_m \cdot \left[1 - \exp(-t/\theta)\right] \cdot t, \qquad (9.2)$$

where: r – flyer plate element radius of curvature; and m – mass of the element of the plate square unit.

When the plate moves by inertial forces, we divide the flyer plate surface into flat areas in contact with the target plate surface and cylindrical areas, which form the bending radius of the flyer plate (Ogorodnikov 2004; Ogorodnikov 2018a) (Figure 9.2).

When solving the equilibrium equation of an element at the bending radius of the flyer plate, together with the plasticity condition, we obtain:

$$\rho \cdot \frac{d^2\delta}{dt^2} = -\frac{d\sigma_r}{dr} + \frac{\sigma_r - \sigma_\varphi}{r} + \frac{\sigma_r}{\delta} \cdot \frac{d\delta}{dr} \qquad (9.3)$$

where: σ_r and $\sigma\varphi$ – respectively, the radial and latitudinal stresses in the flat element of flyer plate; and δ – element thickness.

The solution of this equation has the form:

$$\sigma_r = -P_m + \beta \cdot \sigma_S \cdot f + \rho \cdot \left(\overset{\infty}{r} \cdot \overset{c2}{r} + r \right) \cdot f + \rho \cdot \left(\frac{r^2 \cdot \overset{c2}{r}}{2b^2} - \frac{\overset{c2}{r}}{2} \right) \qquad (9.4)$$

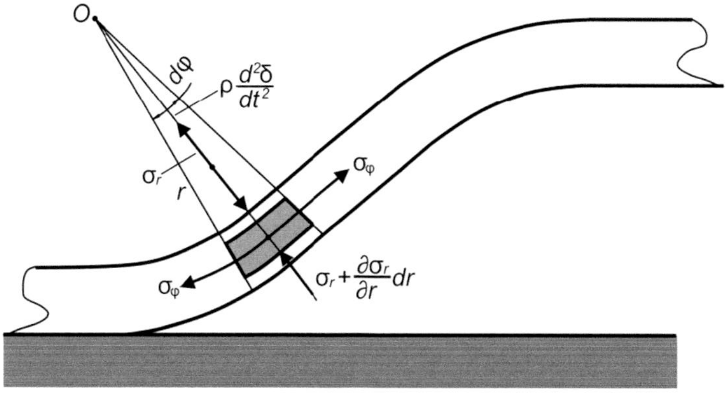

Figure 9.2 Equilibrium of the flyer plate element $\sigma_r + (\partial \sigma_r / \partial r) dr$ on the bending radius.

where: β – Lode–Nadai coefficient; σ_S – yield strength; b – radius of the flyer plate outer surface ($b = r + \delta$); $f = \ln(r/b)$.

9.3 DEVELOPMENT OF THE MATHEMATICAL MODEL: THE SHOCK WAVE PEAK PRESSURES CHANGE DURING THE EXPLOSION WELDING

The pressure P_m is found from the following solutions. The flyer plate elements equation of motion under the action of the pressure of the detonation products can be written in the form (Dragobetskii 2015, 2019):

$$Dm\frac{d^2 x_1(t)}{dt^2} + \rho c \frac{dx_1(t)}{dt} = P(t), \tag{9.5}$$

$$P(t) = P_m \exp(-t/\theta), \tag{9.6}$$

The solution of (equation 9.5) when neglecting the second term has the form:

$$\dot{x}_1(t) = \frac{P_m \theta}{m}\left(1 - \exp(-t/\theta)\right) + v_o, \tag{9.7}$$

where: v_0 – initial velocity of the flyer plate element at the moment of the shock wave release to the free boundary.

To determine P_m and $\dot{x}_1$, the Chapman–Jouguet condition (Dragobetskii 2015; Vorobyov 2017) for the pressure at the detonation front is used:

$$P_m = \frac{\rho_0 D^2}{k+1} \tag{9.8}$$

where: ρ_0 – initial density of the explosive; D – detonation velocity of the explosive; and k – indicator of the explosive adiabatic.

The pressure on the plate to the pressure of the ratio of the detonation products is determined by the equation:

$$P_x / P_0 = \left[0.5(3k-1)k^{-1}\right]^{2k/(k-1)} \tag{9.9}$$

At the moment of the shock wave release to the free surface, the pressure of the detonation products on the plate is determined by the expression:

$$P_x = \left[0.5(3k-1)k^{-1}\right]^{2k/(k-1)} P_0 \left[H \cdot D^{-1}(H/D + t_o)\right] \tag{9.10}$$

where: $t_0 = \delta_M c_M$ – the time of a shock wave passing on the flyer plate; c_M – shock wave velocity in the plate; and δ_M – flyer plate thickness:

$$c_M = a + \lambda u_x \tag{9.11}$$

where: u_x – mass velocity of the flyer plate; a – speed of sound; and λ – shock compression coefficient of the plate material.

The shock wave velocity at the moment of release on the flyer plate free surface is determined by the ratio:

$$c_M = P_o \cdot (\rho_0 \cdot u_x)^{-1}. \tag{9.12}$$

The system of equations for calculating P_x, u_x, and c_M at the moment of the shock wave release on the plate free surface has the form:

$$\begin{cases} P_x = \left[0.5(3k-1) \cdot k^{-1}\right]^{\frac{2k}{k-1}} P\left[H \cdot D^{-1}(H/D+t_0)^{-1}\right]^k \\ c_M = a + \lambda u_x \\ t_0 = \delta_M / (a + \lambda u_x) \\ P_x = \rho_0 \cdot u_x \cdot c_M \end{cases} \tag{9.13}$$

At the explosive charge blast, the maximum pressure at the front of the shock wave is determined by the empirical formula (Dragobetskii 2015, 2019):

$$P_m = \frac{1.08}{(r_o)^{1.08}} \cdot 10^4, \left(0.0773 \cdot 10^{-4} \leq r \leq 1.082 \cdot 10^{-4}\right) \tag{9.14}$$

where: $r_o = h / \sqrt{q}$ – reduced explosion distance; h – explosion standoff distance (air gap size), m; and q – explosion energy per length unit, J/m.

The change in pressure, depending on the location of the shock wave front point, is approximated by:

$$P_m = \left(31.14 + 89.86\lambda + 380.69\lambda^2\right) \cdot 10^5 \tag{9.15}$$

where: λ is the length of the arc along the shock wave front from a given front point assigned to the length of the arc from the axis of charge to the front point equidistant from both ends.

According to the foregoing, the field of the shock wave peak pressures can be described with sufficient accuracy by expression:

$$P_m = 0.1241 \cdot 10^7 \left(\frac{q}{q_0}\right)^{0.572} \cdot r_0^{-1.144} + \\ + \left[0.1166 \cdot 10^8 \cdot \left(\frac{q}{q_0}\right)^{0.4} \cdot r_0^{-0.805} - 0.1241 \cdot 10^7 \left(\frac{q}{q_0}\right)^{0.572} \cdot r_0^{-1.144}\right] \cdot \lambda \tag{9.16}$$

9.4 MATHEMATICAL MODEL OF THE CLAD STRESS STATE AT THE COLLISION MOMENT

The solution of the equilibrium equation of a flyer plate flat element (Figure 9.2), taking into account the mass character of inertia forces with the adoption of the plate cross-sectional velocity linear relationship along its thickness, has the form:

$$\sigma_r(\delta) = \sigma_{max}^{DYN} \cdot \frac{\delta}{\delta_0} = \sigma_{max}^{DYN} \frac{d\delta}{\delta}, \tag{9.17}$$

$$\frac{d\sigma_r}{d\delta} - \frac{\sigma_r}{\delta} = -\frac{\sigma_{max}^{DYN}}{\delta}. \qquad (9.18)$$

The solution of this equilibrium equation has the form:

$$\sigma_r = \sigma_{max}^{DYN} \cdot \frac{\delta}{\delta_0}\left[\left(\frac{1}{\delta} - \frac{1}{\delta_0}\right) \cdot \delta_0 + \frac{\sigma_r(0)}{\sigma_{max}^{DYN}}\right], \qquad (9.19)$$

where: $\sigma_r(0)$ – stress acting in a flyer plate at the moment of collision with the target plate and equal to:

$$\sigma_r(0) = P_m + 2 \cdot \beta \cdot \sigma_S \cdot \ln\frac{h+\delta}{h}; \qquad (9.20)$$

where: σ_{max}^{DYN} – maximum stress at collision;

$$\sigma_{max}^{DYN} = 0.75 \cdot \rho \cdot V^2 \cdot a \qquad (9.21)$$

Taking into account that the displacement velocity of the flyer plate section based on its thickness at the distance δ_0 from the outer surface varies linearly (Del 1975), i.e., $\frac{d\delta_0}{dt} = V_0 \frac{\delta_0}{\delta}$, we obtain:

$$\frac{d^2V}{dt^2} = \frac{d}{dt}\left(V_0 \cdot \frac{\delta_0}{\delta}\right) = \frac{V_0^2 \cdot \delta_0}{\delta_0^2} \qquad (9.22)$$

With this in mind, we get:

$$\sigma_r = \rho \cdot \frac{V_0^2}{\delta_0^2} \cdot \left(\delta^2 - \delta_0^2\right) + \sigma_r(0) \qquad (9.23)$$

$$\frac{d\sigma_r}{d\delta} - \frac{\sigma_r}{\delta} = -\rho \cdot \frac{V_0^2}{\delta_0^2} \cdot \delta \qquad (9.24)$$

The solution of this equation has the form:

$$\sigma_r = e^{\left(-\int_{\delta_0}^{\delta}\frac{1}{\delta}d\delta\right)} \cdot \left[\int_{\delta_0}^{\delta} -\rho\frac{V_0^2}{\delta_0^2} \cdot \delta \cdot e^{\int_{\delta_0}^{\delta}\frac{1}{\delta}d\delta} \, d\delta + C\right] = \rho\frac{V_0^2}{\delta_0^2} \cdot \left(\delta^2 - \delta_0^2\right) + \sigma_r(0) \qquad (9.25)$$

where: C corresponds to the boundary condition; at $b = r + \delta$, it has value $C = \sigma_r(0)$.

When the flyer (clad) plate element strikes the rigid surface of the target (base) plate, its kinetic energy transforms into the potential energy of deformation (Ogorodnikov 2018b).

The value of the specific potential energy W under the action of inertial forces in the clad element with the length dy at the distance z can be expressed by the following:

$$dW = \sigma_{\max}^{\text{DYN}} \cdot \frac{z}{\delta} d\varepsilon_i \qquad (9.26)$$

where: $d\varepsilon_i$ is the increment of strain intensity.

For the flat deformed state, the increment of the strain intensity will be equal to (Dragobetskii 2018; Trotsko 2018):

$$d\varepsilon_i = \frac{\sqrt{2}}{3}\sqrt{\left(\varepsilon_z - \varepsilon_y\right)^2} = \frac{2}{3}\frac{dz}{\delta} \qquad (9.27)$$

where: $\varepsilon_z = dz/\delta$ and $\varepsilon_y = -\varepsilon_z = -dz/\delta$.

Then the clad element specific strain energy value will be equal to:

$$T = 0.5 \cdot \rho \cdot V \qquad (9.28)$$

In addition, the specific kinetic energy margin of the clad element value will be equal to:

$$dW = \sigma_{\max}^{DYN} \cdot \frac{z}{\delta} d\varepsilon_i \qquad (9.29)$$

When neglecting the energy loss on local collapse at collision, deformations of the target plate, and friction of the flyer plate on the target plate, we take $W = T$.

According to the above, the flyer plate displacement velocity at the moment of collision with the target plate surface will be found from the solution of equation (9.1) and system of equation (9.13):

$$V = \frac{P_m \cdot \theta}{m} \cdot \left(1 - \exp^{-t/\theta}\right) + u_x \qquad (9.30)$$

The value of the time constant of the pressure decline in the shock wave is calculated from the empirical dependence (Dragobetskii 2019; Ogorodnikov 2018b):

$$\theta = 0.115 \cdot Q^{0.5} \cdot \left(\frac{h}{Q^{0.5}}\right)^{0.43} \cdot 10^{-3} \qquad (9.31)$$

where Q is the charge mass per plate length meter, kg.

9.5 RESULTS AND DISCUSSION

To study the strain state across the thickness of bimetallic compositions under pulse compression, in particular, under explosion welding, the grid method, the "inserts" method, and the metallographic method are currently used. The metallographic method is based on the study of the hardening of individual grains of the deformed metal microstructure (Del 1974; Dragobetskii 2018; Trotsko 2018).

The above methods for determining the strain state have some drawbacks. The grid method requires a high class of accuracy and surfaces finishing, which provides a zero gap between the surfaces to preserve the dividing grid during the explosive welding. The "inserts" method determines the direction and magnitude of the metal layers' shear displacements, but does not give the picture of the plastic deformation values realized in the contact layers. Therefore, the listed methods are suitable for the study of shear deformations only in the areas located at a certain distance from the joint line, in which the strains are homogeneous.

At present, there are no methods for studying the stress–strain state in the surface layers of colliding plates. Probably the most promising is the application of the method of determining the stress–strain state by hardness distribution, which is based on the assumption of the appearance of a monosemantic functional relationship between the deformed metal hardness and the stress state intensity. This method can be successfully used to find the boundaries of the deformation zone if it occupies only a localized area of the deformed body volume, as well as to determine the lines of different strain intensities (ε_i) and stress intensities (σ_i) over the entire material volume (Kukharchuk 2017; Polishchuk 2018, 2019).

Moreover, the hardening degree is different in each sublayer. Thus, in the field of plastic deformations, the dependence $\sigma_i(\varepsilon_i)$ in each sublayer has its own parameters. This is caused by the heterogeneity of the structural, mechanical and geometric nature as a result of the bimetal production technological cycle.

When modeling the process of layered blanks pulsed deformation, the deformed metal layer is divided into three characteristic zones:

I. the zone of intense deformation of the colliding plates' surface layers. Plastic deformation in this area provides activation and interaction between the initial elements of the composition.
II. the zone of deformation of the colliding plates' surface layers. In case of the wave-like profile of the connection line, its characteristic feature is the plastic deformation non-homogeneity in the layers, which are parallel to the weld zone.
III. the zone of deformation over the entire depth of welded plates, a characteristic feature of which is the deformation homogeneity of the metal in the layers, which are parallel to the weld zone.

Experimental verification of the numerical and analytical calculations results was carried out on the hardness distribution in deformed plates according to the methods of G. Del' and V. Ogorodnikov (Del 1974, 1975). For the target plate (normal quality structural carbon steel Ст3пс (based on GOST codification); US analogues – carbon steels A284Gr.D and A57036, EU analogues – carbon steels Fe37-3FN, Fe37-3FU and S235), a calibration chart was taken under the dynamic compression conditions. For the flyer plate (corrosion- and heat-resistant steel 08X18H10T (in GOST codification); US analogue – stainless steel AISI 321, EU analogues – stainless steels 1.4541, X10Cr-NiTi18-10 and X6CrNiTi18-10), a calibration chart was taken under dynamic stretching, compression, and bending. The hardness measurement was carried out through 0.5 mm in the cross section, starting from the joint zone. As strain gauge transducers, standard sensors of 49.4–49.49 Ω were applied. Measurement of the sensor resistance of the as-exposed to loading was carried out by the potentiometric method. The DC source was connected to the strain gauge through an additional ballast resistance. The signal appearing on the sensor as a result of deformation was amplified by a strain amplifier with a transmission frequency from 0 to 700 Hz and recorded by an oscilloscope.

Table 9.1 Stresses and strains intensities

Distance from the joint zone (mm)	Material (GHOST codification)	Stress intensity (Num. analysis) (MPa)	Hardness (HV)	Stress intensity ($\sigma_i = f$ (HV) method (MPa)
0	Ст3	1,060	460	1,020 ± 120
0.25	Ст3	1,040	440	1,010 ± 142
0.5	Ст3	996	350	980 ± 111
0.75	Ст3	784	260	800 ± 92
1.0	Ст3	560	180	590 ± 64
1.5	Ст3	560	160	540 ± 62
2.0	Ст3	560	160	540 ± 62
0	08Х18Н10Т	1,090	460	1,120 ± 126
0.25	08Х18Н10Т	1,090	455	1,120 ± 126
0.5	08Х18Н10Т	1,090	375	1,090 ± 118
0.75	08Х18Н10Т	1,090	374	1,090 ± 118

Solving together the equations (9.1), (9.21), (9.30), and (9.31) at $Q=4$ kg/m, we obtain the collision velocity value of 148.12 m/s and the maximum stress values by numerical method. Such stresses exceed the yield strength of the clad material and cause an intense flow of metal. The obtained values of HV, σi, and εi by both methods are presented in Table 9.1.

The accuracy of the mathematical description of the strains change in time with the experimental data was confirmed by the F-test (Fisher criterion). The error margin calculated from the accuracy variance was ±14.2% with a confidence level of 0.95 and a significance level of 0.05.

Comparison of the values of stress intensity, determined numerically and found from the graph, testifies to the reliability of the results obtained and the perfection of the techniques. The obtained results are not limited in application only by explosion welding processes. They can also be used for stress evaluating in profiled and sheet blanks calibrating and in the process of step strain by explosive forming.

9.6 CONCLUSIONS

Based on the solution of the equilibrium equations during the collision of the flyer plate (clad) with the target plate (base) in the explosion welding processes, the dependences of the dynamic compression stresses from the explosive loading pressure arising are obtained. This makes it possible to specify the mechanism of the clad shape change under the pulse load before and during the collision with the base plate.

REFERENCES

Del', G.D., Ogorodnikov, V.A., Nakhaichuk, V.G. 1975. Criterion of deformability of pressure shaped metals. *Izv Vyssh Uchebn Zaved Mashinostr* 4: 135–140.
Del', G.D., Ogorodnikov, V.A., Spiridonov, L.K. 1974. Plasticity of metal subjected to complex loading. *Izv Vyssh Uchebn Zaved Mashinostr* 12: 22–26.

Dragobetskii, V., Shapoval, A., Mos'pan, D., Trotsko, O., Lotous, V. 2015. Excavator bucket teeth strengthening using a plastic explosive deformation. *Metallurgical and Mining Industry* 4: 363–368.

Dragobetskii, V., Shapoval, A., Naumova, O., Shlyk, S., Mospan, D., Sikulskiy, V. 2007. The technology of production of a copper – aluminum – copper composite to produce current lead buses of the high – voltage plants; *Proc. IEEE International Conference on Modern Electrical and Energy Systems*: 400–403.

Dragobetskii, V., Zagirnyak, M., Naumova, O., Shlyk, S., Shapoval, A. 2018. Method of determination of technological durability of plastically deformed sheet parts of vehicles. *International Journal of Engineering & Technology* 7(4.3): 92–99.

Dragobetskii, V., Zagirnyak, V., Shlyk, S., Shapoval, A., Naumova, O. 2019. Application of explosion treatment methods for production items of powder materials. *Przeglad Elektrotechniczny* 95(5): 39–42.

Kukharchuk, V.V., Bogachuk, V.V., Hraniak, V.F., Wójcik, W., Suleimenov, B., Karnakova, G. 2017. Method of magneto-elastic control of mechanic rigidity in assemblies of hydropower units, *Proc. SPIE 10445, Photonics Applications in Astronomy, Communications, Industry, and High Energy Physics Experiments 2017*, 104456A (7 August 2017).

Ogorodnikov, V.A., Dereven'ko, I.A., Sivak, R.I. 2018. On the influence of curvature of the trajectories of deformation of a volume of the material by pressing on its plasticity under the conditions of complex loading. *Materials Science* 54(3): 326–332.

Ogorodnikov, V.A., Grechanyuk, N.S., Gubanov, A.V. 2018. Energy criterion of the reliability of structural elements in vehicles. *Materials Science* 53(5): 645–650.

Ogorodnikov, V.A., Savchinskij, I.G., Nakhajchuk, O.V. 2004. Stressed-strained state during forming the internal slot section by mandrel reduction. *Tyazheloe Mashinostroenie* 12: 31–33.

Polishchuk, L.K., Kozlov, L.G., Piontkevych, O.V., Gromaszek, K., Mussabekova, A. 2018. Study of the dynamic stability of the conveyor belt adaptive drive. *Proc. SPIE 10808, Photonics Applications in Astronomy, Communications, Industry, and High-Energy Physics Experiments 2018*, 1080862: 1117–1122.

Polishchuk, L.K., Kozlov, L.G., Piontkevych, O.V., Horbatiuk, R.M., Pinaiev, B., Wójcik, W., Sagymbai, A., Abdihanov, A. 2019. Study of the dynamic stability of the belt conveyor adaptive drive. *Przeglad Elektrotechniczny* 95(4): 98–103.

Trotsko, O., Sergii Shlyk, S. 2018. Development of the sheet blanks forming mathematical model for calculation using simulation in ANSYS software. *Proc. IEEE International Scientific and Technical Conference on Computer Sciences and Information Technologies (CSIT)*: 169–172.

Vorobyov V., Pomazan M., Vorobyova L., Shlyk, S. 2017. Simulation of dynamic fracture of the borehole bottom taking into consideration stress concentrator. *Eastern-European Journal of Enterprise Technologies* 3/1(87): 53–62.

Chapter 10

Tensor models of accumulation of damage in material billets during roll forging process in several stages

V. Matviichuk, I. Bubnovska, V. Mykhalevych, M. Kovalchuk,
W. Wójcik, A. Tuleshov, S. Smailova, and B. Imanbek

CONTENTS

10.1 Introduction 111
10.2 Analysis 111
10.3 Results of the study 112
10.4 Conclusions 118
References 118

10.1 INTRODUCTION

The development of cold rolling forging processes is hindered by the lack of information on the mechanics of forming curvilinear blanks, the stress–strain state, and the materials' deformability (Kukhar 2018).

As a result of the study of the mechanics of forming blanks during roll forging, we have come up with a method for manufacturing curvilinear blanks by two-stage roll forging. In the suggested method, in the first stage, roll forging is performed on a flat roll body, and in the second stage, roll forging of the billet is performed in gauges of cylinder rolls. In this case, in the second stage, the billet is rotated by 90°, and rolling forging is then performed in gauges of cylinder rolls.

In the process of manufacture of curvilinear blanks by cold roll forging, the matter of evaluating fractures of the billet material becomes particularly relevant. The limiting degree of compression and the mechanical characteristics of the material of the products depend on this. It is clear that at different stages of rolling forging, the stress–strain state of the particles of the billet material varies and is not monotone, which causes additional difficulties in the process of fracture evaluation of the billet material.

The aim of the research is to develop models of fracture evaluation of the billet material for cold roll forging in schemes in two or more stages.

10.2 ANALYSIS

To evaluate fractures of the billet material in roll forging, it is necessary to choose an approach to the calculation of equivalent plastic strain at the fracture point. Judging by the hundreds and most likely thousands of papers on this topic, the number of which increases annually, the most widely used approach is based on the damage summation

DOI: 10.1201/9781003225447-10

theory. So, the preface to the proceedings of the forum on the forming technology notes: "A more widely used approach is the identification of fracture strains in the function of stress triaxiality as proposed by Johnson and Cook. This in combination to damage evolution approaches is better able to model the experimental outcomes" (Hora 2018).

It is enough to point out that almost all of the reports submitted to this forum use said approach to model the fracture. However, within the framework of this approach, a large number of damage summation models have been proposed. An idea of these models can be obtained from earlier works (Cockcroft 1968; Rice 1969). In the works (Del 1978; Rene 1976), the more complicated principle of summing damages is substantiated. The authors (Oh 1979), along with the stress triaxiality, discuss the use of invariant indexes in the form of the ratio of principal stresses. Clift et al. (1990) considered a simplified model. A comparative analysis of various fracture models is carried out periodically, for example (Bao 2004b; Golling 2017). Some key features of fracture models for sheet materials are considered in (Bai 2008; Bao 2004a, 2005). In (Gese 2007; Hooputra 2004), revolutionary proposals were made to account for the type of fracture for sheet materials. In (Del 1975a; Ogorodnikov 2018a), a fracture criterion was developed, which takes into account additional features of the strain trajectory, and also some of its properties are investigated. In (Grushko 2012; Mikhalevich 1994), results were obtained justifying the use of new fracture models. These models relate primarily to the initially isotropic body or to equivalent plastic strain at the fracture point. One more important feature of these models is the scalar description for damage particulates.

Models based on the tensor description of the material particulate damage were proposed in (Del 1983; Ilyushin 1967; Mikhalevich 1996). However, over the past decades, the use of tensor models is clearly inferior to the popularity of scalar models. For example, in (Hooputra 2004), the tensor model is mentioned but not discussed.

Scalar fracture models for an initial anisotropic body in the framework of the approach under consideration were proposed in (Ilyushin 1967; Park 2018; Yoon 2017). The paper (Basak 2019; Mikhalevich 1993) is one of the few in which a tensor model was constructed for an initially anisotropic body (Polishchuk 2016a, 2016b, 2018).

10.3 RESULTS OF THE STUDY

By means of experimental and analytical studies and the use of simulation modeling in the DEFORM 3D software package, it was found that for particles of the material on the free surface of the billet, for which the stress triaxiality changes from $\eta = 1$ at the first stage to $\eta = -1$ at the end of roll forging, at the second and third stages, a linear law is valid for the strain path (Kozlov 2019; Ogorodnikov 2018b; Polishchuk 2019):

$$\bar{\varepsilon} = -k \cdot \eta + b, \ k, b > 0 \tag{10.1}$$

where: $\bar{\varepsilon}$ – equivalent plastic strain; and η – stress triaxiality:

$$\eta = \frac{3 \cdot \sigma_m}{\sigma_v} \tag{10.2}$$

where: σ_m – hydrostatic stress; and σ_v – equivalent stress (von Mises).

Equation (10.1) can be represented as:

$$\eta = \frac{\bar{\varepsilon} - \bar{\varepsilon}_0}{\bar{\varepsilon}_1 - \bar{\varepsilon}_0} \tag{10.3}$$

where: $\bar{\varepsilon}_0, \bar{\varepsilon}_1$ – values of the equivalent plastic strain on the path strain at $\eta = 0$ and $\eta = 1$.

Those particles of material that, in the first stage, were deformed under "hard" stress–strain state schemes, in the second stage, will be in "soft" conditions of deformation. In this case, there is non-monotonic deformation. Therefore, to evaluate the fracture of a material, it is necessary to use models that take into account the directional nature of damage accumulation and their dependence on the type of stress state. Such models are tensor models.

Despite the simplicity of relation (1), an attempt to apply a tensor model to describe the process of changing the material's stress–strain state, which leads to the specified strain path, showed the need to adopt a number of assumptions to obtain unambiguous results (Dragobetskii 2015; Ogorodnikov 2004, 2018c).

Let us suggest a deformation model of certain particulates in the billet material in the form of a deformation process within four separate stages. For this case, using a tensor-linear model, we obtain:

$$\psi_{ij}\left((\bar{\varepsilon})_k\right) = n \cdot \sum_{k=1}^{4} \int_{(\bar{\varepsilon})_{k-1}}^{(\bar{\varepsilon})_k} \frac{\bar{\varepsilon}^{n-1}}{\bar{\varepsilon}_{fs}^n [\eta(\bar{\varepsilon})]} \cdot \beta_{ij}(\bar{\varepsilon}) \cdot d\bar{\varepsilon} \tag{10.4}$$

where: $\bar{\varepsilon}_{fs}(\eta)$ – curve of the limit equivalent plastic strain to fracture under stationary deformation; $(\bar{\varepsilon})_{k-1}, (\bar{\varepsilon})_k$ – equivalent plastic strain, respectively, at the beginning and the end of the k-th deformation stage $((\bar{\varepsilon})_0 = 0, (\bar{\varepsilon})_k < \bar{\varepsilon}_f)$; and β_{ij} – directional tensor for increment strain:

$$\beta_{ij} = \sqrt{\frac{2}{3}} \cdot \frac{d\varepsilon_{ij}}{d\bar{\varepsilon}} \tag{10.5}$$

where: $d\varepsilon_{ij}$ – strain increment tensor; $d\bar{\varepsilon}$ – plastic strain increment intensity; and $n > 0$ – material constant.

In the future, we assume that the properties of the material of the billet will be described by the defining relations of flow theory (Del 1975b; Vorobyov 2017; Kukharchuk 2017a):

$$d\varepsilon_{ij} = \frac{3}{2} \cdot \frac{\bar{\varepsilon}}{\sigma_u} \cdot s_{ij} = \frac{3}{2} \cdot \frac{\bar{\varepsilon}}{\sigma_u} \cdot \left(\sigma_{ij} - \delta_{ij} \cdot \sigma\right) \tag{10.6}$$

where: σ_{ij}, s_{ij} – stress tensor and stress deviator, respectively; and δ_{ij} – Kronecker symbol:

$$\delta_{ij} = \begin{cases} 1, i = j, \\ 0, i \neq j. \end{cases} \tag{10.7}$$

114 Mechatronic Systems 2

In this case, both the directional tensor for increment strain and the directional tensors for the stress coincide:

$$\beta_{ij} = \sqrt{\frac{2}{3}} \cdot \frac{d\varepsilon_{ij}}{d\bar{\varepsilon}} \qquad (10.8)$$

For particles of the billet material that belonged to the free surface of the billet in the deformation process in the first stage, each of the four stages is considered separately (Table 10.1).

Table 10.1 Dependencies of the directional tensor for increment strain under various types of stress state

Stage (k)	Stress-state	Directional tensor for increment strain $\left(\beta_{ij}^{(k)}\right)$		
1	Uniaxial tension	$\begin{bmatrix} \frac{\sqrt{6}}{3} & 0 & 0 \\ 0 & -\frac{\sqrt{6}}{6} & 0 \\ 0 & 0 & -\frac{\sqrt{6}}{6} \end{bmatrix}$	, $\eta = 1$	(10.11)
2	Uniaxial tension + torsion	$\begin{bmatrix} \frac{\sqrt{6}}{3} \cdot \eta(\bar{\varepsilon}) & \sqrt{1-\eta^2(\bar{\varepsilon})} & 0 \\ \sqrt{1-\eta^2(\bar{\varepsilon})} & -\frac{\sqrt{6}}{6} \cdot \eta(\bar{\varepsilon}) & 0 \\ 0 & 0 & -\frac{\sqrt{6}}{6} \cdot \eta(\bar{\varepsilon}) \end{bmatrix}$	, $0 \leq \eta < 1$	(10.12)
3	Uniaxial compression + torsion	$\begin{bmatrix} \frac{\sqrt{6}}{6} \cdot \eta(\bar{\varepsilon}) & \sqrt{1-\eta^2(\bar{\varepsilon})} & 0 \\ \sqrt{1-\eta^2(\bar{\varepsilon})} & -\frac{\sqrt{6}}{6} \cdot \eta(\bar{\varepsilon}) & 0 \\ 0 & 0 & -\frac{\sqrt{6}}{6} \cdot \eta(\bar{\varepsilon}) \end{bmatrix}$	, $-1 \leq \eta < 0$	(10.13)
4	Uniaxial compression + hydrostatic compression	$\begin{bmatrix} -\frac{\sqrt{6}}{3} & 0 & 0 \\ 0 & \frac{\sqrt{6}}{6} & 0 \\ 0 & 0 & \frac{\sqrt{6}}{6} \end{bmatrix}$	, $\eta < -1$	(10.14)

In accordance with the representation (10.1), (10.4), (10.9) for the first stage (Kukharchuk 2017b):

$$\psi_{ij}(\bar{\varepsilon}) = \left(\frac{\bar{\varepsilon}}{\varepsilon_t^*}\right)^n \cdot \beta_{ij}^{(1)}, \quad 0 \leq \bar{\varepsilon} \leq -k+b < \varepsilon_p^* \tag{10.9}$$

where: $\varepsilon_t^* = \bar{\varepsilon}_{fs}(\eta = 1)$ – equivalent plastic strain to fracture under tension.
By analogy, for the second and third stages, we obtain:

$$\psi_{ij}(\bar{\varepsilon}) = \left(\frac{b-k}{\varepsilon_t^*}\right)^n \cdot \beta_{ij}^{(1)} + n \cdot \int_{b-k}^{\bar{\varepsilon}} \frac{\bar{\varepsilon}^{n-1}}{\varepsilon_{fs}^n[\eta(\bar{\varepsilon})]} \cdot \beta_{ij}^{(2)}(\bar{\varepsilon}) \cdot d\bar{\varepsilon}, \quad b-k \leq \bar{\varepsilon} \leq k+b \tag{10.10}$$

The curve of the limit equivalent plastic strain to fracture under stationary deformation is represented by the well-known expression:

$$\bar{\varepsilon}_{fs}(\eta) = \varepsilon_s^* \cdot \left(\frac{\varepsilon_t^*}{\varepsilon_c^*}\right)^{\frac{\eta}{2}} \cdot \left(\frac{\varepsilon_t^* \cdot \varepsilon_c^*}{(\varepsilon_s^*)^2}\right)^{\frac{\eta^2}{2}} \tag{10.15}$$

where: $\varepsilon_s^* = \bar{\varepsilon}_{fs}(\eta = 0)$ – equivalent plastic strain to fracture in shear; $\varepsilon_c^* = \bar{\varepsilon}_{fs}(\eta = -1)$ – equivalent plastic strain to fracture in compression.

As a measure function of damage, the second invariant of the damage deviator was adopted:

$$\psi = \psi_{ij} \cdot \psi_{ij}, \quad 0 \leq \psi \leq 1 \tag{10.16}$$

In the initial state $\psi(\bar{\varepsilon} = 0) = 0$, and the achievement of the limit state is determined by the condition:

$$\psi(\bar{\varepsilon} = \bar{\varepsilon}_f) = 1 \tag{10.17}$$

where: $\bar{\varepsilon}_f$ – equivalent plastic strain to fracture under conditions an arbitrary process of deformation (η=const or $\eta \neq$const; if η=const $\Rightarrow \bar{\varepsilon}_f \equiv \bar{\varepsilon}_{fs}$).

The simulation of damage accumulation was carried out in the environment of the Maple computer mathematics system using a program developed by us. The results of the calculations of the values under study, depending on the stress triaxiality, are presented in Figure 10.1.

The simulation results demonstrate that regardless of the value of the nonlinearity index for the summation of damages n, the damage accumulation rate begins to decrease at $\eta < -0.3$.

For the range of values $n = 1 \div 1.5$, there is a maximum value for the damage accumulation curves when approaching the stressed compression state.

This is a very important, at least theoretical, result. Since after reaching the maximum value, the damage begins to decrease against a monotonous increase in the equivalent plastic strain.

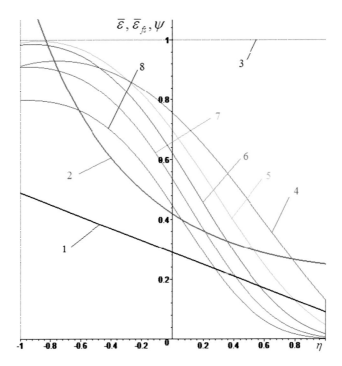

Figure 10.1 Strain path (1 − (1), $k = 0.2$, $b = 0.29$); the curve of the limit equivalent plastic strain to fracture under stationary deformation for steel EP866 ($\varepsilon_t^* = 0.25$, $\varepsilon_s^* = 0.42$, $\varepsilon_c^* = 1.25$) (2 − (15)); 3 − damage intensity limit; 4÷8 − damage accumulation in stages 2, 3 − (10), (16): 4 − $n = 1$; 5 − $n = 1.5$; 6 − $n = 2$; 7 − $n = 2.5$; 8 − $n = 3$.

A similar result in a somewhat different form was discovered earlier in processes accompanied by a discrete change in the direction of plastic strain increment. It is under the given conditions that the indicated effect was experimentally discovered and theoretically described in the papers of Dell (1983). A more detailed theoretical and experimental study of certain aspects of this phenomenon is represented in the papers of V. Mykhalevych and V. Matviichuk (Mikhalevich 1993, 1994).

In this case, the effect under study was found in the processes with continuous changes in the directions of the principal strain increments. The numerical values of this effect are so minor that they do not go beyond the limits of the experimental data scattering. This makes it impossible for us to experimentally substantiate or argue against the discovered theoretical effect. On the other hand, a more detailed theoretical study of this effect may lead to confirming conditions under which it will manifest itself to a greater degree, which, in turn, will create the necessary prerequisites for experimental research.

The graphs in Figure 10.1 clarify that the calculated damage value at the end of the first stage monotonously decreases with an increase in the nonlinearity index of damage summation $n \in [1:3]$. As for the calculated damage value at the end of the third stage, here we get the non-monotonic dependence presented in Figure 10.2.

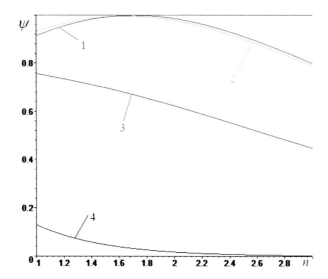

Figure 10.2 Damage at a given point of the deformation path (4) depending on the value of the nonlinearity parameter of damage summation: $1 - \bar{\varepsilon} = 0.49$; $2 - \bar{\varepsilon} = 0.44$; $3 - \bar{\varepsilon} = 0.29$; $4 - \bar{\varepsilon} = 0.09$.

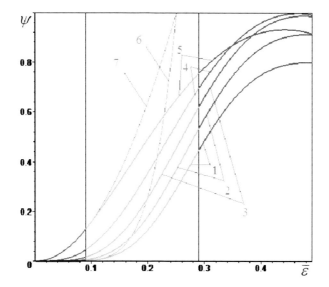

Figure 10.3 Damage at a given point of the deformation path (4) depending on the value of the nonlinearity parameter of damage summation: $1 - \bar{\varepsilon} = 0.49$; $2 - \bar{\varepsilon} = 0.44$; $3 - \bar{\varepsilon} = 0.29$; $4 - \bar{\varepsilon} = 0.09$.

Curves 6 and 7 in Figure 10.3 reflect the pattern of damage accumulation to fracture under tension. As can be seen with $n \geq 1$ and the constant stress state, we get an avalanche accumulation of damage, which is consistent with the classical physical concepts of scattered fracture.

At the same time, by changing the stress state in the process of deformation, the accumulation of damage can be accelerated or slowed down. In this case, the "softening" of the stress state from uniaxial tension to shear reduces the rate of damage accumulation, as can be seen from the curves in Figure 10.3, related to the second stage of the process. Further "softening" of the stress state "shear+compression" – "compression" further slows down the accumulation of damages up to the already noted effect of their partial "healing."

Calculations were carried out for the curve of the limit equivalent plastic strain to fracture at stationary deformation, which is the case for steel EP866. In Figure 10.3, the behavioral regularities of damage accumulation are presented depending on the equivalent plastic strain (Wojcik 1997, 2010).

10.4 CONCLUSIONS

For the first two stages of the deformation trajectory under study, according to the tensor theory of damage accumulation, the most rigorous prediction of limiting deformations corresponds to the linear principle of damage accumulation. An increase in the parameter n, which characterizes the degree of the nonlinear nature of damage accumulation, corresponds to a decrease in the damage accumulation rate. The third stage is characterized by a significant decrease in the rate of damage accumulation and the not monotonic nature of depending on the calculated limit values in the index *n*. For the first time, the effect of partial "healing" of damage in the processes with continuous changes in the directions of the principal strain increments was theoretically discovered.

REFERENCES

Bai, Y., Wierzbicki, T. 2008. A new model of metal plasticity and fracture with pressure and Lode dependence. *International Journal of Plasticity* 24: 1071–1096.

Bao, Y., Wierzbicki, T. 2004. A comparative study on various ductile crack formation criteria. *Journal of Engineering Materials and Technology* 126: 314–324.

Bao, Y., Wierzbicki, T. 2004. On fracture locus in the equivalent strain and stress triaxiality space. *International Journal of Mechanical Sciences* 46(1): 81–98.

Bao, Y., Wierzbicki, T. 2005. On the cut-off value of negative triaxiality for fracture. *Engineering Fracture Mechanics* 72: 1049–1069.

Basak, S., KumarPanda, S. 2019 Failure strains of anisotropic thin sheet metals: Experimental evaluation and theoretical prediction. *International Journal of Mechanical Sciences* 151: 356–374.

Clift, S. E., Hartley, P., Sturgess, C.E.N., Rowe, G.W. 1990. Fracture prediction in plastic deformation processes. *International Journal of Mechanical Science* 32(1): 1–17.

Cockcroft, M. G., Latham, D. J. 1968. Ductility and the Workability of Metals, *Journal of the Institute of Metals* 96: 33–39.

Del, G.D. 1983. Plastichnost deformirovannogo metalla. *Fizika i tehnika vyisokih davleniy* 11: 28–32.

Del, G.D., Ogorodnikov, V.A., Nakhaichuk, V.G. 1975. Criterion of deformability of pressure shaped metals. *Izv. Vyssh. Uchebn. Zaved. Mashinostr.* 4: 135–140.

Del, H.D. 1978. Tekhnolohycheskaia mekhanyka. *Mashynostroenye*: 1–174.

Del', G.D., Ogorodnikov, V.A., Nahaychuk, V.G. 1975. Kriteriy deformiruemosti metallov pri obrabotke davleniem. *Izv. vuzov. Mashinostroenie* 4: 135–137.

Dragobetskii, V., Shapoval, A., Mos'pan, D., Trotsko, O., Lotous, V. 2015. Excavator bucket teeth strengthening using a plastic explosive deformation. *Metallurgical and Mining Industry* 4: 363–368.

Gese, H., Oberhofer, G., Dell, H. 2007. Consistent modeling of plasticity and failure in the process chain of deep drawing and crash with user material model MF-GenYld + CrachFEM for LS-Dyna. *LS-DYNA Anwenderforum*, Frankenthal: 592–598.

Golling, S., Östlund, R., Schill, M. 2017. A comparative study of different failure modeling strategies on a laboratory scale test component. *Proc. of 6th International Conference Hot Sheet Metal Forming of High-Performance Steel CHS2*: June 4–7 2017, Atlanta: 37–46.

Grushko, A.V., Sheykin, S.E., Rostotskiy, I.Y. 2012. Contact pressure in hip endoprosthetic swivel joints. *Journal of Friction and Wear* 33(2): 124–129.

Hooputra, H., Gese, H., Dell, H., Werner, H. 2004. A comprehensive failure model for crashworthiness simulation of aluminium extrusions, *International Journal of Crashworthiness* (9)5: 449–464.

Hora, P. (ed.) 2018. Experimental and numerical methods in the FEM based crack prediction. *Proc. 11th Forming Technology Forum Zurich*: 1–94.

Ilyushin, A.A. 1967. Ob odnoy teorii dlitelnoy prochnosti. *Mehanika tverdogo tela* 13: 21–25.

Kozlov, L.G., Polishchuk, L.K., Piontkevych, O.V., Korinenko, M.P., Horbatiuk, R.M., Komada, P., Orazalieva, S., Ussatova, O. 2019. Experimental research characteristics of counter balance valve for hydraulic drive control system of mobile machine. *Przeglad Elektrotechniczny* 95(4): 104–109.

Kukhar, V.V., Grushko, A.V., Vishtak, I.V. 2018. Shape indexes for dieless forming of the elongated forgings with sharpened end by tensile drawing with rupture. *Solid state phenomena*, 284: 408–415.

Kukharchuk, V.V., Bogachuk, V.V., Hraniak, V.F., Wójcik, W., Suleimenov, B., Karnakova, G. 2017. Method of magneto-elastic control of mechanic rigidity in assemblies of hydropower units. *Proc. SPIE 10445, Photonics Applications in Astronomy, Communications, Industry, and High Energy Physics Experiments 2017*, 104456A: 1221–1232.

Kukharchuk, V.V., Kazyv, S.S., Bykovsky, S.A. 2017. Discrete wavelet transformation in spectral analysis of vibration processes at hydropower units. *Przeglad Elektrotechniczny* 93(5): 65–68.

Mikhalevich, V. M. 1996. Tensor models of rupture strength. Report no. 3. Criterional relations for loading with a change in stress state and the directions of the principal stresses. *Strength of Materials* 28(3): 238–246.

Mikhalevich, V. M., Matvijchuk, V.A., Egorov, V.P., Kornet, I.F. 1994. Isothermal blades rolling. *Kuznechno-Shtampovochnoe Proizvodstvo* 3: 6–9.

Mikhalevich, V.M. 1993. Models of defects accumulation for solids with original and strain-induced anisotropy. *Izvestia Akademii nauk SSSR. Metally* 5: 144–151.

Ogorodnikov, V.A., Savchinskij, I.G., Nakhajchuk, O.V. 2004. Stressed-strained state during forming the internal slot section by mandrel reduction. *Tyazheloe Mashinostroenie* 12: 31–33.

Ogorodnikov, V.A., Zyska, T., Sundetov, S. 2018. The physical model of motor vehicle destruction under shock loading for analysis of road traffic accident. *Proc. SPIE 10808, Photonics Applications in Astronomy, Communications, Industry, and High-Energy Physics Experiments 2018*, 108086C: 1015–1021.

Ogorodnikov, V.A., Dereven'ko, I.A., Sivak, R.I. 2018. On the influence of curvature of the trajectories of deformation of a volume of the material by pressing on its plasticity under the conditions of complex loading. *Materials Science* 54(3): 326–332.

Ogorodnikov, V.A., Dereven'ko, I.A., Sivak, R.I. 2018. On the influence of curvature of the trajectories of deformation of a volume of the material by pressing on its plasticity under the conditions of complex loading. *Materials Science* 54(3): 326–332; doi:10.1007/s11003-018-0188-x.

Oh, S.I., Chen, C.C., Kobayashi, S. 1979. Ductile fracture in axisymmetric extrusion and drawing. Part 2. Workability in extrusion and drawing. *ASME Journal of Engineering for Industry* 101: 36–44.

Park, N., Huh, H., Yoon, J.W. 2018. Anisotropic fracture forming limit diagram considering non-directionality of the equibiaxial fracture strain. *Int J Solids Struct* 151: 181–194.

Polishchuk, L., Bilyy, O., Kharchenko, Y. 2016. Prediction of the propagation of crack-like defects in profile elements of the boom of stack discharge conveyor. *Eastern-European Journal of Enterprise Technologies* 6(1): 44–52.

Polishchuk, L., Kharchenko, Y., Piontkevych, O., Koval, O. 2016. The research of the dynamic processes of control system of hydraulic drive of belt conveyors with variable cargo flows. *Eastern-European Journal of Enterprise Technologies* 2(8): 22–29.

Polishchuk, L.K., Kozlov, L.G., Piontkevych, O.V., Gromaszek, K., Mussabekova, A. 2018. Study of the dynamic stability of the conveyor belt adaptive drive. *Proc. SPIE 10808, Photonics Applications in Astronomy, Communications, Industry, and High-Energy Physics Experiments* 1080862: 1109–1114.

Polishchuk, L.K., Kozlov, L.G., Piontkevych, O.V., Horbatiuk, R.M., Pinaiev, B., Wójcik, W., Sagymbai, A., Abdihanov, A. 2019. Study of the dynamic stability of the belt conveyor adaptive drive. *Przeglad Elektrotechniczny* 95(4): 98–103.

Renne, I.P., Ogorodnikov, V.A., Nakhaichuk, V.G. 1976. Plotting plasticity diagrams by testing cylindrical samples in combined tension and torsion. *Strength of Materials*, 8(6): 733–737.

Rice, J.R., Tracey, D.M. 1969. On the ductile enlargement of voids in triaxial stress fields. *Journal of the Mechanics and Physics of Solids* 3: 201–217.

Vedmitskyi, Y.G., Kukharchuk, V.V., Hraniak, V.F. 2017. New non-system physical quantities for vibration monitoring of transient processes at hydropower facilities, integral vibratory accelerations. *Przeglad Elektrotechniczny* 93(3): 69–72.

Vorobyov, V., Pomazan, M., Vorobyova, L., Shlyk, S. 2017. Simulation of dynamic fracture of the borehole bottom taking into consideration stress concentrator. *Eastern-European Journal of Enterprise Technologies* 3/1(87): 53–62.

Wojcik, J, Wojcik, W, Janoszczyk, B. 1997. Optical fibre system for flame monitoring in energetic boilers. Technology and Applications of Light Guides Book Series: *Proc. SPIE* 3189: 74–82.

Wojcik, W., Kisala, P. 2010. The method for the recovery of the apodization function of the fiber Bragg gratings on the basis of its spectra. *Przeglad Elektrotechniczny* 86(10): 127–130.

Yoon, J.W., Zhang, S., Stoughton, T.B. 2017. Anisotropic fracture criterion and its calibration, *Proc. of 6th International Conference Hot Sheet Metal Forming of High-Performance Steel CHS2*. June 4–7 2017, Atlanta: 26–27.

Chapter 11

Synergetic aspects of growth in machining of metal materials

E. Posviatenko, N. Posvyatenko, O. Mozghovyi,
R. Budyak, A. Smolarz, A. Tuleshov, G. Yusupova,
and A. Shortanbayeva

CONTENTS

11.1 Introduction 121
11.2 Research results 123
11.3 Discussion and practical significance of the research results 126
11.4 Conclusion 127
References 127

11.1 INTRODUCTION

The term "growth" means the formation on a surface of one solid of another substance when they are in contact (Kuznetsov 1956). Concerning the machining of metal materials by pressure, cutting, etc., this can be the formation of the processed material on the working surface of the punch or cutting wedge. The growth is in the state of complete compression and firmly held on this surface of the instrument.

The first experimental research of the growth formation was performed by the physicist Usachov (1952). An optical microscope, artificial thermocouples, a method of stopping the processing, and metallography were used in his research. In 1952, Y.G. Usachov found that the growth is formed from the processed material and serves as an additional cutting wedge. Later research (Kuznetsov 1956) was intended to show that the body of the growth is formed with the involvement of additional environmental factors. As a result, the growth must have a different chemical composition than the material being processed. However, our research (Posvyatenko 1996) showed the falsity of such a physical nature of the growth. Thus, at the current level of experimental technology, the basic provisions of Y.G. Usachov were improved.

The aim of the research was to determine the cause of the growth formation, the positive and negative role of this formation in the machining of metal materials as well as the practical use of the growth formation.

The theoretical part of the research is based on the principles of synergetics (Hacken 1980; Ivanova 1986; Prigogine 1986). A new area of science that studies the patterns of self-organization of ordered structures of diverse origin, focusing on irreversible processes in thermodynamically unbalanced systems, was dubbed "synergetics" (Ivanova 1994; Kanarchuk 2002).

This term coined by Hermann Haken means a concerted, joint action. The principles of synergetics can be successfully applied in the study of technological systems,

DOI: 10.1201/9781003225447-11

particularly, in determining of the interaction of its components, and therefore, when solving the problem of the formation of the work determined in this paper.

This derives from the methodology of synergetics, as unbalanced thermodynamics studies, irreversible processes that lead to a decrease in entropy through the self-organization of ordered or dissipative structures, which occurs in open systems that exchange energy and matter with the surrounding medium.

For nonequilibrium systems, the main characteristic is bifurcations, the mechanism of which was investigated by I. Prigogine. According to his conclusions, the system near the bifurcation points selects one or more options for further development by random fluctuations, losing stability. The alternation of stability and instability is a general phenomenon in the evolution of any open system, which, after passing the bifurcation, can no longer return to its original state.

In the process of our previous studies (Polishchuk 2016b; Posvyatenko 2017), it has been established that from the positions of synergetics in technological systems (TS), the detail is dominant (Figure 11.1).

Due to its shape, size, surface properties, and material, it affects the kinematics and cutting physics, which includes process modes. At the same time, the interaction of the "detail-process" can be considered already stabilized. The chain of some feedback, that is, the self-organization of the TS, is the influence of the physics of the process on the physical–mechanical and geometric properties of the surface of the part. The second most important component of the TS is the cutting tool (CT). Its direct connection with the detail influences the shape and size of the latter based on the type of instrument and structural material (KM) on the instrumental (MI).

The last link is also retroactive and very powerful, since today material science is developing intensively. The geometric parameters of CT and MI influence the physics of the cutting process through contact phenomena, and the effect is reversed due to wear and stability of the CT (Polishchuk 2016a).

The tool acts very strongly on the machine through the MI and the physics of the cutting process, as can be seen from the following examples. The application of tools from high-speed steel at the beginning of the twentieth century increased the speed of cutting by 3–5 times, which could not but affect the design of the machine and the cutting process. So, only 3 years after the invention of a high-speed steel, Nicholson (USA), using a specially created three-component dynamometer caliper, investigated the components of the cutting forces of a tool from the new MI in order to enhance the design of the lathe. Thus, thanks to the high-speed steel, the machines became more high-speed, rigid, and massive (Kozlov 2019; Polishchuk 2019).

Thus, the ranking of the components of the technological system for the processing of materials by cutting in the context of strong direct impact should be considered as follows: the part–tool–process–machine. At the same time, in the interaction of individual components of the parts of the TS, there is traceable and slightly weaker feedback. For the effective development of the TS, it should be synthesized and investigated only in the interconnection of the components (Ogorodnikov 2018, 2019).

It should be noted that the buildup, as the main component of the structural material of the part, deserves special attention, since it is an important characteristic of this material at low and medium cutting speeds (broaching, screw cutting and gear cutting, drilling, planing, slotting, etc.).

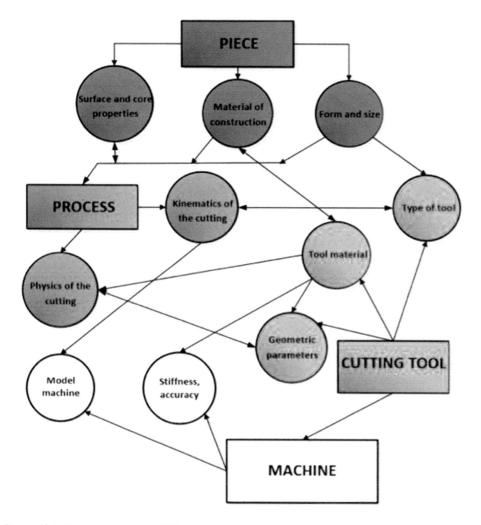

Figure 11.1 Scheme of mutual influence of the components of the technological system in the processing of material cutting.

Our previous studies (Posvyatenko 2018) also showed that the self-organization of the structure of the formation of growth during the mechanical treatment of metal deformable materials can be effectively influenced by the pre-cold plastic deformation and the medium (Dragobetskii 2015; Ogorodnikov 2004).

11.2 RESEARCH RESULTS

The following were used in the work: micro-volume PMT-3; MIM-7 and Neophot-21 microscopes; block profile-profilometer VEI – "Caliber"; REM-1064 electronic microscope and "Camscan 4-DV" scanning electron microscope; Dataletty 150 microdurometer

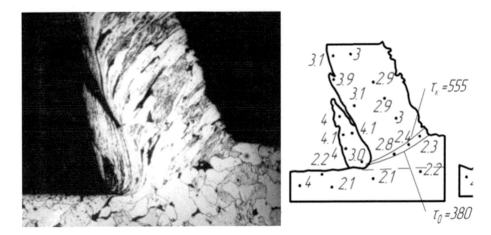

Figure 11.2 Results of the research by methods of microstructural analysis (×50) and microhardness of the chip formation zone during free orthogonal cutting of steel 10 after cold strain hardening (e = 0.25; HV 2.1 GPa): v = 0.15 m/s, Sz = 0.15 mm; cutter – steel P6M5; γ = 15°; ρ = 0.008 mm; coolant liquid – sulfoprezol; dimensional characteristics of the taut-distorted: $H\mu$ – GPa; $\tau 0$, τk-MPa.

(Shimadzu); Talisuri-5 profilograph-profilometer; Taliscan high-speed device for three-dimensional scanning of machined surface. An EWEL web camera and a latest-generation PC were also used in the experiments.

The results of the research are presented in Figure 11.2. Cutting modes: velocity v = 0.15 m/s; depth S = 0.15 mm; front angle γ = 15°; the radius of rounding of the cutting edge ρ = 0.008 mm; coolant liquid – sulfoprezol; microhardness throughout the cutting area – Hm, GPa; tangential stresses – $\tau 0$, τk, MPa.

The formation of the body of the growth occurs at the initial cutting area from a certain volume of the processed material, which passes into the plastic state via the deformation of the shear and compression. Practically, the formation ends after the complete inclusion of the tooth in the work. A wedged chip formation zone forms at the initial cutting area. In this case, the taut-distorted condition of the processed material near the cutting edge is such that the material, turning into a growth, acquires deformation of the relative shear e = 20–50 regardless of the degree of strengthening. The indicated deformation values are higher for the initial displacement deformations (e = 1.5–5). The microhardness of the growth significantly exceeds the microhardness of the chips and does not depend on the degree of hardening of the material being processed. So, when cutting steel 10, this reaches 30%–35% (Del 1975; Vorobyov 2017).

In the area of stable cutting, the body of the growth is elastic. By contrast, in the contact layer of the chips and the surface layer of the parts that adjoin the growth, extensive plastic deformations occur. This is proved by the texture and the increase in microhardness. Thus, in the zone of secondary deformation, the value of microhardness of the chips is close to the value of the microhardness of the growth.

The processed material turns into a growth under the action of extensive plastic deformations, but retains its structure. Thus, the growth has the same structure and chemical composition as the processed material in its original state. Exclusive to

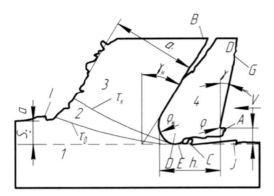

Figure 11.3 General outline of a chip-formation zone on a body of a growth: 1 – processed steel in the unstable state; 2 – chip-formation zone; 3 – chips; 4 – body of the growth.

ordinary steels, the physico-mechanical and cutting properties of the growth have an exclusively deformational nature (Kukharchuk 2017a; Vedmitskyi 2017).

In Figure 11.3, we have provided a general outline of a chip-formation zone on a body of a growth. In the chip-formation zone 2, the processed steel transfers from the elastic into the elastoplastic state.

The zone is constrained by curvilinear areas with the initial circumferential stress $\tau 0$ and the final one τk. The body of the growth 4 is found working in conditions of all-round compression. From one side, the body is restricted by the leading surface of the cutting wedge. From the other side, there is a sliding movement of the chips on the body of the growth 3. And finally, the flank surface of the growth is in contact with the processed surface. Characteristic parameters of the body of the growth and the chips are as follows. (1) part of a growth protruding over the back of the instrument; (2) cutter layer of the chips in contact with the body of the growth; (3) part of the growth remaining on the processed surface; (4) microstructure of the body of the growth at its cutting edge; (5) microstructure of the transition zone from "growth to processed surface"; (6) microstructure of the body of the growth at the cut area; (7) characteristic increase of OP after processing the growth formation (Kukharchuk 2017b; Wojcik 2010).

In Figure 11.4, we have provided a micrograph of the flank surface of the growth 1 and the processed surface 2 with a part of the growth 3 still remaining on the latter. The thickness of the textured layer on the processed surface is tenths of a millimeter in size.

Abrasive protrusions, occasionally left by the growth, break almost in half, and their underside becomes the tip of the ridge on the processed surface with hardness that matches the intense hardness of the growth. At the same time, the top part, which is also quite solid, becomes free and performs its abrasive functions on the processed surface.

In Figure 11.5, the zones of contact of the base of the body of the upright 1 with the chips 2 and the treated surface 3 are shown.

The levels of the texture of the chip 2 and the upright 1 indicate different degrees of cold strain and, accordingly, strengthening.

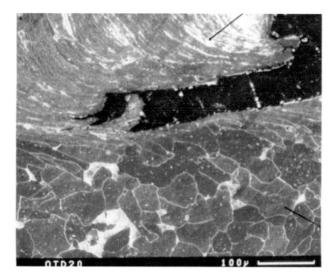

Figure 11.4 Micrograph of the back surface of the topside 1 and the treated surface 2 with the remaining particle on the last (region C in Figure 11.3). Experimental conditions are given in the caption of Figure 11.2.

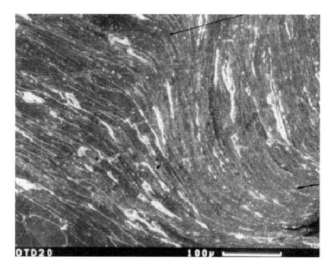

Figure 11.5 Microstructure of the contact zone of the base of the upright 1 and chips 2. The conditions of the experiment are presented in the caption of Figure 11.2.

11.3 DISCUSSION AND PRACTICAL SIGNIFICANCE OF THE RESEARCH RESULTS

The positive role of the growth is to protect the working surface of the tool against wear and hardening of the treated surface. The negative properties of the growth include the growth of roughness of this surface due to the appearance of scallops (Posvyatenko

2019b, 2019c). Growth formation can be used to construct the flow curves of plastic materials. This is due to the fact that in one experiment, it is possible to obtain a wide range of hardnesses of the processed material from the initial to the maximum (hardness of the growth; Posvyatenko 2005; Wojcik 1997).

Such curves are constructed as follows. First, we carry out the processing of plastic material by way of free orthogonal cutting in the zone of the intense formation of growths. Then the process is stopped by the cutter method, which falls, that is, the "root of the chip" is obtained.

The latter is processed by the method of preparation of microsections. Then the microhardness of an experimental sample of the "root of the chip" is studied in the zones: that corresponds to the initial state of the material; in the beginning and at the end of the wedge-shaped section of the chip formation; in an array of chips; at the contact point of a chip with a cutting tool; and finally, in the body of the growth.

These microhardness values (according to Vickers) are known to be related to the values of normal or tangential stresses by simple dependences, by which one can construct an exact yield locus, and to the strain strength of the material.

In the latest studies using the method of studying the distribution of dislocations using an electron microscope, the decisive influence of the dislocation mechanism on the change in the properties of the material being processed under cold plastic deformation is shown (Posviatenko 2019a).

11.4 CONCLUSION

It was established that the mechanical treatment of steels is accompanied by the formation of a "material – tool" in accordance with the laws of the self-organization of the system (synergetic). The growth has the same structure and chemical composition as the treated material. Exclusive for conventional structural and low-carbon steels, the physical–mechanical and cutting properties of the growth have an exclusively deformation nature.

When selecting materials for parts of machines that should work in conditions of long cyclic loads, one should prefer those materials in which the difference between the values of the tensile strength and yield strength is not less than 40%–50%, that is, plastic materials.

The article proposed an original method for constructing a yield locus and determining the limit of cold strain hardening of plastic materials, which includes obtaining a "root of the chip," investigating the microhardness of the latter inclusive and the zone of growth, and recalculating the microhardness into normal and tangential stresses.

REFERENCES

Del', G.D., Ogorodnikov, V.A., Nakhaichuk, V.G. 1975. Criterion of deformability of pressure shaped metals. *Izv Vyssh Uchebn Zaved Mashinostr* 4: 135–140.

Dragobetskii, V., Shapoval, A., Mos'pan, D., Trotsko, O., Lotous, V. 2015. Excavator bucket teeth strengthening using a plastic explosive deformation. *Metallurgical and Mining Industry* 4: 363–368.

Hacken, G. 1980. *Synergetics*. Mir. Moscow: 1–406.

Ivanova, V.C. 1986. Mechanics and synergy of fatigue failure. *Physical and chemical mechanics of materials* 1: 62–68.
Ivanova, V.S. 1994. *Synergetics and Fractals in Materials Science: Monograph*. Nauka. Moscow: 1–383.
Kanarchuk, V.Y., Posvyatenko, E.K., Shechenko, V.I. 2002. Synergetic aspects of cyclic destruction of steel. *Bulletin of National Transport University and Transport Academy of Ukraine. NTU* 7: 7–13.
Kozlov, L.G., Polishchuk, L.K., Piontkevych, O.V., Korinenko, M.P., Horbatiuk, R.M., Komada, P., Orazalieva, S., Ussatova, O. 2019. Experimental research characteristics of counter balance valve for hydraulic drive control system of mobile machine. *Przeglad Elektrotechniczny* 95(4): 104–109.
Kukharchuk, V.V., Bogachuk, V.V., Hraniak, V.F., Wójcik, W., Suleimenov, B., Karnakova, G. 2017. Method of magneto-elastic control of mechanic rigidity in assemblies of hydropower units, *Proc. SPIE* 10445: 1044–1056.
Kukharchuk, V.V., Kazyv, S.S., Bykovsky, S.A., 2017. Discrete wavelet transformation in spectral analysis of vibration processes at hydropower units. *Przeglad Elektrotechniczny* 93(5): 65–68.
Kuznetsov, V.D. 1956. *The Growth During Cutting and Friction*. Mir. Moscow: 1–284.
Ogorodnikov, V.A., Dereven'ko, I.A., Sivak, R.I. 2018. On the influence of curvature of the trajectories of deformation of a volume of the material by pressing on its plasticity under the conditions of complex loading. *Materials Science* 54(3): 326–332.
Ogorodnikov, V.A., Savchinskij, I.G., Nakhajchuk, O.V. 2004. Stressed-strained state during forming the internal slot section by mandrel reduction. *Tyazheloe Mashinostroenie* 12: 31–33.
Ogorodnikov, V.A., Zyska, T., Sundetov, S. 2018. The physical model of motor vehicle destruction under shock loading for analysis of road traffic accident. *Proc. SPIE* 10808: 1080–1086.
Polishchuk, L., Bilyy, O., Kharchenko, Y. 2016. Prediction of the propagation of crack-like defects in profile elements of the boom of stack discharge conveyor. *Eastern-European Journal of Enterprise Technologies*. 6(1), 44–52.
Polishchuk, L., Kharchenko, Y., Piontkevych, O., Koval, O. 2016. The research of the dynamic processes of control system of hydraulic drive of belt conveyors with variable cargo flows. *Eastern-European Journal of Enterprise Technologies* 2(8): 22–29.
Polishchuk, L.K., Kozlov, L.G., Piontkevych, O.V., Horbatiuk, R.M., Pinaiev, B., Wójcik, W., Sagymbai, A., Abdihanov, A. 2019. Study of the dynamic stability of the belt conveyor adaptive drive. *Przeglad Elektrotechniczny* 95(4): 98–103.
Posvyatenko, E.K. 1996. Mechanics of the process of cutting plastic materials after cold strain hardening. *Cutting and tool in technological systems* 50: 149–154.
Posviatenko, E.K., Aksom, P.A. 2019. On the nature of the influence of strain hardening on the processing of austenitic steels. *Bulletin of NTU* 3(45): 113–121.
Posvyatenko, E.K., Aksom, P.A., Posvyatenko, N.I. 2018. The effect of pre-cold deformation and plant lubricating and cooling Liquids on the physico-mechanical and technological properties of austenitic steels. *Cutting and tool in technological systems*. 88: 172–178.
Posvyatenko, E.K., Budiak, R.V., Melnyk, O.V., Nikitin, V.G. 2019. *Physical methods for studying the properties of materials: Textbook*. NTU: 1–176.
Posvyatenko, E.K., Melnyk, O.V., Ivanov, Yu.M. 2005. Express method for determining the mechanical properties of materials and the limits of their deformation strengthening. *Proc. of 61 scientific conference of faculty and university students, NTU*: 1–6.
Posvyatenko, E.K., Mozghovyi, O., Posvyatenko, N.I., Budiak, R.V. 2019. Synergetic aspects of build-up formation in machining. Prospects for the development of mechanical engineering and transport. *Proc. of I International scientific and technical conference, Vinnytsia, 13–15 May*. PP TD Edelveys and Ko: 95–96.

Posvyatenko, E.K., Tveritnikova, O.E., Posvyatenko, N.I., Miller, T.V. 2017. *Historical affinity of development of applied technical sciences*. NTU "KhPI". Kharkiv: 1–224.

Prigogine, I., Stingers, I. 1986. *The Order of Chaos*. Mir. Moscow: 1–430.

Usachyov, Y.G., Panchenko, K.P. 1952. *Phenomena that Occurs when Cutting Metals*. Mashgiz. Moscow: 356–384.

Vedmitskyi, Y.G., Kukharchuk, V.V., Hraniak, V.F. 2017. New non-system physical quantities for vibration monitoring of transient processes at hydropower facilities, integral vibratory accelerations. *Przeglad Elektrotechniczny* 93(3): 69–72.

Vorobyov, V., Pomazan, M., Vorobyova, L., Shlyk, S. 2017. Simulation of dynamic fracture of the borehole bottom taking into consideration stress concentrator. *Eastern-European Journal of Enterprise Technologies* 3/1(87): 53–62.

Wojcik, W. 1997. Optical fiber system for flame monitoring in energetic boilers, technology and applications of lightguides, *Proceedings of SPIE*: 3189: 74–82.

Wojcik, W., Kisala, P. 2010. The method for the recovery of the apodization function of the fiber Bragg gratings on the basis of its spectra. *Przeglad Elektrotechniczny* 86(10): 127–130.

Chapter 12

Theoretical and experimental studies to determine the contact pressures when drawing an axisymmetric workpiece without a blank flange collet

R. Puzyr, R. Argat, A. Chernish, R. Vakylenko,
V. Chukhlib, Z. Omiotek, M. Mussabekov,
G. Borankulova, and B. Yeraliyeva

CONTENTS

12.1 Introduction 131
12.2 Analysis of the methods of the drawing of an axisymmetric workpiece
 without a blank flange collet 131
12.3 Results and discussion 136
12.4 Conclusions 138
References 139

12.1 INTRODUCTION

The drawing down of axisymmetric parts is characterized not only by the loss of flange part stability, but also by the localized deformation in the zone of the transition of the wall to the bottom and the destruction of the workpiece at a given location (Golovlev 1995). The destruction and localization of deformations can be eliminated by reducing the value of the tensile stresses, and when drawing down with a collet, this problem is solved by reducing the collet force, using highly effective lubricants, and also by polishing the die body contact surfaces and the blank holder (Kalyuzhnyi 2015).

12.2 ANALYSIS OF THE METHODS OF THE DRAWING OF AN AXISYMMETRIC WORKPIECE WITHOUT A BLANK FLANGE COLLET

The drawing down without a workpiece flange by pressing is limited by the formation of an uncompressed workpiece part, and bottom separation can occur if the process of workpiece retraction into the die body hole continues, ignoring the folds on the flange. As has been shown in the papers (Luo 2011; Yan 2014), it is possible to extend the interval of non-collet forging by increasing the resistance of the flange, loss of stability or by increasing tensile stresses at the initial stage of deformation. The blank flange practically does not have contact with the surface of the die body, since, when the

DOI: 10.1201/9781003225447-12

punch is lowered, the flange portion rises above the die body and forms a conical surface with an angle α at the apex (Popov 1977; Puzyr 2015). Therefore, the contact of the workpiece with the die body occurs along its toroidal surface with a certain rounded radius. The focus is the theoretical definition of contact stresses on the torus-like part of the die body and the factors that allow you to control the magnitude of these stresses in order to find conditions for expanding the possibilities of drawing down without a blank holder.

To determine the contact stresses, we start from the equilibrium equations for the torus (Puzyr 2015), but taking into account the surface load:

$$\left. \begin{array}{l} \dfrac{\partial R_2 N_1}{\partial \phi} + \dfrac{\partial \left((R_1 \sin\theta) T\right)}{\partial \theta} - N_2 \dfrac{\partial R_2}{\partial \phi} + T \dfrac{\partial R_1 \sin\theta}{\partial \theta} = 0, \\[2mm] \dfrac{\partial \left((R_1 \sin\theta) N_2\right)}{\partial \theta} + R_2 \dfrac{\partial T}{\partial \phi} - N_1 \dfrac{\partial (R_1 \sin\theta)}{\partial \theta} + T \dfrac{\partial R_2}{\partial \phi} = 0, \\[2mm] \dfrac{N_1}{R_1} + \dfrac{N_2}{R_2} = q. \end{array} \right\} \quad (12.1)$$

where: N_1, N_2, T – internal meridional, tangential, and shear stresses; and q – contact pressure.

We take into account the fact that the deformation is axisymmetric, and therefore the terms of the system (12.1) contain $\partial \phi$, as well as the shearing force T, and that $R_1 = \dfrac{R}{\sin\theta} = \dfrac{a(1 + k\sin\theta)}{k\sin\theta}$, $R_2 = a$, $k = \dfrac{a}{R}$, so we will have:

$$\left. \begin{array}{l} N_2 a \cos\theta + \dfrac{\partial N_2}{\partial \theta} \dfrac{a}{k}(1 + k\sin\theta) - N_1 a \cos\theta = 0, \\[2mm] \dfrac{N_1 \sin\theta}{R} + \dfrac{N_2}{a} = q. \end{array} \right\} \quad (12.2)$$

Next, we calculate according to the procedure (Volmir 1967) with the only difference being that the pressure on the workpiece when it slides along the edge of the die body is applied from within half a torus. Assuming $q = const$, we exclude N_1 from equations (12.2), which gives us the differential equation:

$$\dfrac{dRN_2 \sin\theta}{d\theta} = qaR\cos\theta \quad (12.3)$$

Integrating the resulting equation from 0 to θ, we have:

$$N_2 = -\dfrac{qa}{2} \dfrac{R_d + r_m + R}{R} \quad (12.4)$$

where: R_d – radius of the drawing part; r_m – radius of the die body input edge; $R = R_Д$ $a\sin\theta$; and $a = r_m$.

Substituting (12.3) into the second equation (12.2), we obtain:

$$N_1 = \frac{qa}{2} \tag{12.5}$$

When the torus is affected by a surface load with loss of stability in the form of bulge formation, the expression for q has been obtained by the authors of (Volmir 1967) as $k \rightarrow 0$:

$$q = \frac{E}{12(1-\mu^2)} \left(\frac{s}{a}\right)^3 (n^2 - 1) \tag{12.6}$$

where: μ – Poisson's ratio; E – Young's modulus; and n – number of emerging half-waves with loss of stability.

Regarding our drawing conditions, it should be slightly modified. For further calculation, we use the cylindrical rigidity of the shell corresponding to the secant modulus $D \approx D' = \frac{E_c s^3}{9}$, we sort out the value of n for which $(n^2 - 1) = 1$. Further, since $E_c = \frac{\sigma_i}{\varepsilon_i}$, we accept $\sigma_i = \sigma_{cp}$, and $\varepsilon_i = \frac{1}{2}\varepsilon_\theta$ (Ogrodnikov 2018b; Puzyr 2018b). The draw ratio is $m = \frac{D}{D_0}$, hence $\varepsilon_\theta = 1 - m = 1 - \frac{D}{D_0}$, where D and D_0 are the diameter of the workpiece and the diameter of the blank, respectively. Then, expression (12.6) will have the form:

$$q = \frac{2\sigma_s}{9\left(1 - \frac{D}{D_0}\right)} \left(\frac{s}{r_m}\right)^3 \tag{12.7}$$

We substitute this relation into the formulas for the internal forces (12.4), (12.5) and, passing to the stresses, we get:

$$\sigma_\rho = \frac{\sigma_s}{9\left(1 - \frac{D}{D_0}\right)} \left(\frac{s}{r_m}\right)^2 \tag{12.8}$$

$$\sigma_\theta = -\frac{1 + \frac{k}{2}\sin\theta}{1 + k\sin\theta} \frac{2\sigma_s}{9\left(1 - \frac{D}{D_0}\right)} \left(\frac{s}{r_m}\right)^2 \tag{12.9}$$

Before (Puzyr 2015), formulas were derived for calculating the magnitude of the stress tensor components on the torus-like part of the die body without taking into account the surface load. Using the principle of superposition (Savelov 2015) and carrying out simple transformations, the final dependencies for calculation of meridional and tangential stresses on the rounded radius of the die body at $\theta = 90°$ will look like (Polishchuk 2016a, 2016b):

$$\sigma_p = \sigma_s \left(\ln \frac{R_0}{R_d + r_m} - \frac{r_m}{R_d + r_m} + 0{,}1 \frac{R_0}{R_0 - R_d} \left(\frac{s}{r_m} \right)^2 \right) \tag{12.10}$$

$$\sigma_\theta = -\sigma_s \left(1 - \ln \frac{R_0}{R_d + r_m} + \frac{r_m}{R_d + r_m} + 0{,}22 \frac{1 + \frac{k}{2}}{1 + k} \frac{R_0}{R_0 - R_d} \left(\frac{s}{r_m} \right)^2 \right) \tag{12.11}$$

In contrast to the correlations (Puzyr 2015), in order to calculate the meridional and tangential stresses on the torus-like part of the die body, the dependences (12.10) and (12.11) contain a term that relates the thickness of the blank to the value of the stresses arising while drawing, which describes their distribution on the drawing edge of the die body more accurately (Kozlov 2019; Polishchuk 2016a, 2018).

Using the second system equation (12.2), as well as (12.10) and (12.11) and carrying out simple transformations, we find the surface pressure, which causes the plastic deformation of the blank during drawing without a workpiece collet:

$$\frac{q}{s} = \sigma_s \left(\begin{array}{c} \left(\ln \frac{R_0}{R_d + r_m} - \frac{r_m}{R_d + r_m} \right) \frac{R_d}{(R_d + r_m)^2} + \\ + \frac{R_0}{R_0 - R_d} \left(\frac{s}{r_m} \right)^2 \left(\frac{1}{9(R_d + r_m)} - \frac{2}{9 r_m} \frac{1 + \frac{k}{2}}{1 + k} \right) - \frac{1}{r_m} \end{array} \right) \tag{12.12}$$

Let us estimate the terms of the given expression for given deformation parameters – $R_0 = 50$ mm, $s = 2$ mm, $R_d = 25$ mm, and $r_m = 10$ mm. The first and second terms are of the same order of smallness and can be neglected with an error up to the third decimal place. Then the surface load is the following (Del 1975; Kukharchuk, 2017):

$$q = -\frac{s}{r_m} \sigma_s \tag{12.13}$$

Thus, an expression has been developed for calculating the surface contact pressure during the drawing down of a cylindrical workpiece, which makes it possible to take friction stresses into the rounded radius of the die body, and also to calculate the drawing force. The dependence obtained differs from the existing ones by its simplicity and visibility and can be used at the preliminary stage of choosing equipment for forging.

To verify the obtained dependence by the determination of contact pressures, experiments have been carried out to measure the load cell deformations caused by frictional forces. Experimental and theoretical studies (Comsa 2007; Del 1975; Ogorodnikov 2004; Wang 2009) demonstrate that the frictional stresses between the surfaces to be contacted can be controlled within wide limits, while achieving a significant change in the stress state and the distribution of deformations in the volume of the blank, which is of great importance for the production of high-quality workpieces by means of drawing down.

Since the flange of the workpiece practically does not contact the mirror of the die body while drawing down without the blank holder, friction occurs mainly along the rounded radius of the latter, which is of great interest in determining the frictional stresses and studying their effect on the occurrence of the loss of flange stability.

The experiments were carried out on a UME-10 TM tensile machine with a force of 10 tf, which is intended for testing metal and plastic samples across a wide range of deformation rates during static stretching, compression and transverse bending, and also under cyclic loading with any asymmetry coefficient in the limits of the machine's load capacity for specified load limits or deformation under normal and elevated temperatures. The machine is equipped with an electronic measuring instrument and a chart apparatus for recording the stress–strain curve across a large scale (Puzyr 2016; Puzyr 2018a; Renne 1976).

The experimental equipment was a special die body with an inlet of 50 mm and a set of punches (Figure 12.1) with diameters of 49.3, 46.4, and 49 mm, providing the drawing of blanks without thinning the stiffening plate of 08 kp steel with a thickness of 0.15 mm, aluminum A2 with thickness of 1.4 mm, and copper M4 of 0.25 mm correspondingly. The die body was installed on the lower traverse of the tensile machine, the exchangeable punches were on the upper traverse, and both traverses were equipped with clamping elements (Kukharchuk 2017a, 2017b; Vedmitskyi 2017).

In the course of the experiment, the following was also recorded: the drawing force, the punch stroke, the number of corrugations emerging, and the height of the half-wave, depending on the displacement of the punch.

The special equipment was a die body with an insert for the drawing radius, which can be displaced in the direction of movement of the punch during drawing and a punch of a conventional design (Figure 12.2) (Borisevich 2009; Mori 2016; Zagirnyak 2015).

The equipment for measuring the load cell deformations was a MCP606-I/P strain gauge operational amplifier with an ADC E14–440, an AX-1803D power supply unit, and Pentium 4 CPU 2.40 GHz 1.0 GB RAM personal computer. Load cell deformations caused by frictional forces during drawing determined the strain gauge deformation, which is included in a quarter-bridge scheme without compensation of temperature stresses. This, in turn, caused a change in the resistance of the resistor and the current

Figure 12.1 Experimental equipment: (a) die body without top overlays; (b) changeable punches.

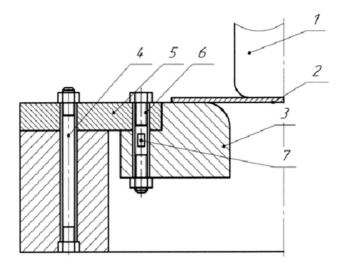

Figure 12.2 Equipment for measuring frictional stresses at the rounded radius of the exhaust die body: 1– punch; 2 – blank; 3 – movable insert; 4 – fixing bolt; 5 – connecting ring; 6 – load cell; 7 – strain gauge.

in the circuit. The current oscillations were amplified by an operational amplifier and transmitted to the ADC. On the computer monitor, these current readings were recorded for a predetermined period of time. A KF 5P1–5–200-A-12 strain gauge was used, with an operating resistance of R = 199.7±0.2 Ohms and a base of 5mm. The label of the resistors was produced with cyacrine glue according to the technology described in the paper (Puzyr 2016).

12.3 RESULTS AND DISCUSSION

The contact pressure between the hemispherical surface of the die body and the workpiece was measured. Measurements were performed for a minimum of 16 blanks for each metal and alloy with different coefficients of drawing, the load cell rounded radius r_m = 4mm and r_m = 1.5mm, without lubrication of the surfaces of the workpiece and the tool. At the same time, the depth of the punch's progress was recorded, such as force increase; a sharp force increase meant the occurrence of corrugations that made it difficult to draw a flat workpiece into the hole in the die. Based on the measurement results, typical ADC voltage vs. time charts are shown in Figure 12.3. Also, the experimental data are used to verify the accuracy of the mathematical model of the distribution of contact pressures on the die body rounded radius when drawing without a collet.

For greater accuracy of contact pressure measurements, the first lot of blanks was chosen in such a size that their diameter was equal $D_0 = D + 2r_m$, where D is the diameter of the drawing workpiece. This was to minimize the effect of blank flange deformation and reduce the effect of the bending moment on the torus-like section. Since the tangential stresses act in the latitudinal direction along the conical surface, they form a moment toward the die body edge. The given moment balances the bending

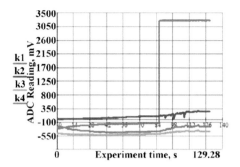

Figure 12.3 Typical schedule of measurement of load cell deformations: k1 – first load cell (breakage); k2 – second load cell; k3 – third load cell; k4 – fourth load cell.

moment, which changes the curvature of the middle surface of the elements upon their transition to the rounded edge of the die body. The remaining two lots of blanks were selected with dimensions $D_0 = D + 4r_m$ and $D_0 = D + 6r_m$ correspondingly. Figure 12.4 shows the blanks from the metals researched having different sizes.

Similarly, some semifinished products after deformation (Figure 12.5).

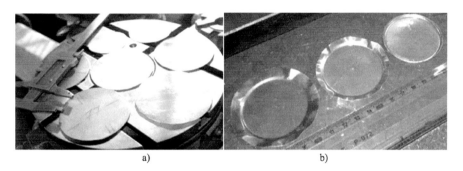

Figure 12.4 Results of experimental studies: (a) blanks for drawing; (b) drawing of steel without a workpiece collet, $k = 1.04$; $k = 1.08$; $k = 1.12$ from right to left.

Figure 12.5 Extraction of steel and copper without a blank holder, $k = 1.04$.

Table 12.1 Results of experimental studies on contact pressure measurements

Material	Deformation parameters					Contact pressure	Estimated contact pressure	Δ (%)
	s (mm)	D_0 (mm)	k	r_m (mm)	h (mm)	q (MPa)	q (MPa)	
Steel 08kp	0.15	58/52	1.16/1.04	4/1.5	8.0/6.0	58.04/38.55	17.25/46.0	71.3/17.2
		66/54	1.32/1.08	4/1.5	8.0/6.02	94.34/55.33		81.7/16.9
		74/56	1.48/1.12	4/1.5	8.1/6.01	133.30/57.02		87.1/19.4
Aluminum A2	1.4	58/52	1.16/1.04	4/1.5	8.2/6.1	27.25/46.26	42/56	35.1/17.4
		66/54	1.32/1.08	4/1.5	8.3/6.02	37.54/48.93		10.6/12.6
		74/56	1.48/1.12	4/1.5	8.0/6.04	46.08/51.76		8.7/7.5
Copper M4	0.25	58/52	1.16/1.04	4/1.5	8.0/6.05	21.14/17.47	10/26.6	52.3/34.3
		66/54	1.32/1.08	4/1.5	7.9/6.05	34.24/21.18		70.5/20.3
		74/56	1.48/1.12	4/1.5	7.95/6.0	43.15/24.81		76.7/6.7

The results of the experimental data were subjected to statistical processing (Ogorodnikov 2018a; Sosenushkin 2015), then averaged and summarized in Table 12.1. In the table, the numerator represents values for $r_m = 4.0$ mm; the denominator – for $r_m = 1.5$ mm; Δ, % represents deviations of the experimental data from the theoretical ones, calculated from the dependence (12.3).

12.4 CONCLUSIONS

The results of the experimental studies show that the greatest coincidence with theory is 6.7% for drawing with k = 1.12 and $r_m = 1.5$ mm for copper, and the greatest discrepancy in calculations is up to 87% for steel with an $r_m = 4.0$ mm rounded radius of the die body. The discrepancy between the results of the experimental studies and calculated data in general can be explained by the imperfection of the mathematical model, where all factors of constructive and technological nature are not taken into account, as well as hardening of the metal during plastic deformation (Del 1975; Guo 2000; Harpell 2000). In particular, the greatest discrepancy in the results is observed with the drawing blanks of steel and copper with a rounded radius of the die body of 4 mm, which is obviously connected with the choice of the rounded radius of the die body. For the metal thicknesses used in the experiments, it is excessive, since it goes beyond the recommendation of the technical literature, which is $r_m = 8$–$10\,s$. Here, at the very beginning of the deformation process, the formation of folds on the flange begins, which prevent the metal from being drawn into the die hole and dramatically increase the required force. The best coincidence between the results of the theoretical and experimental studies, which does not exceed 20% (with the exception of copper drawing, k = 1.04, $r_m = 1.5$ mm), showed data for all metals with $r_m = 1.5$ mm. However, as can be seen from Table 12.1, the increase in the flange width results in an increase in the contact pressure, and the best coincidence of the results is observed precisely with the drawing of blanks with an enlarged flange. This pattern does not reflect the analytical relationship (12.3), but it provides a more accurate match of the data. Therefore, the completion of the dependence is obvious, but in the existing form, it can be used to calculate the drawing process with the rational geometry of the tool.

REFERENCES

Borisevich, V.K., Zahirnyak, M.V., Drabobetsky, V.V. 2009. Choosing the optimal parameters for extracting sheet blanks. *Kuznechno-shtampovochnoye proizvodstvo. Obrabotka materialov davleniyem.* 2: 38–41.

Comsa, D.S., Banabic D. 2007. Numerical simulation of sheet metal forming processes using a new yield criterion. *Key Engineering Materials.* 344: 833–840.

Del', G.D., Ogorodnikov, V.A., Nakhaichuk, V.G. 1975. Criterion of Deformability of Pressure Shaped Metals. *Izv. Vyssh. Uchebn. Zaved. Mashinostr.* 4: 135–140.

Del', G.D., Tomilov, F.K., Ogorodnikov, V.A. 1975. *Stressed state during the hot extrusion of steel.* Steel USSR 5(4): 223–224.

Golovlev, V.D. 1995. The initial stage of drawing sheet metal. *Kuznechno-shtampovochnoye proizvodstvo* 7: 4–10.

Guo, Y.Q. 2000. Recent developments on the analysis and optimum design of sheet metal forming parts using a simplified inverse approach. *Comput. Struct.* 78: 133–148.

Harpell, E.T., Worswick, M.J., Finn, M., Jain, J., Martin, P. 2000. Numerical prediction of the limiting draw ratio for aluminum alloy sheet. *Journal of Materials Processing Technology* 100: 131–141.

Kalyuzhnyi, O.V., Kalyuzhnyi, V.L. 2015. *Intensification of forming processes of cold sheet stamping.* Sik Group, Kyiv, Ukraine LLC: 1–315.

Kozlov, L.G., Polishchuk, L.K., Piontkevych, O.V., Korinenko, M.P., Horbatiuk, R.M., Komada, P., Orazalieva, S., Ussatova, O. 2019. Experimental research characteristics of counter balance valve for hydraulic drive control system of mobile machine. *Przeglad Elektrotechniczny* 95(4): 104–109.

Kukharchuk, V.V., Bogachuk, V.V., Hraniak, V.F., Wójcik, W., Suleimenov, B., Karnakova, G. 2017. Method of magneto-elastic control of mechanic rigidity in assemblies of hydropower units. *Proc. SPIE 10445*: 44–52.

Kukharchuk, V.V., Kazyv, S.S., Bykovsky, S.A. 2017. Discrete wavelet transformation in spectral analysis of vibration processes at hydropower units. *Przeglad Elektrotechniczny* 93(5): 65–68.

Luo, J.C. 2011. Study on stamping-forging process and experiment of sheet metal parts with non-uniform thickness. Wuhan: Huazhong University of Science & Technology 51: 49–54.

Mori, K., Nakano, T. 2016. State-of-the-art of plate forging in Japan. *Production Engineering* 10 (1): 81–91.

Ogorodnikov, V.A., Dereven'ko, I.A., Sivak, R.I. 2018. On the influence of curvature of the trajectories of deformation of a volume of the material by pressing on its plasticity under the conditions of complex loading. *Materials Science* 54(3): 326–332.

Ogorodnikov, V.A., Grechanyuk, N.S., Gubanov, A.V. 2018. Energy Criterion of the Reliability of Structural Elements in Vehicles. *Materials Science.* 53(5): 645–650.

Ogorodnikov, V.A., Savchinskij, I.G., Nakhajchuk, O.V. 2004. Stressed-strained state during forming the internal slot section by mandrel reduction. *Tyazheloe Mashinostroenie.* 12: 31–33.

Polishchuk, L., Bilyy, O., Kharchenko, Y. 2016. Prediction of the propagation of crack-like defects in profile elements of the boom of stack discharge conveyor. *Eastern-European Journal of Enterprise Technologies* 6(1): 44–52.

Polishchuk, L., Kharchenko, Y., Piontkevych, O., Koval, O. 2016. The research of the dynamic processes of control system of hydraulic drive of belt conveyors with variable cargo flows. *Eastern-European Journal of Enterprise Technologies* 2(8): 22–29.

Polishchuk, L.K., Kozlov, L.G., Piontkevych, O.V., Gromaszek, K., Mussabekova, A. 2018. Study of the dynamic stability of the conveyor belt adaptive drive. *Proc. of SPIE* 10808: 1117–1121.

Popov, E.A. 1977. *Fundamentals of the Theory of Sheet Punching.* Mashinostroyeniye. Moskva: 1–278.

Puzyr, R., Savelov, D., Argat, R., Chernish, A. 2015. Distribution analysis of stresses across the stretching edge of die body and bending radius of deforming roll during profiling and drawing of cylindrical workpiece. *Metallurgical and Mining Industry* 1: 27–32.

Puzyr, R., Haikova, T., Majerník, J., Karkova, M., Kmec, J. 2018. Experimental study of the process of radial rotation profiling of wheel rims resulting in formation and technological flattening of the corrugations. *Manufacturing Technology*. 18(1): 106–111.

Puzyr, R., Haikova, T., Trotsko, O., Argat, R. 2016. Determining experimentally the stress-strained state in the radial rotary method of obtaining wheels rims. *Eastern-European Journal of Enterprise Technologies* 4/1(82): 52–60.

Puzyr, R., Savelov, D., Shchetynin, V., Levchenko, R., Haikova, T., Kravchenko, S., Yasko, S., Argat, R., Sira, Y., Shchipkovskyi, Y. 2018. Development of a method to determine deformations in the manufacture of a vehicle wheel rim. *Eastern-European Journal of Enterprise Technologies. Engineering Technological Systems* 4/1(94): 55–60.

Renne, I.P., Ogorodnikov, V.A., Nakhaichuk, V.G. 1976. Plotting plasticity diagrams by testing cylindrical samples in combined tension and torsion. Strength of Materials 8(6): 733–737.

Savelov, D., Dragobetsky, V., Puzyr, R., Markevych, A. 2015. Peculiarities of vibrational press dynamics with hard-elastic restraints in the working regime of metal powders molding. *Metallurgical and Mining Industry* 2: 67–74.

Sosenushkin, E.N., Yanovsky, E.A., Sosenuushkin, A.E., Emelyanov, V.V. 2015. Mechanics of nonmonotonic plastic deformation. Russian Engineering Research 35(12): 902–906.

Vedmitskyi, Y.G., Kukharchuk, V.V., Hraniak, V.F. 2017. New non-system physical quantities for vibration monitoring of transient processes at hydropower facilities, integral vibratory accelerations. *Przeglad Elektrotechniczny* 93(3): 69–72.

Volmir, A.S. 1967. *Stability of Deformable Systems*. Nauka. Moskva: 1–984.

Wang, X.Y., Ouyang, K., Xia, J.C. 2009. FEM analysis of drawing-thickening technology in stamping-forging hybrid process. *Forging & Stamping Technology* 34(4): 73–78.

Yan, G.X., Wang, X.Y., Deng, L. 2014. A study of hole flanging-upsetting process. *Advanced Materials Research* 939: 291–298.

Zagirnyak, M.V., Drahobetskyi, V.V. 2015. New methods of obtaining materials and structures for light armor protection. *Military Technologies (ICMT), International Conference*: 1–6.

Chapter 13

Modification of surfaces of steel details using graphite electrode plasma

*V. Savulyak, V. Shenfeld, M. Dmytriiev, T. Molodetska,
V. M. Tverdomed, P. Komada, A. Ormanbekova, and
Y. Turgynbekov*

CONTENTS

13.1 Introduction 141
13.2 Preparation of the equipment and selection modes alloying 142
13.3 Analysis and research study of the microstructure of samples.......................143
13.4 Measuring the hardness of samples by depth 145
13.5 Conclusion 147
References 148

13.1 INTRODUCTION

Improving the reliability and wear resistance of machine parts working with significant loads and in abrasive environments is a relevant task. Examples of such details are crankshafts, gear transmissions, working bodies of soil-working and earth-moving machines, etc. Various methods for modifying working surfaces are used to improve these properties: using a laser (Engel 1981, Komanduri & Hou 2001, Schaaf 2003), an electron beam (Bataev et al. 2014a, 2014b), a plasma arc (Edenhofer et al. 2001, Maliska et al. 2003), steel modifications with nanostructured compositions (Klein et al. 2013, Kondrat'ev et al. 2014). The disadvantages of the vast majority of these surface modification methods are the significant costs of electricity, additives, time, and finances.

There is an increase in the use of plasma machining methods for the surfaces of machine parts in order to improve their properties (Bendo et al. 2011, Lamima et al. 2015), which allows them to realize their inherent advantages: saving time and consumable materials during processing (thanks to the high reaction speeds), as well as considerable decrease in ecologically dangerous emissions.

The paper (Ivanaysky et al. 2011) presents the results of studying the technology of plasma coagulation at low temperatures (500°C and 700°C) in a glow discharge for 3 and 6 hours. The technology allows obtaining a polycrystalline cementite layer. Increasing the time of the process from 3 to 6 hours leads to a doubling of the cementite layer's thickness. The disadvantage of this method of surface coagulation is a considerable process performance time expenditure.

In the work (Kirgizov et al. 2010), the induction surfacing technology was proposed for steel parts made from the PG-C27 alloy, followed by the melting of the applied coating using a graphite electrode. Using the additive material of the specified type involves the preservation of the structural components that are already present in it. The impact of the high temperature on the powder additive material destroys the available wear-proof components. The effectiveness of using such technology is somewhat doubtful, involves significant financial costs for materials, energy, and technology implementation.

In the work (Shevchenko 2004), a carbonization method of the surface layer using a graphite electrode is considered. The essence of the offered method consists of the following. At the momentary breaking of temporary contact of a graphitic electrode with the detail causes a spark discharge and carbon from an electrode passes into the main material, leading to the creation of a cementite layer on its surface, resulting in significantly higher hardness of the main material.

The disadvantages of this carbonization method are: the complexity and significant cost of the required equipment; a low coagulation thickness; no opportunity to build upon worn surfaces.

The work (Savulyak & Shenfeld 2016) presents the results of studying a technological process, where, in the first stage, an iron–chromium coating is applied using the electrolytic method, and in the second stage that coating is carbonized using diffusion carburization technology in paste-like carburizers. As the first and second stages of the technological process require considerable expenses of time and energy. However, although the possibility of obtaining fine-grained coatings of sufficiently high hardness is shown, the reliability of the coat-based system is limited by the insufficient strength of their predominantly adhesive interaction. Additionally, attention must be paid to such a disadvantage as a known limitation on the productivity of both electrolytic deposition and diffusion carburization. Therefore, the layer strengthening is limited to a thickness up to 0.5mm that is not always satisfactory to technologists or operators performing finishing machining, due to the fast degradation of the coating.

The object of consideration in the proposed work is to study the possibility of increasing the productivity through processes of coagulation of working surfaces of details that are exposed to intense wear (Vedmitskyi et al. 2017), by treating the surface with a plasma stream saturated with ionized carbon atoms. At the same time, creating the possibility of alloying the surface with high-carbon compounds synthesized refractory metals.

13.2 PREPARATION OF THE EQUIPMENT AND SELECTION MODES ALLOYING

Implementation and research of surface coagulation technology and its modification were carried out on a modernized UD 209M installation on 10mm thick plates made of 40X steel. An EGP brand graphite electrode was used to excite the plasma (Figure 13.1).

The research was carried out in two stages. In the first stage, an arch was lit between the sample and a graphite electrode, and there was a plasma stream sated with carbon ions. This flow activated the surface of the samples and partially melted it. Carbon ions were sated through fusion and the activated steel surface.

The second stage of research involved studying the possibility of modifying a sample's surface by a graphite electrode plasma arc with simultaneous alloying with

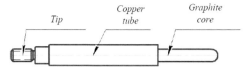

Figure 13.1 Modernized EGP brand graphite electrode.

refractory metals. During this surface treatment in the metal layers, solid structural components were synthesized, mainly carbides.

Before conducting research, the samples were carefully prepared by degreasing and purifying from dirt and rust. After the samples were prepared, the modification process was carried out in the following modes: the speed of the electrode displacement was 11, 17, 23 m/h; graphite electrode diameter was 5 mm; current was 100 A; polarity – straight; Volt–Ampere characteristic (VAC) is rigid. A VD-306M1 welding rectifier was used for feeding the arc.

Studying the parameters of the stable arc burning modes and maintaining a stable plasma column allowed us to recommend the operation at a current of about 100 A and the speed of the electrode moving relative to a steel sample, which is processed by plasma, of about 11 m/h. Work in other modes does not ensure the stability of the plasma column and is followed by fluctuations concerning the direction of movement.

The transfer of carbon ions from the electrode to the steel was studied by changing the structure of the plasma surface treated by the microstructure obtained by optical microscopes and by changing their microhardness.

For studying the influence of the impurity of alloying, such as molybdenum, vanadium, and chromium on the structure and hardness of the modified layer of steel samples, 20 mm plates in 40X steel were used. In the first series of experiences, the alloying element chosen was molybdenum with concentration in a fusion bath near 0.5%. Molybdenum and vanadium, the mass fraction of which was 0.5% and 0.3%, respectively, were used in the second series of experiments for alloying. Alloying of the third series of samples was carried out using molybdenum, vanadium, and chromium, containing 0.5%, 0.3%, and 2%, respectively. The depth of the weld bath in established modes was 5 mm. For fixing of the alloying impurity on the surface of the metal, special glue mix was used. For all samples, electrode distance from a detail was 2–3 mm.

13.3 ANALYSIS AND RESEARCH STUDY OF THE MICROSTRUCTURE OF SAMPLES

Research of the microstructure of the modified coats was performed on micro slips manufactured according to standard techniques using optical microscopes. The capture of the image was done by a webcam with special software for recording on a computer.

An analysis of the microstructure hardened surfaces at an electrode transfer speed of 23 and 17 m/h showed a cementite mesh on the surface with perlite grains in its cells. The formation of a cementitious grid shows a significant increase in the concentration of carbon due to its transfer in plasma from the electrode and a relatively high cooling rate.

Figure 13.2 Microstructures of the first series samples reinforced by plasma arc processing using a graphite electrode without alloying (X100): (a) surface layer; (b) the middle layer; (c) base.

At the electrode transfer speed of 11 m/year, it was found that the strengthened surface has a layered structure. A reinforced layer can be divided into two sublayers by depth.

The surface sublayer also has a cementite mesh (Figure 13.2a), and the middle layer is granular perlite (Figure 13.2b), which is explained by the higher energy input per unit area of the hardened surface compared with the coatings obtained at the electrode transfer velocities of 23 and 17 m/h and, accordingly, its cooling is slower. Figure 13.2b shows the structure of the base metal from 40X steel with a ferrite–perlite structure of pre-eutectoid steel.

Studying of samples with alloying mixes showed that these are processed by plasma arc coverage at a speed of 11 m/h and alloying molybdenum up to 0.5%, while a mix of molybdenum of 0.5% and vanadium of 0.3% has no significant structural changes in comparison when processed using plasma without alloying. Structural changes will become apparent when more than 1% of components by weight are added to the alloying complex.

Figure 13.3 shows the microstructures of a series of samples with an alloying mass of 0.5% molybdenum, 0.3% vanadium, and 2% chromium, where two zones with different structures can be observed. The first zone that is superficial has the structure of the ledeburite (Figure 13.3a). The structure with considerable release of small complex carbides (Fe, V, Mo, Cr) xC is characteristic of the second area (Figure 13.3b). This second coating area is formed directly on the metal base with a ferrite–pearlites structure (Figure 13.3c). Considering the 11 m/h (3 mm/s) transfer speed of the graphite electrode and the diameter of the stack plasma column of 10 mm, it is possible to calculate the residence time of the metal surface under the exposure to energy flux and carbon ions, which will be about 3.3 s. During this time, the surface is warmed, activated, saturated with carbon, melted, and crystallized according to the concentration of carbon, the alloying elements, and other kinetic conditions. During crystallization, carbides and other high-carbon compounds in the shape and size of crystals are formed, which ensures the corresponding physical and mechanical properties and the surface wear resistance. The transition between the second zone and the base metal is smooth and is evidence of the diffusive nature of carbon penetration into the base metal.

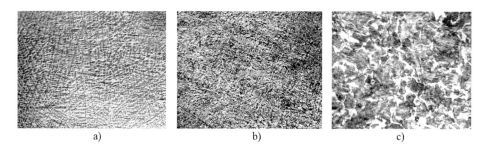

Figure 13.3 Microstructure of the surface layers, which was processed by the graphite electrode plasma arc and alloyed with a mixture of molybdenum, vanadium, and chromium (×100).

13.4 MEASURING THE HARDNESS OF SAMPLES BY DEPTH

After analysis of the microstructure of surface layers, Rockwell's hardness (HRC) was determined, as well as microhardness of metal by depth for Vickers (HV).

Let us consider the hardness of the samples that were modified by graphite electrode plasma arc processing and alloyed by molybdenum, vanadium, and chromium (rice 4). The highest hardness was shown by a series of samples whose alloying process used a mix of molybdenum, vanadium, and chromium. The hardness on the coat's surface was established within the limits of 49–52 HRC, which corresponds to a ledeburite microstructure. Under the surface layer, a layer with complex carbides and slightly lower hardness was formed (Figure 13.4).

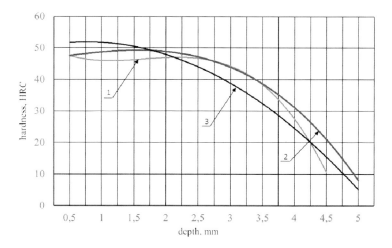

Figure 13.4 The hardness of the coat by depth for different compositions of alloying components (1 – molybdenum 0.5%, 2 – molybdenum 0.5% and vanadium 0.3%, 3 – molybdenum 0.5%, vanadium 0.3%, and chromium 2%).

When molybdenum and vanadium were used for alloying, the hardness was reduced compared with samples treated with plasma, but without alloying, slightly, within 10%.

To study the change in the coating hardness by depth, microhardness measurements were performed using a microhardness meter (Figures 13.5–13.7). After analyzing the dependence of hardness on the depth of a series of samples (Figure 13.5), which was applied with a suspension of 0.5% ferromolybdenum, it was established that it was possible to increase the hardness of the surface layer in a range from 720 to 960 MPa and part of the middle layer to a depth up to 2.5 mm. This makes it possible to carry out machining of the surface without loss of hardness. In the transition zone from the coat to the base metal, a gradual reduction of hardness toward a basis metal is observed, which is an average of 280 MPa. The maximum depth of the impact of graphite electrode plasma arc processing on the specified modes with alloying is up to 4 mm. The analysis of the impact of a processing combination of plasma with alloying showed increases in hardness in a range of 100–150 MPa from the highest average hardness of specimens strengthened by the graphite electrode plasma arc processing.

Studying of the coverage at the samples alloyed by 0.5% molybdenum and 0.3% vanadium (Figure 13.6), it was possible to establish that the hardness of the strengthened coating, namely: superficial and almost all centered, fluctuating within on average from 500 to 650 MPa at a depth of up to 2.6 mm. It follows that alloying with the addition of 0.3% of vanadium is leveled by hardness at the depth of the carbonization layer, but at the same time, the general average hardness decreases. There is also an increase in the transition zone between the middle layer and the base metal and an increase in the harness of the base metal in the thermal influence zone

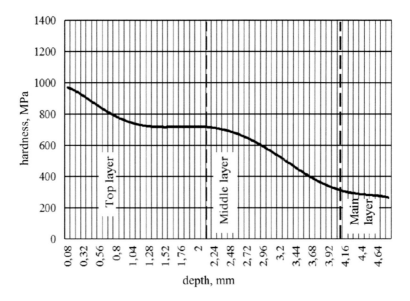

Figure 13.5 The hardness of coating for samples processed by plasma with molybdenum alloying by weight 0.5%.

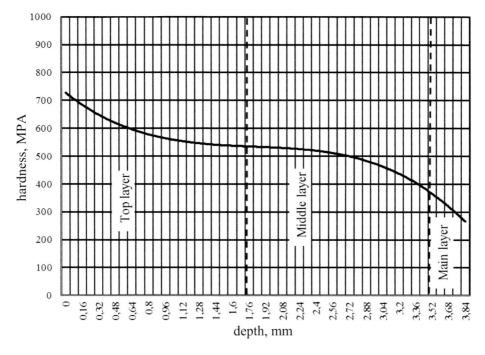

Figure 13.6 The hardness of coating for the samples processed by plasma with alloying by weight with 0.5% of molybdenum and 0.3% of vanadium.

Figure 13.7 shows the results of studying the coating produced by the surface treatment of a steel sample by a plasma arc of a graphite electrode and alloying with 0.5% molybdenum, 3% vanadium 0.3%, and 2% chromium by weight. The study shows that the top layer has a hardness within 644–1,448 MPa at a depth of 1 mm. In addition, the perlite structure was found below the upper layer, which had a hardness within 927–1,448 MPa along the entire depth of 1–2.2 mm, to finally sharply pass into the center layer having an average hardness of 700 MPa. The transition zone between the middle layer and the base metal also decreases steadily as well as in the previous sample, moving to the values of the hardness of the base metal.

The overall effect of plasma processing and alloying appears at a depth up to 4.5 mm. Thanks to the addition of chromium in the alloyed complex, it was possible to not only successfully raise the hardness to 1,448 MPa in separate structural components, but also to extend it to a depth of 3.2 mm.

13.5 CONCLUSION

1 The plasma of the electric arc between the detail and the graphite electrode allows to effectively transfer of the carbon and saturation of the steel surface and the resulting melt.

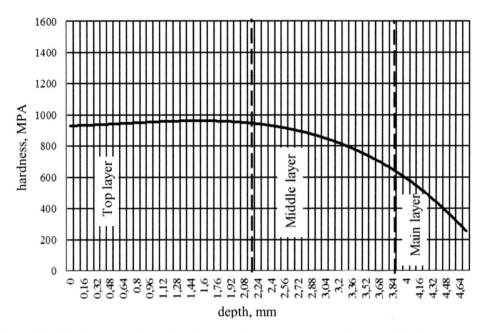

Figure 13.7 The hardness of coating for samples, processed by plasma with alloying by weight 0.5% molybdenum, 0.3% vanadium, and 2% chromium.

2. The treatment of the steel surfaces by the graphite electrode plasma arc allows modification of the surface with the formation of high-density structures characterized by high hardness and wear resistance.
3. The treatment of steel surfaces with a pre-applied layer of alloying elements using a graphite electrode plasma arc allows not only to increase the hardness of the surface but also to compress the structure and improve the physical and mechanical characteristics of the details.
4. The proposed method shows its universality and suitability for application in various types of production and usage conditions.
5. In terms of productivity, the method of strengthening with the use of a graphite electrode plasma arc surpasses other methods of carbonization. During an hour it is possible to strengthen the surface of a steel detail on the area of more than 1,000 cm² at a depth up to 5 mm.
6. The processing combination of plasma and alloying with a complex of refractory metals and bonds in the studied concentration allows increasing surface hardness by more than 50 units according to the Rockwell scale without additional heat treatment.

REFERENCES

Balanovskii, A.E., Grechneva, M.V., Van Huy, V. & Zhuravlev, D.A. 2017. New plasma carburizing method. *IOP Conference Series: Earth and Environmental Science*: 092003.

Bataev, I.A., Golkovskii, M.G., Bataev, A., Losinskaya, A., Dostovalov, R., Popelyukh, A. & Drobyaz, E. 2014. Surface hardening of steels with carbon by non-vacuum electron-beam processing. *Surface & Coatings Technology* 242: 164–169.

Bataev, I.A., Golkovskii, M.G., Losinskaya, A.A., Bataev, A.A., Popelyukh, A.I., Hassel, T. & Golovin, D.D. 2014. Non-vacuum electron-beam carburizing and surface hardening of mild steel. *Applied Surface Science* 322: 6–14.

Bendo, T., Pavanati, H.C., Klein, A.N., Martinelli, A.E. & Maliska, A.M. 2011. Plasma nitriding of surface Mo-enriched sintered iron. *ISRN Materials Science* 2011: 121464.

Dragobetskii, V., Shapoval, A., Mos'pan, D., Trotsko, O. & Lotous, V. 2015. Excavator bucket teeth strengthening using a plastic explosive deformation. *Metallurgical and Mining Industry* 4: 363–368.

Edenhofer, B.U., Grafen, W. & Müller-Ziller, J. 2001. Plasma-carburising — a surface heat treatment process for the new century. *Surface and Coatings Technology* 2001: 142–144.

Engel, S.L. 1981. Basic of laser heat treating. In Metzbower, E.A. (ed.), *Source Book on Applications of the Laser in Metalworking*: 149–171, American Society for Metals Metals Park.

Ivanaysky, V.V., Krivochurov, N.T., Shaykhudinov, A.S. & Ishkov, A.V. 2011. Joint induction-arc surfacing of the working bodies of agricultural machinery. *Altai State Agrarian University Bulletin* 3: 35–38.

Kirgizov, V.E., Shishkin, G.M., Baldanov, K.P., Andreev, S.V. & Gom, P.E. 2010. Increasing the durability of plow plowshares when restoring the welding with a carbon electrode. *Vestnik IrGSKHA* 38: 65–71.

Klein, A.N., Cardoso, R.P., Pavanati, H.C., Binder, C., Maliska, A.M., Hammes, G., Fusão, D., Seeber, A., Brunatto, S.F. & Muzart, J.L.R. 2013. DC Plasma technology applied to powder metallurgy: An overview. *Plasma Sci. Technol.* 15(1): 70–81.

Komanduri, R. & Hou, Z.B. 2001. Thermal analysis of the laser surface transformation hardening process. *International Journal of Heat and Mass Transfer* 44(15): 2845.

Kondrat'ev, V.V., Balanovskii, A.E., Ivanov, N.A., Ershov, V.A. & Kornyakov, M.V. 2014. Evaluation of the effect of modifier composition with nanostructured additives on grey cast iron properties. *Metallurgist* 58(5–6): 377.

Kozlov, L.G., Polishchuk, L.K., Piontkevych, O.V., Korinenko, M.P., Horbatiuk, R.M., Komada, P., Orazalieva, S. & Ussatova, O. 2019. Experimental research characteristics of counter balance valve for hydraulic drive control system of mobile machine. *Przegląd Elektrotechniczny* 95(4): 104–109.

Kukharchuk, V.V., Hraniak, V.F., Vedmitskyi, Y.G., Bogachuk, V.V., Zyska, T., Komada, P., Sadikova, G. 2016. Noncontact method of temperature measurement based on the phenomenon of the luminophor temperature decreasing. *Proc. SPIE* 10031: 100312F.

Kukharchuk, V.V., Kazyv, S.S. & Bykovsky, S.A. 2017. Discrete wavelet transformation in spectral analysis of vibration processes at hydropower units. *Przegląd Elektrotechniczny* 93(5): 65–68.

Lamima, T.S., Bernardellia, E.A., Binder, C., Kleina, A.N. & Maliskaa, A.M. 2015. Plasma carburizing of sintered pure iron at low temperature. *Materials Research* 18(2): 320–327.

Maliska, A.M., Pavanati, H.C., Klein, A.N. & Muzart, J.L.R. 2003. The influence of ion energy bombardment on the surface porosity of plasma sintered iron. *Materials Science and Engineering A*. 352(1–2): 273–278.

Medvedev, S.I., Nezhivlyak, A.E., Grechneva, M.V., Balanovsky, A.E. & Ivakin, V.L. 2015. Optimization of plasma hardening conditions of the side surface of rails in PUR-1 experimental equipment. *Welding International* 29(8): 643–651.

Ogorodnikov, V.A., Dereven'ko, I.A. & Sivak, R.I. 2018. On the influence of curvature of the trajectories of deformation of a volume of the material by pressing on its plasticity under the conditions of complex loading. *Materials Science* 54(3), 326–332.

Ogorodnikov, V.A., Savchinskij, I.G. & Nakhajchuk, O.V. 2004. Stressed-strained state during forming the internal slot section by mandrel reduction. *Tyazheloye Mashinostroyeniye* 12: 31–33.

Ogorodnikov, V.A., Zyska, T. & Sundetov, S. 2018. The physical model of motor vehicle destruction under shock loading for analysis of road traffic accident. *Proc. SPIE* 10808: 108086C.

Polishchuk, L.K., Kozlov, L., Piontkevych, O., Horbatiuk, R., Pinaiev, B., Wójcik, W., Sagymbai, A. & Abdihanov, A. 2019. Study of the dynamic stability of the belt conveyor adaptive drive. *Przegląd Elektrotechniczny* 95(4): 98–103.

Savulyak, V.I. & Shenfeld, V.Y. 2016. *Welding of High-Carbon Wear-Resistant Coatings*. Vinnitsia: VNTU.

Schaaf, P. 2003. Laser nitriding and laser carburizing of surfaces. *Proc. SPIE* 5147: 404.

Shevchenko, O.V. 2004. Compositional plasma plasters on the basis of powder rolls. *Abstract of the scientific degree of candidate of technical sciences*. Kiev.

Vedmitskyi, Y.G., Kukharchuk, V.V. & Hraniak, V.F. 2017. New non-system physical quantities for vibration monitoring of transient processes at hydropower facilities, integral vibratory accelerations. *Przegląd Elektrotechniczny* 93(3): 69–72.

Vorobyov, V., Pomazan, M., Vorobyova, L. & Shlyk, S. 2017. Simulation of dynamic fracture of the borehole bottom taking into consideration stress concentrator. *Eastern-European Journal of Enterprise Technologies* 3/1(87): 53–62.

Chapter 14

Complex dynamic processes in elastic bodies and the methods of their research

B. I. Sokil, A. P. Senyk, M. B. Sokil, A. I. Andrukhiv,
O. O. Koval, A. Kotyra, P. Droździel, M. Kalimoldayev,
and Y. Amirgaliyev

CONTENTS

14.1 Introduction 151
14.2 Construction of mathematical model of dynamics of researched elastic body ...152
14.3 Conclusions 161
References 161

14.1 INTRODUCTION

Operation of machines and mechanisms is accompanied by various dynamic actions – external and internal. Their consequence is the change in quantitative and, in some cases, qualitative characteristics of motion. Investigating them in a general is a complex task, therefore, when solving many (applied) problems, identify the most relevant elements and conduct a thorough analysis of the influence of these or other factors. Such elements are most often elastic bodies (beams, shafts, rotors, etc.), which carry out complex oscillations (a combination of longitudinal, bending, and twisting). By interacting with one another, as well as with other bodies, the indicated oscillations can lead to a significant increase in their dynamic loads. These loads of magnitude considerably increase in so-called resonance cases, namely when the frequencies of oscillations (bending, longitudinal, torsion) are interconnected or one of them is related to the frequency of external perturbation by rational relations (Bogolyubov & Mitropolsky 1974) (we are talking about oscillations in bodies with nonlinear elastic properties of their material). To prevent them, and therefore to extend the life of the exploitation of elastic bodies, it can only be based on the adequate dynamic process of mathematical models and quantitative analysis on their basis of influence of certain parameters. Such tasks are the subject of consideration of the article.

Analytical methods for investigating the resonance phenomena for mechanical systems with lumped masses are developed relatively thoroughly for quasilinear (Bogolyubov & Mitropolsky 1974, Valeev 1977, Grebennikov 1984) and simplest nonlinear mathematical models (Grubel et al. 2015, Pelicano et al. 2001, Khaustov 2016). With regard to the analytical study of mathematical models of systems with distributed parameters, the use of the principle of single frequency oscillations in conjunction with the basic ideas of asymptotic integration of systems with "small nonlinearity" has allowed to create the basic principles (Mitropolsky & Moseyenkov 1976) for solving many practical problems. Their basic position in combination with the wave theory of motion is generalized to new classes of dynamic systems – systems that are characterized by a constant component of the velocity of longitudinal motion (Sokil 1983, Rogatinsky et al. 2014, Lyashuk et al. 2017) or strongly nonlinear systems (Lyashuk et al. 2018). However, the study of complex nonlinear oscillations of elastic bodies did not find proper development through purely mathematical problems. This is primarily about the mutual influence of some forms of oscillation on others (including internal resonances) (Ogorodnikov et al. 2014, Lukovsky et al. 2013). Such tasks are the subject of consideration of work. For a partial solution to this problem, a methodology based on the use of partial information about one of the forms of oscillation of the elastic body (Gulyaev & Borsch 2007, Pirogova & Taranenko 2015) and its consideration in the "refined" mathematical model of others, and hence the construction of its analytical solution, are discussed. As a rule, it is most appropriate to experimentally obtain information on the basic parameters of oscillation of a smaller amplitude. This, first, allows for the refined mathematical model to adapt the existing analytical methods for nonlinear systems with low perturbation; and second, leads to a decrease in the inaccuracy of the analytical description of the complex oscillations of elastic bodies.

14.2 CONSTRUCTION OF MATHEMATICAL MODEL OF DYNAMICS OF RESEARCHED ELASTIC BODY

Along the elastic body, which rotates around the vertical axis, a continuous homogeneous flow of zero-rigid medium travels through twisting and bending oscillations (wells for drilling, a conveyor screw for moving friable or viscous media, turbine rotor, etc.). Partial information of the dynamics of the specified system is known based on experimental data, for example, discrete information about the amplitude and frequency (period) of small bending oscillations. Our aim is to study the influence of the main parameters of bending vibrations of an elastic body, the motion of a continuous flow of the medium, physical and mechanical characteristics of the body on its torsional oscillations, in particular to find the conditions for the existence of external and internal resonances and the peculiarities of passing through them (Vedmitskyi et al. 2017, Kukharchuk et al. 2016, 2017b).

To partially solve this problem, first we need to build a mathematical model of the investigated elastic body. To do this, choose a moving coordinate system XOYZ where the OZ axis coincides with the axis of rotation. In case when relative bending oscillations occur in the XOZ plane by processing experimental data, they can be described by the relations $u(z,t) = b\sin\frac{k\pi}{l}z\cos\psi_k$, $\psi_k = \wp_k t + \psi_{0k}$, $k = 1, 2, \ldots$ where l denotes the distance between the bearings in which the elastic body rotates. The basis for determining the parameters b, ω_k, ψ_{0k} can be a sequence of magnitudes and periods (half-life) of oscillations of an elastic body (Senyk & Sokil 1977). If you denote the angle of the curl of the body (portable movement) body section with coordinate z at an arbitrary time $t - \theta(z,t)$, then from the equation of "dynamic equilibrium" for arbitrarily selected item of small lengths, it follows

$$I_z \frac{\partial^2 \theta(z,t)}{\partial t^2} - GJ_z \frac{\partial^2 \theta}{\partial z^2} - 2(m_1+m_2)\Omega u \frac{\partial u}{\partial t} - 2m_2\Omega V u \frac{\partial u}{\partial x} - F\left(\theta, \frac{\partial \theta}{\partial t}, \frac{\partial \theta}{\partial x}, \varphi\right) = 0 \quad (14.1)$$

We denote by I_z the running moment of inertia relative to the axis of rotation of the body, together with the liquid, G – shear modulus, J_z – the equatorial moment of the cross section, $F\left(z, \theta, \frac{\partial \theta}{\partial t}, \frac{\partial \theta}{\partial z}, \varphi\right)$ – nonlinear with respect to variables $\theta, \frac{\partial \theta}{\partial t}, \frac{\partial \theta}{\partial x}$ periodic by $\varphi = \eta t + \varphi_0$ function, which describes the distribution along the length of the body of moments of external forces relative to the axis of rotation (including moments of resistance), η – frequency of periodic perturbation; m_1 – mass of unit of elastic body length, m_2 – mass of the unit length continuous fluid flow, which moves along the elastic body at a relative speed V, $2m_1\Omega u \frac{\partial u}{\partial t} dz$ – moment relative to the axis of rotation of the Coriolis force of the inertia of the element of the elastic body, $2m_2\Omega u\left(\frac{\partial u}{\partial t} + V\frac{\partial u}{\partial z}\right)dz$ – the moment, relative to the axis of rotation, of the Coriolis force of inertia of a continuous flow of medium with a mass, which at this time is "located" in the selected element, Ω – the value of the angular velocity. In addition, it is considered that the maximum value of moments of Coriolis forces of inertia is a small value in comparison with the maximum value of the restoring moment of elastic forces of the body. This allows rewriting equation (14.1) in the form

$$\frac{\partial^2 \theta(z,t)}{\partial t^2} - \frac{G}{\rho}\frac{\partial^2 \theta}{\partial z^2} = \varepsilon\left\{2(m_1+m_2)\Omega u \frac{\partial u}{\partial t} + 2m_2\Omega V u \frac{\partial u}{\partial x} + F\left(\theta, \frac{\partial \theta}{\partial t}, \frac{\partial \theta}{\partial x}, \varphi\right)\right\}, \varepsilon = \frac{1}{I_z}$$
(14.2)

where $\rho = \frac{J}{I_z}$ is the density of the material of the elastic body, together with the continuous flow of the medium, and the maximum value of the right part of equation (14.2) is a small value in comparison with the maximum value of the second term of the left part of it. The given equation for simplicity will be considered under homogeneous boundary conditions (Polishchuk et al. 2016b, Polishchuk et al. 2016a, 2018)

$$\theta(z,t)\big|_{z=0} = \theta(z,t)\big|_{z=l} = 0 \quad (14.3)$$

Before proceeding to construct a solution to equation (14.2), at the above boundary conditions, briefly let's dwell on the moments of the Coriolis forces of inertia relative to the axis of rotation, which are described by the expression $2(m_1+m_2)\Omega u \frac{\partial u}{\partial t} + 2m_2\Omega V u \frac{\partial u}{\partial x}$. If we replace functions and their derivatives with expressions, which are consistent with the relative motion of the elastic body (bending oscillations), then we will get

$$2(m_1+m_2)\Omega u \frac{\partial u}{\partial t} + 2m_2\Omega V u \frac{\partial u}{\partial x}$$
$$= -(m_1+m_2)\Omega \omega_k \, b^2 \sin^2 \frac{k\pi}{l} z \sin 2\psi_k + m_2\Omega V \frac{k\pi}{l} b^2 \sin \frac{2k\pi}{l} \cos^2 \psi_k$$
(14.4)

Thus, the Coriolis forces of inertia of the relative motion of a continuous medium along the elastic body and its bending oscillation create an additional periodic action relative to the axis of rotation with period $\dfrac{\pi}{\wp_k}$. On the other side, the frequency of bending oscillations of an elastic body, which rotates around a fixed axis with angular velocity Ω in the k-th form of "dynamic equilibrium," depends on the value of the latter and is defined by

$$\wp_k = \sqrt{\dfrac{EI}{m_1+m_2}\left(\dfrac{k\pi}{l}\right)^4 - \Omega^2} \qquad (14.5)$$

Hence, the "extra moment" of the forces of inertia caused by bending oscillations is determined not only by the geometric and physical–mechanical properties of the elastic body, but also by its angular velocity of rotation. All this creates, as will be shown below, not only difficulties but also some features of the dynamics for the proposed system class. They can be identified and described by the solution to the boundary value problem (14.2) (14.3). The right part of the differential equation (14.2), as follows from (14.4), (14.5) is a π periodic function of the argument $\psi_k = \wp_k t + \psi_{0k}$, and 2π – periodic by argument Θ, so for a given equation we have to consider the following cases: nonresonant and resonant. As for the latter, it can be predetermined as an external periodic perturbation with η frequency, and bending oscillations that create periodic action for an argument ψ_k and the period of this action is $\dfrac{\pi}{\wp_k}$. First, consider a simple, nonresonant case where there is no rational dependence between the above frequencies and frequency of its own oscillations: $p\eta \neq q\dfrac{s\pi}{k}\sqrt{\dfrac{G}{\rho}}$, $p\dfrac{s\pi}{k}\sqrt{\dfrac{G}{\rho}} \neq q\dfrac{\wp_k}{2}$, p, q – are mutually simple numbers, and $\dfrac{s\pi}{k}\sqrt{\dfrac{G}{\rho}}$ – is the own frequency of torsional oscillations of the undisturbed motion in the s-th form of dynamic equilibrium (Polishchuk et al. 2019, Kozlov et al. 2019).

First, let's consider a simpler nonresonant case of fluctuations of the studied system: the elastic body – a continuous flow of medium. For this construct boundary problem solution (1)–(2) provided that $p\eta \neq q\dfrac{s\pi}{k}\sqrt{\dfrac{G}{\rho}}$, $p\dfrac{s\pi}{k}\sqrt{\dfrac{G}{\rho}} \neq q\dfrac{\wp_k}{2}$. Its first approximation in the s-th form of a dynamic equilibrium we will present in the form

$$\theta(z,t) = a\sin\dfrac{s\pi}{l}z\sin\vartheta_s + \varepsilon U(a,z,\vartheta_s,\gamma,\phi_k)$$
$$\phi_k = \wp_k t + \phi_{0k}, \vartheta_s = \dfrac{s\pi}{l}\sqrt{\dfrac{G}{\rho}}t + \vartheta_{0s} \qquad (14.6)$$

where $U(a,z,\vartheta_s,\gamma,\phi_k) - 2\pi$ is a periodic with respect to its own (ϑ_s), forced (γ) and flexural oscillation (ϕ_k) function, which satisfies the boundary condition, followed from (14.3) and does not contain in its schedules the basic mode of oscillation (ϑ_s), that is satisfying the condition

$$\int_0^{2\pi} U(a,z,\vartheta_s,\gamma,\phi_k)\begin{Bmatrix}\sin\vartheta_s\\ \cos\vartheta_s\end{Bmatrix} d\vartheta_s = 0 \tag{14.7}$$

As for the parameters a and ϑ_s, then they will be variables in time and for the considered approximation are bound relations $\dfrac{da}{dt} = \varepsilon A_1(a)$ and $\dfrac{d\vartheta_s}{dt} = \omega_s + \varepsilon B_1(a)$. Unknown functions $A_1(a)$ and $B_1(a)$ are determined in a such way that satisfies the original equation (14.2) with the accuracy considered asymptotic representation of the solution in the form (14.6), if on the place of the parameters a and ϑ_s substitute functions of time determined by the above relations. Performing a known procedure of asymptotic integration of the boundary-value problem under consideration, we obtain a differential equation that binds unknown functions

$$\begin{aligned}
&\frac{\partial^2 U}{\partial \vartheta_s^2}\left(\frac{s\pi}{l}\right)^2 \frac{G}{\rho} + \frac{\partial^2 U}{\partial \gamma^2}\eta^2 + \frac{\partial^2 U}{\partial \psi_k^2}\wp_k^2 + 2\frac{\partial^2 U}{\partial \vartheta_s \partial \gamma}\frac{s\pi}{l}\eta \\
&+ 2\frac{\partial^2 U}{\partial \vartheta_s \partial \psi_k}\frac{s\pi}{l}\wp_k + 2\frac{\partial^2 U}{\partial \gamma \partial \psi_k}\eta\wp_k - \frac{G}{\rho}\frac{\partial^2 U}{\partial z^2} \\
&= \tilde{F}(a,z,\vartheta_s,\gamma,\psi_k) - 2\omega_s \sin\frac{s\pi}{l}z\left(A_1(a)\cos\vartheta_s + 2aB_1(a)\sin\vartheta_s\right)
\end{aligned} \tag{14.8}$$

where

$$\begin{aligned}
\tilde{F}(a,z,\vartheta_s,\gamma,\psi_k) &= \bar{F}(a,z,\vartheta_s,\gamma) - (m_1+m_2)\Omega\wp_k\, b^2 \sin^2\frac{k\pi}{l}z \sin 2\psi_k \\
&+ m_2\Omega V \frac{k\pi}{l}b^2 \sin\frac{2k\pi}{l}\cos^2\psi_k
\end{aligned}$$

and $\bar{F}(a,z,\vartheta,\gamma)$ corresponds to the value of the function $F\left(\theta,\dfrac{\partial\theta}{\partial t},\dfrac{\partial\theta}{\partial x},\varphi\right)$ provided that $\theta(z,t)$ and its derivatives are determined in accordance with the main value shown in (14.6). Restrictions imposed on the function $U(a,z,\psi_s,\gamma,\phi_k)$ allow the relation (14.8) to get (after the procedure of averaging over the phases of own and forced oscillations) a system of ordinary differential equations to describe the basic parameters of torsional oscillations of an elastic body in nonresonant case (Ogorodnikov et al. 2018).

$$\frac{da}{dt} = -\frac{\varepsilon}{8\omega_s l \pi^3}\int_0^l\int_0^{2\pi}\int_0^{2\pi}\int_0^{2\pi} \tilde{F}(a,z,\vartheta_s,\gamma,\psi_k)\sin\frac{s\pi}{l}z\cos\vartheta_s\, d\gamma\, d\vartheta_s\, d\psi_k\, dz$$

$$\frac{d\vartheta_s}{dt} - \omega_s = -\frac{\varepsilon}{8a\omega_s l \pi^3}\int_0^l\int_0^{2\pi}\int_0^{2\pi}\int_0^{2\pi} \tilde{F}(a,z,\vartheta_s,\gamma,\psi_k)\sin\frac{k\pi}{l}z\sin\vartheta_s\, d\gamma\, d\vartheta_s\, d\psi_k\, dz \tag{14.9}$$

$$= -\frac{\varepsilon}{4a\omega_s l \pi^2}\int_0^l\int_0^{2\pi}\int_0^{2\pi} \bar{F}(a,z,\vartheta_s,\gamma,)\sin\frac{s\pi}{l}z\sin\vartheta_s\, d\gamma\, d\vartheta_s\, dz$$

It follows from (14.9) that for the considered approximation in nonresonant case, small bending oscillations do not affect on the law of variation in the amplitude and frequency of torsional oscillations, but only partially replace their form.

Resonance case. As noted above, resonance oscillations in the system of an elastic body that rotates around the axis are a continuous stream of media moving along the body, possible resonant oscillations are caused by external periodic action and bending oscillations. As for resonant torsional oscillations that are caused by its bending oscillations and the rotation of the elastic body, then the angular velocity for which the abovementioned resonant oscillations are determined by the dependence

$$p\frac{s\pi}{l}\sqrt{\frac{G}{\rho}} = q\sqrt{\frac{EI}{m_1+m_2}\left(\frac{k\pi}{l}\right)^4 - \Omega^2} \rightarrow \Omega_{ks}^{pq} = \sqrt{q^2\frac{EI}{m_1+m_2}\left(\frac{k\pi}{l}\right)^4 - p^2\left(\frac{s\pi}{l}\right)^2\frac{G}{\rho}}$$

Let's consider initially more simple resonant oscillations that are caused by external periodic perturbation, that is the case $p\eta \approx q\frac{s\pi}{k}\sqrt{\frac{G}{\rho}}$. The value of the amplitude of oscillations during the passage of resonance essentially depends on the phase difference of its own and forced oscillations. Therefore, formally entering the specified parameter $\varphi_{qp} = q\theta_s - p\gamma \Rightarrow \theta_s = \frac{p}{q}\gamma + \frac{1}{q}\varphi_{qp}$ in equation (14.9) after averaging only by the phase of forced oscillations, we get

$$\frac{da}{dt} = \frac{-\varepsilon}{2\omega_s l\pi}\int_0^l\int_0^{2\pi} \overline{F}\left(a,z,\frac{p}{q}\gamma+\frac{1}{q}\varphi_{qp},\gamma\right)\sin\frac{s\pi}{l}z\cos\left(\frac{p}{q}\gamma-\frac{1}{q}\varphi_{qp}\right)dx\,d\gamma$$

$$\frac{d\varphi_{qp}}{dt} = q\omega_s - p\eta - \frac{\varepsilon}{2\omega_s al\pi}\int_0^l\int_0^{2\pi} \overline{F}\left(a,z,\frac{p}{q}\gamma+\frac{1}{q}\varphi_{qp},\gamma\right)\sin\frac{s\pi}{l}z\sin\left(\frac{p}{q}\gamma+\frac{1}{q}\varphi_{qp}\right)dx\,d\gamma$$

(14.10)

Thus, resonant twist oscillations caused by external periodic perturbation to the proposed approximation are described by dependencies $\theta(z,t) = a\sin\frac{s\pi}{l}z\sin\vartheta_s + \varepsilon U(a,z,\vartheta_s,\gamma,\phi_k)$, $\phi_k = \wp_k t + \phi_{0k}$, $\vartheta_s = \frac{s\pi}{l}\sqrt{\frac{G}{\rho}}t + \vartheta_{0s}$ where the parameters a, φ_{qp} are connected by differential equations (14.10).

For research the resonant phenomenon caused by flexural oscillations of the elastic body and the motion of a continuous medium stream, more precisely, "moments of the Coriolis forces of inertia" elastic body, and the continuous flow of the medium simplify the right part of the relations (14.9). For this purpose, we will present an interrelation that determines the given moments of forces in the form

$$2(m_1+m_2)\Omega u\frac{\partial u}{\partial t} + 2m_2\Omega Vu\frac{\partial u}{\partial x} = -(m_1+m_2)\Omega\wp_k b^2\sin^2\frac{k\pi}{l}z\sin 2\psi_k$$
$$+ \frac{1}{2}m_2\Omega V\frac{k\pi}{l}b^2\sin\frac{2k\pi}{l}z(1+\cos 2\psi_k)$$

(14.11)

If taking into account that

$$\delta_{ks} = \int_0^l \sin^2\frac{k\pi}{l}z \sin\frac{s\pi}{l}z\,dz = \begin{cases} 0 & \text{if } s \text{ even number} \\ \dfrac{l}{\pi} - \dfrac{l}{2\pi}\dfrac{2s}{s^2 - 4k^2} & \text{if } s \text{ odd number} \end{cases}$$

$$\Delta_{ks} = \int_0^l \sin\frac{2k\pi}{l}z \sin\frac{s\pi}{l}z = \frac{l}{\pi}\begin{cases} 0 & \text{if } s \neq 2k \\ \dfrac{\pi}{2} & \text{if } s = 2k \end{cases}$$

then likewise, as for the resonance caused by external periodic perturbation, basic equations describing resonant torsional oscillations caused to bending oscillations take the form

$$\frac{da}{dt} = -\frac{\varepsilon}{4\omega_s l\pi^2}\int_0^l\int_0^{2\pi}\int_0^{2\pi}\bar{F}(a,z,\vartheta_s,\gamma,)\sin\frac{s\pi}{l}z\cos\vartheta_s\,d\gamma\,d\vartheta_s\,dz$$

$$-\frac{(m_1+m_2)b^2}{2Il\omega_s\pi}\Omega\wp_k\delta_{ks}\left[\cos\frac{\bar{\phi}_{qp}}{q}\int_0^\pi\sin\xi\cos\frac{p}{q}\xi\,d\xi - \sin\frac{\bar{\phi}_{qp}}{q}\int_0^\pi\sin\xi\sin\frac{p}{q}\xi\,d\xi\right]$$

$$+\frac{b^2}{4Il\hbar\omega_s}k\pi m_2\Omega V\Delta_{ks}\left[\cos\frac{\bar{\phi}_{qp}}{q}\int_0^\pi\cos\xi\cos\frac{p}{q}\xi\,d\xi - \sin\frac{\bar{\phi}_{qp}}{q}\int_0^\pi\cos\xi\sin\frac{p}{q}\xi\,d\xi\right] \quad (14.12)$$

$$\frac{d\phi_{qp}}{dt} - q\omega_s + 2p\wp_k = -\frac{\varepsilon}{4a\omega_s l\pi^2}\int_0^l\int_0^{2\pi}\int_0^{2\pi}\bar{F}(a,z,\vartheta_s,\gamma,)\sin\frac{s\pi}{l}z\sin\vartheta_s\,d\gamma\,d\vartheta_s\,dz$$

$$-\frac{(m_1+m_2)b^2}{2Il\omega_s\pi a}\Omega\wp_k\delta_{ks}\left[\sin\frac{1}{q}\bar{\phi}_{qp}\int_0^\pi\sin\xi\cos\frac{p}{q}\xi\,d\xi + \cos\frac{1}{q}\bar{\phi}_{qp}\int_0^\pi\sin\xi\sin\frac{p}{q}\xi\,d\xi\right]$$

$$+\frac{b^2}{4Il a\pi\omega_s}k\pi m_2\Omega V\Delta_{ks}\left[\sin\frac{\bar{\phi}_{qp}}{q}\bar{\phi}_{qp}\int_0^\pi\cos\xi\cos\frac{p}{q}\xi\,d\xi + \frac{1}{2}\cos\frac{\bar{\phi}_{qp}}{q}\int_0^\pi\cos\xi\sin\frac{p}{q}\xi\,d\xi\right]$$

where $\bar{\phi}_{qp}$ is the phase difference torsional and bending oscillation of elastic body

$$\bar{\varphi}_{qp} = q\theta_s - p2\psi_k \quad \Rightarrow \quad \frac{1}{q}\bar{\varphi}_{qp} + 2\frac{p}{q}\psi_k = \theta_s$$

As for the first improved approximation of the solution, then it can be found without difficulty by the expansion of the unknown and right part of equation (14.8) in multiples of Fourier series followed by a comparison of the coefficients with the same harmonics.

At first opinion, the ratios obtained are too complicated for practical implementation. Let's show that this is not so on the example of a complex shaft oscillation under the conditions: (1) nonlinear-elastic properties of its material are described by a nonlinear technical law; (2) the strength of the resistance depends on the speed; (3) harmonious external periodic perturbation; (4) empirical information on bending oscillations, which, by mathematical treatment, can lead to appearance $\vartheta(z, \Xi_1 t + \vartheta_{01}) = b_1 \sin\frac{\pi}{l} x \cos(\wp_1 t + \vartheta_{01})$. In this case, the right part of the differential equation (14.2) becomes a form

$$Q(x,t,\theta,\theta_t,\theta_x,\varphi) - 2m\Omega u_t(z,t)u(z,t)$$
$$= \lambda\theta_z^2\theta_{zz} - \delta\theta_t^{2r+1} + h\sin\varphi + m\Omega\Xi_s b_s^2 \sin 2\varphi_s \sin^2\frac{s\pi}{l}x \qquad (14.13)$$
$$+ \frac{1}{2}m_2\Omega V \frac{k\pi}{l} b^2 \sin\frac{2k\pi}{l} z(1+\cos 2\psi_k)$$

where λ, δ, h are the known constants. As for the nonresonant case, then, as follows from the differential equation (14.9), amplitude frequency characteristics fading oscillations in the form close to the "first" dynamic are described by the equations

$$\frac{da}{dt} = -\frac{\varepsilon\Gamma^2\left(\frac{2r+3}{2}\right)}{2\pi\Gamma^2(r+2)}\left(\frac{G}{\rho}\right)^r a^{2r+1}, \qquad (14.14)$$

For a resonant case at a frequency close to the main ($k=1$) frequency of its own torsional oscillations, the main parameters of the dynamic process are determined by dependencies

$$\frac{da}{dt} = -\frac{\varepsilon\Gamma^2\left(\frac{2r+3}{2}\right)}{2\pi\Gamma^2(r+2)}\left(\frac{G}{\rho}\right)^r a^{2r+1} + \frac{h}{2}\sqrt{\frac{\rho}{G}}\cos\vartheta_1$$
$$\qquad (14.15)$$
$$\frac{d\vartheta_1}{dt} = \frac{\pi}{l}\sqrt{\frac{G}{\rho}} - \mu + \frac{3\lambda\varepsilon a^2}{64}\sqrt{\frac{\rho}{G}}\left(\frac{\pi}{l}\right)^3 - \frac{h}{2a}\sqrt{\frac{\rho}{G}}\sin\vartheta_1$$

Somewhat more difficult equations describe resonant oscillations, which are caused by bending oscillations of the elastic body and the motion along its continuous medium flow. Then, from the differential equation (14.11), we get to the main internal resonance oscillation ($k=s=1$) system of equations

$$\frac{da}{dt} = -\frac{\varepsilon\Gamma^2\left(\frac{2r+3}{2}\right)}{2\pi\Gamma^2(r+2)}\left(\frac{G}{\rho}\right)^r a^{2r+1} + \frac{2(m_1+m_2)b^2}{3I\omega_1\pi^2}\Omega_{11}\wp_1\sin\bar\phi_{11}$$
$$\qquad (14.16)$$
$$\frac{d\bar\phi_{11}}{dt} - \omega_1 + 2\wp_1 = \frac{3\lambda\varepsilon a^2}{64}\sqrt{\frac{\rho}{G}}\left(\frac{\pi}{l}\right)^3 - \frac{2(m_1+m_2)b^2}{3I\omega_2\pi^2 a}\Omega_{11}\wp_1\cos\bar\phi_{11}$$

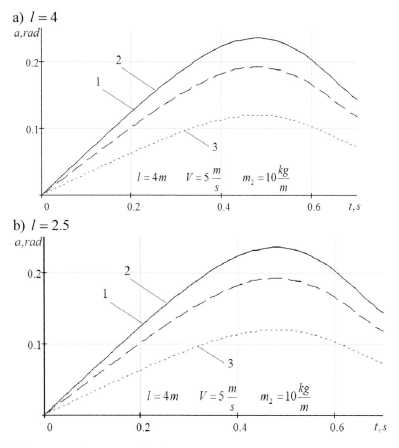

Figure 14.1 Changing the amplitude of resonant torsional oscillations caused by the main flexural oscillations.

In Figure 14.1, there is given the change in the amplitude of torsional oscillations when passing through resonance caused by the main mode of bending oscillations.

In addition to the abovementioned resonance in the studied system, possible resonances caused by other modes of bending oscillations, in particular the second (k = 2, s = 1) and bending taking into account the motion of a continuous flow of medium (for example, when k = 1, s = 2). The indicated oscillations are described, respectively, by the equations

$$\frac{da}{dt} = -\frac{\varepsilon \Gamma^2 \left(\frac{2r+3}{2}\right)}{2\pi \Gamma^2 (r+2)} \left(\frac{G}{\rho}\right)^r a^{2r+1} + \frac{16(m_1+m_2)b^2}{15I\omega_2\pi^2} \Omega_{21}\wp_2 \sin\bar{\phi}_{21},$$

$$\frac{d\bar{\phi}_{21}}{dt} - \omega_1 + 2\wp_2 = \frac{3\lambda\varepsilon a^2}{64}\sqrt{\frac{\rho}{G}}\left(\frac{\pi}{l}\right)^3 - \frac{16(m_1+m_2)b^2}{15I\omega_1\pi^2 a}\Omega_{21}\wp_2 \cos\bar{\phi}_{21} \qquad (14.17)$$

and

$$\frac{da}{dt} = -\frac{\varepsilon\Gamma^2\left(\frac{2r+3}{2}\right)}{2\pi\Gamma^2(r+2)}\left(\frac{G}{\rho}\right)^r a^{2r+1} - \frac{b^2}{16I\omega_2}m_2\pi\Omega_{12}V\cos\bar{\varphi}_{12} \qquad (14.18)$$

$$\frac{d\bar{\varphi}_{12}}{dt} - \omega_2 + 2\wp_1 = \frac{3\lambda\varepsilon a^2}{64}\sqrt{\frac{\rho}{G}}\left(\frac{2\pi}{l}\right)^3 + \frac{b^2}{16Ia\omega_2}k\pi m_2\Omega_{12}V\sin\bar{\varphi}_{12}$$

Below, for different lengths of elastic body, it presents the change of amplitude torsional oscillations over time at transition through resonance at a frequency close to the frequency of the main mode of torsional oscillations caused by bending oscillations and the motion of a continuous flow of medium (the second mode) (Figure 14.2).

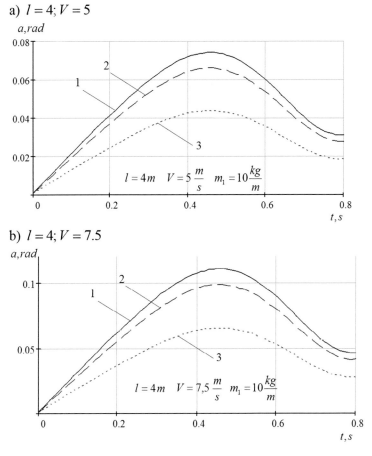

Figure 14.2 Change in time of the amplitudes of torsional oscillations at transition through resonance caused by bending oscillations and the motion of a continuous flow of medium.

14.3 CONCLUSIONS

This is the overview research methodology of complex processes in elastic bodies in the case of torsional and bending oscillations provided that along it moves with a constant relative velocity a continuous flow of nonelastic medium. It has been obtained with the mathematical model of the system of the elastic body – a continuous flow of medium in the case where the basic parameters of bending body oscillations are known. It has been shown that the methodology is particularly effective in cases where components of moving an elastic body caused by some forms of oscillation are much smaller in comparison with others. In particular, in the paper it is established that:

- in the studied system the elastic body – a continuous flow of medium there can be resonant phenomena caused not only to external periodic perturbations but also to the mutual influence of some oscillations to others including the motion of a continuous medium along the body;
- amplitude of torsional oscillations during transition through resonance at small angular speeds of rotation is larger for smaller values of the relative amount of motion of the medium, but for the greater ones – on the contrary;
- as regards the resonance phenomena caused by bending oscillations of the elastic body, then this may take place on odd modes of torsional oscillations (caused to the moment of the Coriolis forces of inertia relative to the axis of rotation of the elastic body) and paired mods of twisting oscillations, and this caused Coriolis forces of inertia of the moving oscillations (more precisely, its moment relative to the axis of rotation). As for the magnitudes of the amplitudes of the passage of the resonance, then for the case of resonance on the main frequency of their own torsional oscillations, the great value of the angular velocity of the elastic body rotation corresponds to the greater value of the amplitude of the transition through the resonance; greater value of the relative amount of motion of the continuous medium corresponds to a smaller value of the amplitude of the transition through the resonance.

The results obtained in the work can be the basis for determining the operational parameters of machines, the elements of which are in a complex movement.

REFERENCES

Bogolyubov, N.N. & Mitropolsky, Yu.A. 1974. *Asymptotic Methods in the Theory of Nonlinear Oscillations*. Moscow: Nauka.
Grebennikov, E.A. 1984. *On Some Problems of the Theory of Resonant Systems. Proc. IX International conf. on nonlinear oscillations*. Kiev: Naukova Dumka.
Grubel, M.G., Nanivsky, R.A. & Falcon, M.B. 2015. Resonant oscillations of the sprung part of wheeled vehicles when moving along the ordered system of inequalities. *Bulletin of the Vinnytsia Polytechnic Institute* 1: 155–161.
Gulyaev, V.I. & Borsch, O.I. 2007. Spiral waves in elastic twisted rotating tube rods with internal flows of liquid. *Acoustic bulletin* 10(3): 12–18.
Khaustov, D.Ye. 2016. Conditions internal resonance military tracked vehicles. *Systems of Arms and Military Equipment* 1(45): 73–76.

Kozlov, L.G., Polishchuk, L.K., Piontkevych, O.V., Korinenko, M.P., Horbatiuk, R.M., Komada, P., Orazalieva, S. & Ussatova, O. 2019. Experimental research characteristics of counter balance valve for hydraulic drive control system of mobile machine. *Przegląd Elektrotechniczny* 95(4): 104–109.

Kukharchuk, V.V. et al. 2016. Noncontact method of temperature measurement based on the phenomenon of the luminophor temperature decreasing. *Proc. SPIE* 10031: 100312F.

Kukharchuk, V.V., Bogachuk, V.V., Hraniak, V.F., Wójcik, W., Suleimenov, B. & Karnakova, G. 2017. Method of magneto-elastic control of mechanic rigidity in assemblies of hydropower units. *Proc. SPIE* 10445: 104456A.

Kukharchuk, V.V., Kazyv, S.S. & Bykovsky, S.A. 2017. Discrete wavelet transformation in spectral analysis of vibration processes at hydropower units. *Przegląd Elektrotechniczny* 93(5): 65–68.

Lukovsky, I.A., Solodun, A.V. & Timoha, A.N. 2013. Intrinsic resonance oscillations of the liquid in conical tanks. *Applied Hydromechanics* 15(2): 46–52.

Lyashuk, O., Sokil, M., Vovk, Y., Tson, O. & Dzyura, V. 2017. The impact of the kinematic parameters of bounce and pitch motions of sprung mass on wheeled vehicles handling. *Scientific Journal of Silesian University of Technology* (Series Transport) 97: 81–91.

Lyashuk, O., Sokil, M. Vovk, Y. Tson, A. Gupka, A. & Marunych, O. 2018. Torsional oscillations of an auger multifunctional conveyor's screw working body with consideration of the dynamics of a processed medium continuous flow. *Ukrainian Food Journal* 7(3): 499–510.

Mitropolsky, Yu.A. & Moseyenkov, B.I. 1976. *Asymptotic Solutions of Partial Differential Equations*. Kiev: Vishcha school.

Ogorodnikov, P.I., Svetlitsky, V.M. & Gogol, V.I. 2014. Investigation of the connection between the longitudinal and torsional oscillations of the drill column. *Naftogazovaya branch of Ukraine* 2: 6–9.

Ogorodnikov, V.A., Zyska, T. & Sundetov, S. 2018. The physical model of motor vehicle destruction under shock loading for analysis of road traffic accident. *Proc. SPIE* 10808: 108086C.

Pelicano, F., Fregolent, A., Bertizzi, A. & Vestroni, F. 2001. Primary parametric resonances of a power transmission belt: theoretical and experimental analysis. *Journal of Sound and Vibration* 224(4): 669–684.

Pirogova, N.S. & Taranenko, P.A. 2015. Calculated and experimental analysis of natural and critical frequencies and forms of the high-speed rotor of a microgas turbine installation. *SUSU Journal (Series Engineering)* 15(3): 47.

Polishchuk, L., Bilyy, O. & Kharchenko, Y. 2016. Prediction of the propagation of crack-like defects in profile elements of the boom of stack discharge conveyor. *Eastern-European Journal of Enterprise Technologies* 6(1): 44–52.

Polishchuk, L., Kharchenko, Y., Piontkevych, O. & Koval, O. 2016. The research of the dynamic processes of control system of hydraulic drive of belt conveyors with variable cargo flows. *Eastern-European Journal of Enterprise Technologies* 2(8): 22–29.

Polishchuk, L.K. et al. 2019. Study of the dynamic stability of the belt conveyor adaptive drive. *Przegląd Elektrotechniczny* 95(4): 98–103.

Rogatinsky, R.M., Gevko, I.G. & Dyachun, A.Y. 2014. *Scientific and Applied Foundations of Guided Transport and Technological Mechanisms*. Ternopil: Ternopil Ivan Puluj National Technical University.

Senyk, P.M. & Sokil, B.I. 1977. On the determination of the parameters of a nonlinear oscillatory system from the amplitude-frequency characteristic. *Mathematical Methods and Physical Fields* 7: 94–99.

Sokil, B.I. 1983. Asymptotic representation of the solution of a nonlinear system with resonance. *Ukrainian Mathematical Journal* 35(3): 339–341.

Valeev, K.G. 1977. *The Construction of Periodic Solutions in Complex Resonance Cases.* Kiev: Publishing House of the Institute of Mathematics.

Vedmitskyi, Y.G., Kukharchuk, V.V. & Hraniak, V.F. 2017. New non-system physical quantities for vibration monitoring of transient processes at hydropower facilities, integral vibratory accelerations. *Przegląd Elektrotechniczny* 93(3): 69–72.

Chapter 15

Analysis of the character of change of the profilogram of micro profile of the processed surface

*N. Veselovska, S. Shargorodsky, V. Rutkevych,
R. Iskovych-Lototsky, Z. Omiotek, O. Mamyrbaev,
and U. Zhunissova*

CONTENTS

15.1 Introduction 165
15.2 Analysis and problem statement 166
15.3 Materials and research methods 168
15.4 Research results and discussion 169
15.5 Conclusion 172
References 173

15.1 INTRODUCTION

Processing by face milling is widespread and described in reference literature (Guzeev et al., 2005; Pukhalsky & Gavrilov, 2012; Yalçın et al., 2009). In the course of face milling, various cutting parameters can be set: cutting depth t, mm; milling width B, mm; serve on the mill tooth S_z, mm/rotation; cutting speed V, m/min; rotation frequency n, minutes^{-1}. The cutting parameters are set by either the technologist from cutting mode standards (Guzeev et al., 2005) or reference books by cutting tool manufacturers, for example, of Sandvik, Pramet, Seco, and many others (Sandvik Coromant, 2019), or the machine operator a trial and error method.

The cutting parameters in standards (Guzeev et al., 2005) are given in the first case on the basis of the statistical technological transitions, collected from a large number of enterprises. These cutting parameters have to guarantee the uniformity of the carried-out size and the required roughness of the processed surface. As it is known, the wear occurs from the very beginning of processing using a tool (Loladze, 1982; Pimenov, 2103; To Simsiva et al., 2012).

In the work (Isakov, 2013), the wear intensity model of a mill is shown in processing and the geometrical model of microroughness of the processed surface, taking into account the wear of the face mill's teeth, but it does not consider the physical properties of the processed materials. Forecasting models of roughness without taking note of the wear of the tool are provided in scientific articles (Bajić et al., 2012; Grzenda & Bustillo, 2013).

In the works (Kovac et al., 2013; Rosales et al., 2010; Simunovic et al., 2013), the influence of processing parameters on the roughness of processed surfaces is investigated

DOI: 10.1201/9781003225447-15

through face milling, for example, the processed surface quality arising from the milling modes or the cooling method is investigated, studies using the Taguti method, etc., are conducted. However, the abovementioned works do not consider an important component – the change in the roughness of processed flat surfaces in connection with the growth of the size of the buildup of wear on the back surface of face mill teeth. Therefore it is necessary to establish how the wear of face mills influences the roughness of the processed surface.

15.2 ANALYSIS AND PROBLEM STATEMENT

The aim is to investigate the influence of the size of the wear buildup on a back surface of face mill teeth and of the face milling parameters on the processed surface roughness during STEEL 45 (C45) processing.

This chapter (Yalçın et al., 2009) concentrates on the influence of various cooling strategies on the roughness of a surface and wear of the tool during computer milling of materials of soft preparations. The study involved a selection of milling operations in the form of dry milling, cooling with cold air, and cold liquid milling. The air cooling system was developed and manufactured for cooling of final milling tools.

The study (Pukhalsky & Gavrilov, 2012) presents a theoretical model for defining the beating of the face mill cutting edges with the mechanical fastening of the replaceable many-sided plates, considering the valid sizes and the beating of a mandrel determining the tool installation eccentricity. The interrelation of the mutual beating of step mill teeth and their wear is considered.

In the works (Guzeev et al., 2005; Polishchuk et al., 2019; Sandvik Coromant, 2019), the data necessary for defining the parameters of cutting at turning, boring, processing of openings, milling, are given for machines with numerical program control, as well as applications including data on modern CNC machine models.

In the book (Loladze, 1982), the mechanism of cutting tool destruction and wear in various processing conditions and questions of the fragility and plastic durability of a tool's cutting part are considered. The work provides durability calculation methods, as well as the theory concerning adhesive, fatigue, and diffusive wear of tools. As a result, recommendations are made on the increase in tool firmness and increase in the productivity of processing using cutting.

In the work (Kozlov et al., 2019; Pimenov, 2103), features of porosity formation are considered during the hardening of a cast high-manganic steel plate. A model for assessing the influence of this type of defect on the intensity and deformation condition of a plate resulting from strain loads of ore preparation crushers is offered. The finite element method is used for tension calculation in the DEFORM software package and the calculation of equivalent tension in characteristic sections of a plate for various parameters of the macro time of a round section.

The work (Ogorodnikov et al., 2018a; To Simsiva et al., 2012) offers a physical-probabilistic model of back surface wear of a cutting tool during high-speed turning. The authors did not consider the influence of processes of abrasive and diffusive wear.

In the work (Isakov, 2013; Ogorodnikov et al., 2018b), the geometrical model of processed flat surface face milling microroughness considering the wear of the tool is presented. The model considers the height change of microroughness of flat surfaces caused by the tops of teeth connected with the dimensional wear face mills. As a result, the operating values of microroughness height for various serves, radii of rounding, various front and back corners of mill teeth taking into account tool wear on the back surface are obtained.

The work (Grzenda & Bustillo, 2013) concentrates on the initial transformation of data and its influence on forecasting of face milling surface roughness during high torque face milling. In the experiments conducted in industrial conditions, an extensive data set was generated. The data set includes a very broad set of parameters that influence the roughness of a surface: properties of the cutting tool, processing parameters, and the cutting method. Some of these parameters can be potentially connected with the others or can slightly impact the forecasting model. Moreover, depending on the number of available records of the model, machine learning may or may not be able to model some of the primary dependences. Therefore, it is necessary to choose a suitable quantity of input signals and the appropriate configuration of the coordinated prediction model. In this article, the hybrid algorithm that unites a genetic algorithm with neural networks is offered to consider the choice of the corresponding parameters and their corresponding transformation. The algorithm was tested in a number of experiments conducted in the conditions of a master class with data sets of different sizes for research concerning the impact of the available data on the choice of the corresponding data transformations. The data set size directly impacts the accuracy of forecasting models of roughness modeling, as well as the use of individual parameters and the transformed functions. Test results show considerable improvement in the quality of forecasting models constructed using that method. These improvements become obvious when these models are compared with the standard multilayered perseptrona, which were trained with all parameters, and with data that decrease by means of the standard operation of the main components.

In works (Bajić et al., 2012; Kovac et al., 2013; Rosales et al., 2010), the influence of three cutting parameters, components of tool wear, and cutting force in face milling on surface roughness is considered in the technological process as a part of off-line control. Experiments were conducted in order to define the process planning model. The cutting speed, tooth serve, and cutting depth were accepted as influential factors. For experimentally relevant data, two methodologies of modeling were used, namely the regression analysis and neural networks. The results obtained by means of models were compared. Both models have a relative forecasting error lower than 10%. The research showed that when a set of training materials represents a small methodology of neural network modeling, they are comparable of the methodology of regression analysis and can even allow obtaining the best results, in this case with an average relative error of 3.35%. In the theory, the advantages of off-line control are explained by a process that uses process models incorporating these two methodologies of modeling.

In the works (Benardos & Vosniakos, 2002; Elhami et al., 2013; Grzenda et al., 2012), the application of the Taguti method for ANN model optimization developed by the Levenberg–Marquardt algorithm is presented. For the purpose of demonstrating the implementation of the approach, the situational research of modeling the resulting cutting force in the course of a rotation is used. The educational and architectural ANN parameters were located in the orthogonal L18 array, and the predictive productivity of the ANN model was estimated with the use of the offered equation. Using the dispersion analysis (ANOVA) and the analysis of parameters (ANOM), optimal levels of the ANN parameters are defined. The ANN model optimized across Taguti was developed and exhibited high forecasting precision. Analyses and experiments showed that optimum training and the architectural ANN parameters can be defined systematically, thereby avoiding the long procedure of tests and errors (Dragobetskii et al., 2015; Ogorodnikov et al., 2004; Vorobyov et al., 2017).

15.3 MATERIALS AND RESEARCH METHODS

The purpose of the research is to determine the degree of impact that wear of the cutting tool blade and cutting speed have on roughness Rz. The object of the research is the process of milling with a face mill equipped with T15K10 carbide plates of semi-manufacture of 45 steel.

Pilot studies are conducted to assess the roughness of surface processed using face milling in terms of different degrees of wear of the mill teeth on the back surface. For that purpose, processing of a 45 steel detail has been carried out (composition of carbonaceous qualitative structural 45 steel according to GOST 1050–88: carbon C – 0.42 ... 0.5%, silicon Si – 0.17% ... 0.37%, magnesium Mn – 0.5% ... 0.8%, is lame Cr – no more than 0.25%, other Fe iron) (Polishchuk et al., 2019) with sizes "L = 200 mm × B = 75 mm × H = 100 mm" on an SF15 (6S12) vertical milling machine without the use of cooling, using a tool with the following parameters: cutting part material (pentahedral plate) – T5K10 (composition of solid T5K10 alloy of the titano-volframo-cobalt group as per GOST 3882-74: WC tungsten carbide – 85%, titanium carbide TiC – 6%, cobalt Co – 9%) (Loladze, 1982); mill diameter: D = 125 mm; the main angle in the plan: $\varphi = 60°$; the auxiliary angle in the plan: $\alpha_1 = 12°$; forward angle: $\gamma = 15°$; back angle: $\alpha = 8°$; quantity of mill teeth: $z = 1$; tilt angle of the main cutting edge: $\lambda = 0$. Using Brinell's TB 500403 hardness gauge the firmness of detail – HB190 is measured.

The cutting parameters were selected for different stages of processing, according to the reference book (Guzeev et al., 2005), and are given in Table 15.1.

Measurement of roughness Rz was carried out according to indications of the Profilometer Abris PM 7.0 Outline. Instrument readings were taken for basic length $L = 0.4$ mm at the beginning, the middle, and at the end of the operating course of a mill. Thus, in each experiment 3×5 repetitions are carried out ($k = 15$).

After each operating course, macrographs of the back surface of the face mill tooth are taken. In addition, photographs were taken of the processed surface. Pictures were processed on a personal computer and, in the mode of picture magnification, measurements of the degree of wear on the back surface of a mill tooth were conducted through

Table 15.1 The cutting modes for different face milling processing stages

No	Milling stage	Milling depth t (mm)	Serve on tooth S_z (mm/tooth)	Cutting speed V (m/min)	Rotation frequency n (min^{-1})
1	Finishing	1	0.125	392.6	1,000
2	Finishing	1	0.16	392.6	1,000
3	Semi-finishing	1	0.25	392.6	1,000
4	Semi-finishing	1	0.25	247.3	630
5	Draft	1	0.32	196.3	500

comparison with a dimensional ruler. Thus, experimental roughness points of the processed surface for different degrees of wear and the different face milling parameters were obtained. Afterward, statistical processing of experimental data for the set statistical reliability 0.95 was carried out. Average values of the measured size were defined according to the results of five experiments. As not displaced assessment of general dispersion selective dispersion is defined. The uniformity of selective dispersions was checked by Kokhren's criteria (Kozlov et al., 2019; Yuchshenko & Wójcik, 2014).

15.4 RESEARCH RESULTS AND DISCUSSION

Figure 15.1 shows the back surface of a face mill tooth and the flat surfaces of a detail processed by the mill after the first and second pass of the tool. The first pass has an $l_z = 1.31$ mm, the second pass – $l_z = 4.6$ mm.

Wear on the back surface of the face mill tooth is shown on the left and details with the processed flat surfaces on the right with a milling depth of $t = 1.0$ mm; to serve $S_z = 0.125$ mm/tooth; cutting speed $V = 392.6$ mm/min; to the rotation frequency of the mill $n = 1000 \text{min}^{-1}$.

The experimental average roughness values of the processed surface for different degrees of wear and the different face milling parameters are provided in Figure 15.2.

Figure 15.2 shows that roughness Rz increases from 15% to 30% with the growth of the degree of wear l_z from 0 to 3.1–4 mm. At the same time, at a constant speed of cutting $V = 392.6$ m/min, an increase in the serve S_z with 0.125 mm/tooth up to 0.16 mm/tooth leads to an increase in roughness by 7%–16%, up to 0.25 mm/by tooth – for

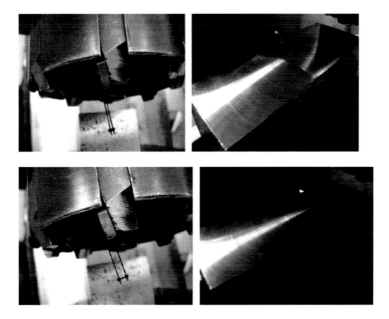

Figure 15.1 Wear on the back surface of the face mill tooth at the left and details with the processed flat surfaces on the right with a milling depth $t = 1.0$ mm; to serve $S_z = 0.125$ mm/tooth; cutting speed $V = 392.6$ mm/min.

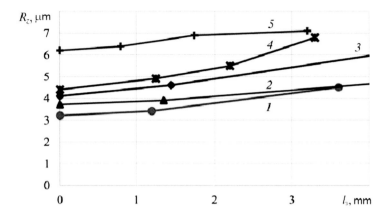

Figure 15.2 Experimental average roughness values of the processed surface Rz for different degrees of wear l_z and different face milling parameters: 1 – $S_z = 0.125$ mm/tooth, $V = 392.6$ m/min; 2 – $S_z = 0.16$ mm/tooth, $V = 392.6$ m/min; 3 – $S_z = 0.25$ mm/tooth, $V = 392.6$ m/min; 4 – $S_z = 0.25$ mm/tooth, $V = 247.6$ m/min; 5 – $S_z = 0.32$ mm/tooth, $V = 196.3$ m/min.

28%–48%, that corresponds with data (Isakov, 2013). An increase in cutting speed V from 247.3 to 392.6 m/min at invariable serve $S_z = 0.25$ mm/tooth, on the other hand, leads to a reduction of roughness by 7%–15%.

Let us show on one of the examples of face milling parameters as the profilogram of a processed surface micro profile changes in the event an increase in wear of a mill tooth on the back surface. Processing using face milling, in this case, is carried out with a depth of $t = 1.0$ mm; serve $S_z = 0.32$ mm/tooth; cutting speed $V = 196.3$ mm/min; spindle rotation frequency $n = 500 \min^{-1}$; quantity of mill teeth: $z = 4$. For evident comparison of roughness parameters of the processed planes at different degrees of wear, a detail with four steps of 1 mm high and 50 mm long each is received (Figure 15.3). Each of the steps is processed using a mill with various degrees of wear on a back surface. After processing the subsequent step, macrographs of the back surface of the face mill tooth are made. The photos were processed on a personal computer and in a photo

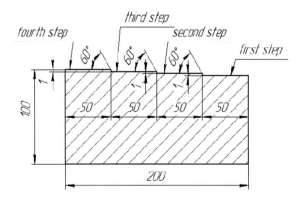

Figure 15.3 A detail with four steps.

magnification mode, and measurement of the degree of wear on the back surface of the mill tooth was carried out by comparison with a dimensional ruler.

After processing the obtained step, the detail micro profile profilogram of a processed surface with a basic length $L=0.8$ mm is measured using a Protonmietprofilometer 130 intended for measuring parameters of a profile and roughness parameters of a surface on the average line system (GOST 2514282). For each step, a profilogram is made for five points; if the number of repetitions is $k=5$, each time there is a change in a micro profile profilogram of the processed surface at the increase in wear of a mill tooth on a back surface, as a result, 20 profilograms are obtained. For the first step $l_z=0$ mm, the second – the 0.18, the third – the 0.45, and the fourth – 0.88 mm. Figure 15.4 shows the profilograms for four steps at the points in the middle of the corresponding process steps.

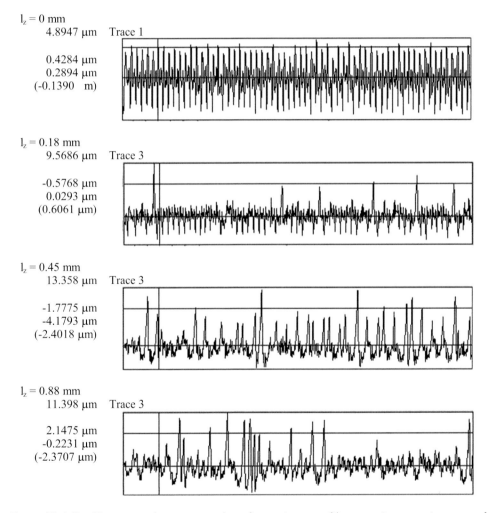

Figure 15.4 Profilograms of a processed surface micro profile at an increase in wear of mill teeth on the back surface.

From Figure 15.4 it is visible that at zero wear (the first step) the profilogram has uniform character with a high frequency and amplitude. At the same time, the sizes of the ledges $H_{imax} = 4.8947\,\mu m$ and hollows $H_{imin} = -5.2874\,\mu m$ are almost equal and the average line passes within a significant shift from the middle at $-0.1390\,\mu m$. The average value of roughness $Rz = 10.2\,\mu m$.

At wear $l_z = 0.18\,mm$ (the second step) character of a profilogram changes: sizes of ledges $H_{imax} = 9.5686\,\mu m$ prevail over the sizes of hollows $H_{imin} = -5.0386\,\mu m$, at the same time the average line is considerably displaced from the middle on $0.6016\,\mu m$. At the same time, the sizes of hollows are greater than the sizes of ledges. The average value of roughness $Rz = 10.9\,\mu m$.

At wear $l_z = 0.45\,mm$ (the third step) the same tendency remains: sizes of ledges $H_{imax} = 13.358\,\mu m$ prevail over the sizes of hollows $H_{imin} = -5.3434\,\mu m$. The average roughness value $Rz = 15.8\,\mu m$.

At wear $l_z = 0.88\,mm$ (fourth step): sizes of ledges $H_{imax} = 11.398\,\mu m$ prevail over the sizes of hollows $H_{imin} = -6.1568\,\mu m$. The average roughness value is $Rz = 16.5\,\mu m$.

In general, at the increasing degree of wear on a back surface, a reduction of frequency of ledges and hollows of a micro profile profilogram is observed. The amplitudes of ledges prevail over the amplitudes of hollows in the process of wear increase from 0 to $0.88\,mm$. This arises from the fact that the degree of wear on the back surface of a mill tooth smooths out the processed surface. In the process of increase in the degree of wear up to $0.88\,mm$, a significant increase in roughness of 15%–30% is observed.

15.5 CONCLUSION

The conducted research shows that roughness Rz increases by 15%–30% for the adopted cutting modes in the case of an increase in the degree of wear from 0 to 3.1–$4\,mm$;

An increase in serve S_z from 0.125 up to 0.16 mm/tooth, at a constant cutting speed of $V = 392.6\,m/min$, leads to an increase in roughness by 7%–16%. An increase in serve S_z up to 0.25 mm/tooth, at a constant cutting speed of $V = 392.6\,m/min$, leads to an increase in roughness by 28%–48%.

An increase in cutting speed V from 247.3 m/min to 392.6 m/min at invariable serve $S_z = 0.25\,mm/tooth$ leads to a reduction of roughness by 7%–15%.

It is shown that the increasing degree of wear on the back surface causes a reduction in the frequency of ledges and hollows on the micro profile profilogram of the processed surface to be observed. At the same time, amplitudes of ledges prevail over the amplitudes of hollows in the process of wear increase.

During the design of face milling operations, it is necessary to consider the change of the roughness parameters, taking into account the wear of mill teeth on the back surface.

Unlike the traditional approach to the design of operations, perhaps, in processing by milling it is necessary to place more emphasis on serve, as well as the process of increasing wear of the mill teeth on the back surface, and the resulting connected increase in roughness, in order to correct the serve toward roughness reduction. This will in turn lead to an increase in productivity.

REFERENCES

Bajić, D., Celent L. & Jozić, S. 2012. Modeling of the influence of cutting parameters on the surface roughness, tool wear and the cutting force in face milling in off-line process control. *Strojniški vestnik - Journal of Mechanical Engineering* 58(11): 673–682.

Benardos, P.G. & Vosniakos, G.C. 2002. Prediction of surface roughness in CNC face milling using neural networks and Taguchi's design of experiments. *Robotics and Computer Integrated Manufacturing* 18(5–6): 343–354.

Dragobetskii, V., Shapoval, A., Mos'pan, D., Trotsko, O. & Lotous, V. 2015. Excavator bucket teeth strengthening using a plastic explosive deformation. *Metallurgical and Mining Industry* 4: 363–368.

Elhami, S., Razfar, M.R., Farahnakian, M. & Rasti, A. 2013. Application of GONNS to predict constrained optimum surface roughness in face milling of high silicon austenitic stainless steel. *The International Journal of Advanced Manufacturing Technology* 66(5–8): 975–986.

Grzenda, M. & Bustillo, A. 2013. The evolutionary development of roughness prediction models. *Applied Soft Computing Journal* 13(5): 2913–2922.

Grzenda, M., Bustillo, A., Quintana, G. & Ciurana, J. 2012. Improvement of surface roughness models for face milling operations through dimensionality reduction. *Integrated Computer Aided Engineering* 19(2): 179–197.

Guzeev, V.I., Batuyev, V.A. & Surkov, I.V. 2005. *The Cutting Modes for Turning and Frezernorastochny Machines with Numerical Program Control: The Reference Book*. Moscow: Mechanical Engineering.

Isakov, D.V. 2013. Geometrical model of height of microroughnesses of the processed surface taking into account wear of teeths of a face mill. *Friction and Wear* 34(4): 382–386.

Kovac, P., Rodic, D., Pucovsky, V., Savkovic, B. & Gostimirovic, M. 2013. Application of fuzzy logic and regression analysis for modeling surface roughness in face milling. *Journal of Intelligent Manufacturing* 24(4): 755–762.

Kozlov, L.G., Polishchuk, L.K., Piontkevych, O.V., Korinenko, M.P., Horbatiuk, R.M., Komada, P., Orazalieva, S. & Ussatova, O. 2019. Experimental research characteristics of counter balance valve for hydraulic drive control system of mobile machine. *Przegląd Elektrotechniczny* 95(4): 104–109.

Loladze, T.N. 1982. *Durability and Wear Resistance of the Cutting Tool*. Moscow: Mechanical engineering.

Ogorodnikov, V.A., Dereven'ko, I.A. & Sivak, R.I. 2018b. On the influence of curvature of the trajectories of deformation of a volume of the material by pressing on its plasticity under the conditions of complex loading. *Materials Science* 54(3): 326–332.

Ogorodnikov, V.A., Savchinskij, I.G. & Nakhajchuk, O.V. 2004. Stressed-strained state during forming the internal slot section by mandrel reduction. *Tyazheloye Mashinostroyeniye* 12: 31–33.

Ogorodnikov, V.A., Zyska, T. & Sundetov, S. 2018a. The physical model of motor vehicle destruction under shock loading for analysis of road traffic accident. *Proc. SPIE* 10808: 108086C.

Pimenov, D.Y. 2013. Dependence of size of the platform of wear on a back surface of teeths of a face mill in processing. *Friction and Wear* 2: 199–203.

Polishchuk, L.K. et al. 2019. Study of the dynamic stability of the belt conveyor adaptive drive. *Przegląd Elektrotechniczny* 95(4): 98–103.

Pukhalsky, V.A. & Gavrilov, G.A. 2012. Influence of beating of the cutting edges of face mills on their wear. *Messenger of mechanical engineering* 11: 65–68.

Rosales, A., Vizán, A., Diez, E. & Alanís A. 2010. Prediction of surface roughness by registering cutting forces in the face milling process. *European Journal of Scientific Research* 41(2): 228–237.

Sandvik Coromant. 2019. The metal-cutting equipment. www.sandvik.coromant.com.

Simunovic, G., Simunovic, K. & Saric, T. 2013. Modelling and simulation of surface roughness in face milling. *International Journal of Simulation Modelling* 12(3): 141–153.

To Simsiva, Z.V., Kutyshkin, A.V. & Simsiva, D.T. 2012. Estimation of adhesive wear of a back surface of the cutting tool at high-speed machining. *STIN* 4: 18–22.

Vorobyov, V., Pomazan, M., Vorobyova, L. & Shlyk, S. 2017. Simulation of dynamic fracture of the borehole bottom taking into consideration stress concentrator. *Eastern-European Journal of Enterprise Technologies* 3/1(87): 53–62.

Yalçın, B., Özgür, A.E. & Koru, M. 2009. The effects of various cooling strategies on surface roughness and tool wear during soft materials milling. *Materials and Design* 30: 896–899.

Yuchshenko, O. & Wójcik, W. 2014. Development of simulation model of strip pull self-regulation system in dynamic modes in a continuous hot galvanizing line. *Informatyka, Automatyka, Pomiary w Gospodarce i Ochronie Srodowiska* 1: 11–13.

Chapter 16

Investigation of interaction of a tool with a part in the process of deforming stretching with ultrasound

N. Weselowska, V. Turych, V. Rutkevych, G. Ogorodnichuk, P. Kisała, B. Yeraliyeva, and G. Yusupova

CONTENTS

16.1 Introduction 175
16.2 Analysis and problem statement 175
16.3 Purpose, objectives, materials, and methods 176
16.4 Research results 176
16.5 Conclusions 182
References 182

16.1 INTRODUCTION

The current development of hydraulic cylinders requires their developers to further improve their technical level, competitiveness and to expand functional capabilities. Requirements for the working surfaces of hydraulic cylinders are increasing – the accuracy of holes in terms of the permissible deviations from axial straightness and non-circularity, the working surface roughness, the surface microrelief, as well as the increase of working pressures. These requirements can be successfully fulfilled by a technological process based on stretching deformation using ultrasound (Kumar 2013, Mashkov et al. 2014, Turych & Rutkevych 2016). In this connection, the task of developing the technology of cylinder sleeves is topical.

16.2 ANALYSIS AND PROBLEM STATEMENT

The problem of improving the accuracy of hollow detail surfaces of machine parts such as sleeves and cylinders is highlighted by many authors, namely Proskuryakov Yu. G., Rosenberg O. A Posvyatenko E. K. (Turych & Rutkevych 2016, Turych et al. 2017). However, from the standpoint of resource conservation, which is extremely important for the Ukrainian economy, these processes are not sufficiently studied. Since the adhesion phenomena in the process of treatment of materials by cold plastic deformation are certainly harmful, a number of process researchers recommend the use of oils with high screening properties, that is, anti-agglutinating materials, in which the fillers are molybdenum disulfide, graphite, and other similar substances, which are able to withstand high contact pressure, provide a reliable separation of the surfaces of parts and tools during processing and low external friction coefficient values (0.07–0.1)

DOI: 10.1201/9781003225447-16

(Kumar 2013). However, this method of dealing with adhesion is unsuitable for cold plastic deformation finishing processes, since it does not allow the possibility of lowering the roughness of the surfaces, obtaining high values of deformation strength, texture, and useful compressive stresses in the surface layer. The problem of increasing the accuracy of surface treatment of machine parts such as hollow-type cylinder liners has been covered by many authors, namely Proskuryakov Y. G., Rosenberg A. A., Posvyatenkom E. K., etc. (Moriwaki 2010, Turych et al. 2017, 2018). However, from the resource standpoint, which is extremely important for the Ukrainian economy, these processes have been insufficiently studied. Since adhesion phenomena in the processing of materials by cold plastic deformation are undoubtedly harmful, some process researchers have recommended the use of oils with high screening properties such as adhesive coatings materials, fillers, including molybdenum disulfide, graphite, and similar substances able to withstand high contact pressure, provide reliable separation of the surfaces of parts and tools during processing and a low external friction coefficient (0.07–0.1) (Romashkyna 2009, Shao-Yi et al. 2016).

Several studies show that during cold plastic deformation using ultrasonic vibrations, i.e., using periodic forced separation of the tool and the parts during processing, the quality is improved greatly and the intensity of operations is reduced, as a result of periodic recess into the surface of the part and immediate termination of contact between surfaces of the tools and parts (Kumar 2013, Titov et al. 2017, Turych et al. 2018, Tymchyk et al. 2018).

The conducted studies are relevant to solve practical problems of using such methods.

16.3 PURPOSE, OBJECTIVES, MATERIALS, AND METHODS

The aim of the study is to improve the processing of hollow machine part surfaces such as sleeves and cylinders by stretching deformation using ultrasound and to define the theoretical dependencies in order to calculate the pulling forces during contact between the tool and the part.

A description of the ultrasonic deforming through stretching with ultrasound has been developed based on the application of rheological models of materials that reflect their actual elastic–plastic properties. Such an approach makes it possible to identify the mechanism of how the ultrasound impacts the process of stretching deformation. A study of force parameters was carried out using a loading chart of a perfect elastic–plastic body. Investigation of the machining method's effect on torque involved the following materials: Steel 10, aluminum alloy ΛK4.

16.4 RESEARCH RESULTS

The use of stretching deformation operations can reduce processing complexity and increase the hydraulic drive reliability and durability by improving the quality of cylinder holes.

The application of ultrasonic technology allows expanding the technological capabilities of stretching deformation, namely: by increasing the hole precision, axial straightness of the cylinder and opening, creating a microrelief on the inner surface to maintain lubrication (Turych & Rutkevych 2016, Turych et al. 2018).

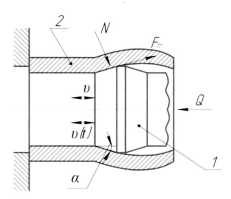

Figure 16.1 Ultrasound stretching schematic.

For the development of the technological process, we consider the contact interaction of the tool with the part during stretching deformation with the use of ultrasound. Figure 16.1 shows a schematic of stretching deformation using ultrasound (Polishchuk et al. 2016, 2018, 2019).

The deforming element 1 oscillates harmonically with an amplitude of ξ and passes through the hole of part 2 with a static force of Q. The equation of motion of the deforming element can be represented as follows

$$u(t) = vt + \xi \sin \omega t \tag{16.1}$$

where v – speed of the tool; t – time; $\omega = 2\pi f$, here f – the frequency of oscillation.

Moving the deforming element leads to a displacement of the contact surfaces of the tool and the part along the normal direction leading to the formation of a working cone tool:

$$u_n(t) = u(t)\sin\alpha \tag{16.2}$$

and to the bias by the tangent:

$$u_\tau(t) = u(t)\cos\alpha \tag{16.3}$$

here α – the inclination angle of the formation of a working cone of the deforming element.

By moving, (16.2) and (16.3) cause normal tension q_n with a resultant N and tangential stresses q_τ with a resultant F_{fr}.

Due to the fact that the tools (16.2), (16.3) are intermittent in nature, the stretching deformation process will take place in three stages, namely: elastic, plastic, and unloading. By analogy with (Rimkeviciene et al. 2009), let's consider the load diagram of ideal elastic–plastic material. The load diagram is shown in Figure 16.2.

The diagram shows: (1) zone of elastic loading; (2) zone of plastic deformation; (3) unloading zone. In view of (16.2), (16.3), the loading characteristics are expressed as follows:

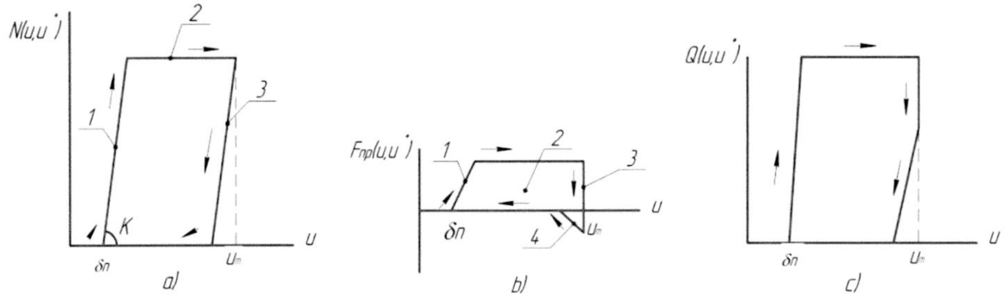

Figure 16.2 The load diagram at $F_{fr} = \eta \cdot N$.

$$N(u_n, \dot{u}_n) = \begin{cases} 0 & u_n \leq \Delta_n & \dot{u} \geq 0 \\ k_n(u_n - \Delta_n)\sin\alpha & \Delta_n \leq u_n \leq \Delta_n + S & \dot{u} \geq 0 \\ N & \Delta_n + S \leq u_n \leq u_{nm} & \dot{u} \geq 0 \\ N - k_n(u_{nm} - u_n)\sin\alpha & u_{nm} - S \leq u_n \leq u_{nm} & \dot{u} \leq 0 \\ 0 & u_n \leq u_{nm} - S & \dot{u} \leq 0 \end{cases} \quad (16.4)$$

where k_n – the rigidity of the section of the detail in the normal direction; N – normal force during plastic deformation; Δ – coordinate of the beginning of surface contact of tools and parts; u_{nm} – the maximum mean of the function within a period (16.1); $S = N/k_n \sin\alpha$ – the movement of the deforming element to achieve the normal force of N, that is appropriate for plastic deformation (Kozlov et al. 2019, Ogorodnikov et al. 2018, Ogorodnikov, Dereven'ko, et al. 2018).

Let us consider that between the working surface of the tool and the treated surface during their relative tangential displacement, a friction force occurs directed toward the opposite direction of movement.

$$|F_{fr}| \leq \eta N \quad (16.5)$$

A sign of equality (16.5) is placed when slipping occurs. In the absence of slipping the friction force is equal to the classic tangential force. Taking into account the above, it is worth considering the following three cases.

Case I. Friction force $|F_{fr}| = \eta N$

The characteristic of a tangential interaction is as follows:

$$F_{fr}(u,\dot{u}) = \eta \cdot N(u,\dot{u}) sign\dot{u} \quad (16.6)$$

The characteristic of the friction force in the direction of the motion has the following form:

$$F_{frv}(u,\dot{u}) = \eta N(u,\dot{u})\cos\alpha \cdot sign\dot{u} \quad (16.7)$$

In (16.7) F_{frv} accepts the following values

$$\eta \cdot N(u,\dot{u})\cos\alpha \cdot \text{sign}\dot{u} = \begin{cases} \eta \cdot N(u,\dot{u})\cos\alpha & \dot{u} \geq 0 \\ -\eta \cdot N(u,\dot{u})\cos\alpha & \dot{u} \leq 0 \end{cases} \tag{16.8}$$

The total characteristic in the direction of movement that is stretching force:

$$Q(u,\dot{u}) = N(u,\dot{u})\sin\alpha + F_{fr}(u,\dot{u})\cos\alpha \tag{16.9}$$

Figure 16.2b shows the dependence (16.8). In Sections 16.1 and 16.2, characteristics of friction force (Figure 16.2b) corresponding to the load result in the sliding of surfaces; in Section 16.3, unloading takes place, followed by elastic negative tension, in Section 16.4 – unloading with slipping. In Figure 16.2c, the solid line shows the total characteristic. The total characteristic describes all possible situations (Figure 16.2c).

The case, as described above, can be considered as a limiting transition since the relative slip of the tool and the part is preceded by a shift of the treated surface together with the tool (Landeta 2015). Calculation using the above method, which does not take into account the previous surface shift, will produce an error, especially at a low instrument oscillation amplitude.

Taking into account the previous shift, we assume that the treated part surface also has a shear stiffness k_τ in the direction of the formed working cone of deforming element (Dragobetskii et al. 2015, Ogorodnikov et al. 2004, Vasilevskyi 2013). Taking the above into account, let us consider the following situations.

Case II. Friction coefficient

$$\eta \leq k_\tau ctg\alpha / k_n \tag{16.10}$$

The characteristic of the tangential interaction for this case is as follows:

$$F_{fr}(u,\dot{u}) = \begin{cases} \eta N & \dot{u} \geq 0 \\ \eta N - k_\tau(u_m - u)\cos\alpha & u_m - \dfrac{2S}{1+m} \leq u \leq u_m \quad \dot{u} \leq 0 \\ -\eta(N - k_n(u_m - u))\sin\alpha & u_m - S \leq u \leq u_m - \dfrac{2S}{1+m} \quad \dot{u} \leq 0 \\ 0 & u_m - S \leq u \quad \dot{u} \leq 0 \end{cases} \tag{16.11}$$

where $m = \dfrac{k_\tau}{k_n \eta} ctg\alpha$.

The total characteristic in the direction of the movement is similar to (16.9).

Figure 16.3a and b show the dependencies (16.4) and (16.11). Sections 16.1 and 16.2 show the characteristics of friction force (Figure 16.3b), corresponding to the load – there is a sliding of surfaces; in Section 16.3 – the elastic unloading with further negative elastic tension due to the friction; in Section 16.4 – unloading with slippage. The solid line in Figure 16.3c shows the total characteristic, and the dashed one – its components: $N(u,\dot{u})$, $F_{fr}(u,\dot{u})$.

If the amplitude of tool oscillations is $\xi \leq S_1/2$, a purely elastic interaction takes place, which is established after the transition process (Section 16.5, Figure 16.3c). Here the material behaves as elastic–plastic with a rigidity coefficient:

$$k = k_n \sin^2\alpha + k_\tau \cos^2\alpha \tag{16.12}$$

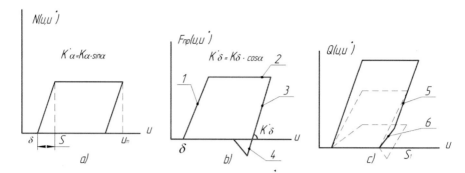

Figure 16.3 Load diagram at $\eta \leq \dfrac{k_\tau}{k_n} \operatorname{ctg}\alpha$.

Note that according to (16.11):

$$S_1 = 2S/(1+m) \qquad (16.13)$$

As shown in (16.12), (16.13), k_τ and η are important in calculating the total rigidity coefficient. For example, when $\alpha = 5°$ $k \approx 0{,}0076 k_n + 0{,}99 k_\tau$.

Under these conditions, all conclusions are valid (Romashkyna 2009), according to which:

$$Q_y = Q - \xi \cdot k \left(\text{at } \xi \leq S_1 \right) \qquad (16.14)$$

where Q_y, Q – stretching forces with ultrasound and without it, respectively.

At large amplitudes, nonlinear distortions begin, associated with the approach to the characteristic branch, marked by line 6 in Figure 16.3c. In this case, there are slippage and friction losses on the contact surfaces of the tool and the part.

Case III. If the friction coefficient is

$$\eta \geq k_\tau \cdot \operatorname{ctg} \cdot \alpha / k_n \qquad (16.15)$$

the characteristic of tangential interaction has the following form:

$$F_{fr}(u,\dot{u}) = \begin{cases} 0 & u \leq \Delta \\ k_\tau(u-\Delta)\cos\alpha & u \leq \Delta + S/m \quad \dot{u} \geq 0 \\ \eta \cdot N & \Delta + S/m \leq u \leq u_m \quad \dot{u} \geq 0 \\ \eta \cdot N(u,\dot{u}) & \dot{u} \leq 0 \quad \dot{u} \geq 0 \end{cases} \qquad (16.16)$$

Figure 16.4b shows the characteristics (16.4) (16.16). Line 1 in Figure 16.4b shows tangent characteristics (16.16), elastic deformation takes place; 2 – slipping; 3 – unloading with slippage. The total characteristic is presented in Figure 16.4c. Here the elastic deformation zone is limited by line 4.

For the three cases at $\xi \geq S_1$, with the help of the impulse theorem, substituting in (17) the relevant characteristics of normal and tangential interactions, the correlation

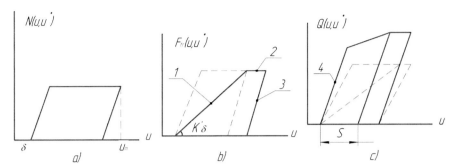

Figure 16.4 Load diagram at $\eta \geq \dfrac{k_\tau}{k_n} c \operatorname{tg}\alpha$.

that ties a constant statistical force with parameters of tool motion and the characteristics of the processed material can be obtained:

$$Q_y = \frac{1}{T} \int_{t_1}^{t_1+T} (N(u,\dot{u})\sin\alpha + F_{fr}(u,\dot{u})\cos\alpha)\,dt \qquad (16.17)$$

To check received dependencies, experiments were conducted involving stretching out of steel 10 sleeves with the dimensions (mm): outer diameter – 30, hole diameter – 10, deforming element tension – 0.2. The sleeves were processed using carbide deforming elements with a 5° angle of inclination. Sulfofresol was used as a lubricant.

The stiffness of the linear section of the detail in the normal direction was determined using the method described in (Turych et al. 2018). For investigated details, it was $k_n = 5.92 \cdot 10^6$ MN/m.

Tangential stiffness was determined as follows. From the processed sleeve, microfloors were made and photographed. The photograph shows that the metal grains have a typical, pronounced texture, that is, grain elongation and inclination in the direction of stretching. The value of grain elongation was measured on several sections of microfloors, and then the average value X_0 was determined. Using the technique, the normal force and friction force were determined (Shao-Yi 2016). The value of the tangential stiffness was determined as the ratio of friction to X_0:

$$k_\tau = F_{fr}/X_0 \qquad (16.18)$$

For the experiments, the value was 6.02×10^5 MN/m.

As the stretching was conducted with a maximum vibration amplitude of 15 mm, and the mean $S_1/2$ was 17 mm, the stretching force was determined by the dependence (14). The experimental and calculated values of the forces were well coincided, the difference amounted to no more than 15%.

The analysis provides a clear picture of the contact interaction between a tool and a product surfaces during stretching deformation using ultrasound and also makes it possible to calculate the stretching force, observed in different conditions of contact interaction.

16.5 CONCLUSIONS

The obtained theoretical dependences for calculating stretching forces, when processing with the use of ultrasound, show that the mechanism of influence of ultrasound on reduction of stretching forces is manifested in the accumulation of small plastic deformations, which increase between periods of fluctuations due to the total vibrational and translational motions of the tool and reduce friction forces by changing the kinematics of sliding.

It was established that the type of contact interaction between the tool with the detail has a decisive impact on how ultrasounds reduce stretching forces. With impulse interaction (with a break between contact of instrument and detail surfaces), stretching force is reduced to a minimum, tending toward zero, and in unceasing (without breaking of contact between surfaces) – by 60%–70% in comparison to stretching force without the use of ultrasound.

The construction of the deforming tool for deformation processing with the use of ultrasound was proposed.

REFERENCES

Dragobetskii, V., Shapoval, A., Mos'pan, D., Trotsko, O. & Lotous, V. 2015. Excavator bucket teeth strengthening using a plastic explosive deformation. *Metallurgical and Mining Industry* 4: 363–368.

Kozlov, L.G., Polishchuk, L.K., Piontkevych, O.V., Korinenko, M.P., Horbatiuk, R.M., Komada, P., Orazalieva, S. & Ussatova, O. 2019. Experimental research characteristics of counter balance valve for hydraulic drive control system of mobile machine. *Przeglad Elektrotechniczny* 95(4): 104–109.

Kumar, J. 2013. Ultrasonic machining-a comprehensive review. *Mach. Sci. and Technol.* 3: 325–379.

Landeta, J.F., Valdivielso, A.F., Lacalle, L.L., Girot F. & Pérez, J.P. 2015. Wear of Form Taps in Threading of Steel Cold Forged Parts. *Journal of Manufacturing Science and Engineering* 137(3): 1–11.

Mashkov, V., Smolarz, A., Lytvynenko, V. & Gromaszek, K. 2014. The problem of system fault-tolerance. *Informatyka, Automatyka, Pomiary w Gospodarce i Ochronie Środowiska* 4: 41–44.

Moriwaki, T. 2010. Development of 2DOF ultrasonic vibration cutting device for ultraprecision elliptical vibration cutting. *Key Engineering Materials* 447–448: 164–168.

Ogorodnikov, V.A., Dereven'ko, I.A. & Sivak, R.I. 2018. On the Influence of Curvature of the Trajectories of Deformation of a Volume of the Material by Pressing on Its Plasticity Under the Conditions of Complex Loading. *Materials Science* 54(3): 326–332.

Ogorodnikov, V.A., Savchinskij, I.G. & Nakhajchuk, O.V. 2004. Stressed-strained state during forming the internal slot section by mandrel reduction. *Tyazheloe Mashinostroenie* 12: 31–33.

Ogorodnikov, V.A., Zyska, T. & Sundetov, S. 2018. The physical model of motor vehicle destruction under shock loading for analysis of road traffic accident. *Proc. SPIE* 108086C: 1–5.

Polishchuk, L., Bilyy, O. & Kharchenko, Y. 2016. Prediction of the propagation of crack-like defects in profile elements of the boom of stack discharge conveyor. *Eastern-European Journal of Enterprise Technologies* 6(1): 44–52.

Polishchuk, L., Gromaszek, K., Kozlov, L.G. & Piontkevych, O.V. 2019. Study of the dynamic stability of the belt conveyor adaptive drive. *Przeglad Elektrotechniczny* 95(4): 98–103.

Polishchuk, L., Kozlov, L.G., Piontkevych, O.V., Gromaszek, K. & Mussabekova, A. 2018. Study of the dynamic stability of the conveyor belt adaptive drive. *Proc. SPIE 1080862*: 1791–1800.

Rimkeviciene, J., Ostasevicius, V., Jurenas, V. & Gaidys, R. 2009. Experiments and simulations of ultrasonically assisted turning tool. *Mechanika (Lietuva)* 75(1): 42–46.

Romashkyna, O.V. 2009. Yssledovanye vlyianyia parametrov ultrazvukovoi obrabotkyna formyrovanye ostatochnikh napriazhenyi pry narezanyy naruzhnikh rezb maloho dyametra. *VestnykSamHTU. Ser. Tekhnycheskyenauky* 2: 113–119.

Shao-Yi, H., Yu-Tuan, C. & Guan-Fan, L. 2016. Analysis of Sheet Metal Tapping Screw Fabrication Using a Finite Element Method. *MPDI Appl. Sci.* 6: 1–15.

Titov, A.V., Mykhalevych, V.M., Popiel, P. & Mussabekov, K. 2017. Statement and solution of new problems of deformability theory. *Proc. SPIE 108085E*: 1611–1617.

Turych, V.V. & Rutkevych, V.S. 2016. Kontaktna vzaiemodiia instrumenta z detalliu v protsesi deformuiuchoho protiahuvannia z ultrazvukom. *Promyslova hidravlika i pnevmatyka* 4(54): 71–76.

Turych, V.V., Rutkevych, V.S., Goncharuk, N. & Ogorodnichuk, G. 2018. Investigation of the process of smoothing with ultrasound. *Eastern–European Journal of Enterprise Technologies. Engineering technological systems* 3(93): 22–33.

Turych, V.V., Weselowskaya, N., Rutkevych, V. & Shargorodskiy, S. 2017. Investigation of the process of thread extrusion using the ultrasound. *Eastern–European Journal of Enterprise Technologies. Engineering technological systems* 6(90): 60–68.

Tymchyk, S.V., Skytsiouk, V.I., Klotchko, T.R., Ławicki, T. & Denisova, N. 2018. Distortion of geometric elements in the transition from the imaginary to the real coordinate system of technological equipment. *Proc. SPIE* 108085C: 1595–1604.

Vasilevskyi, O.M. 2013. Advanced mathematical model of measuring the starting torque motors. *Technical Electrodynamics* 6: 76–81.

Chapter 17

Robotic complex for the production of products special forms with filling inside made from dough

R. Grudetskyi, O. Zabolotnyi, P. Golubkov, V. Yehorov, A. Kotyra, A. Kozbakova, and S. Amirgaliyeva

CONTENTS

17.1 Introduction 185
17.2 Literature overview 186
17.3 Aim of work 186
17.4 Material and research methods 186
17.5 Results and discussion 194
References 195

17.1 INTRODUCTION

As it is known, in every production process, there is continuity, which turns it into reproduction. In it, there is a division into consumed and accumulated parts, for compensation of the expended means of production.

Let us list the factors whose change can generate savings in financial expenses. (1) currently, in the production of goods and products made from dough with filling inside the main task is controlling finished products, with a high human factor percentage; (2) the technological production process consists of defrosting minced meat before modeling and re-freezing it after obtaining the product, which entails additional financial costs. (3) the current manufacturing form of products made from dough with filling inside is traditional. Here we can point to two factors that can be changed to increase the system's efficiency. The first factor is changing the product's form in the direction of complication to prevent counterfeiting. The second factor is improving warehousing, performing density, and transportation. Combining the above factors, we can create new dough products with filling inside in a special form, as well as a technological production process in which the secondary defrosting of ingredients is not required. The optimal and difficult to implement form of dough products with filling inside is the cubic form. This form:

(1) will protect against fake product manufacturers; (2) allows incoming raw materials (meat) to be processed process (minced meat production), and frozen ingredients (minced cubes) to be stored; (3) will allow working with raw materials, without subjecting them to full defrosting and further re-freezing; (4) allows obtaining products with a given shape, also results in a reduction in energy costs during product storage and transportation; (5) minimizes the human factor in production.

DOI: 10.1201/9781003225447-17

17.2 LITERATURE OVERVIEW

In general, obtaining finished products can be achieved by a fairly simple Automated Control Systems (ACS) that implements only the regulatory functions, i.e., stabilization of process variables at their given values. These control functions are, as a rule, implemented based on the simplest typical algorithms – PID control algorithms. This is only possible due to the fact that the recipe of finished and semifinished products and the characteristics of raw materials and ingredients are not changed (Kvaternyuk et al.2018, Lima et al. 2015, Martin et al. 2014, Mashkov et al. 2016). This causes the two most important circumstances for process management. (1) maintenance of the optimal production modes found at the stage of special studies. As a result, there is no need to implement control functions such as optimization and adaptation during the process. (2) The lack of disturbances of raw materials, i.e., the lack of operational process variables, the consequences of which must be compensated by ACS when implementing the regulatory function. Disturbances that persist are associated with fluctuations in ambient temperature and the speed of pneumatic operation, according to specified algorithms. The intensities of these disturbances are rather low, and the task of stabilizing them is fairly simple.

However, when working with raw materials and ingredients of plant and animal origin (biopolymers), in particular minced meat and dough, the situation is completely different. It is characterized by the following factors: (1) the recipe of products and their composition changes due to the presentation of increasingly high product requirements: increased nutritional value, cost reduction, resource base expansion, enrichment with vitamins and additives. (2) the characteristics of the original product, even within the same recipe, always differ from each other (the composition of the dough and the composition of minced meat) and, therefore, they can change significantly during the process. Thus, in comparison with other technological production processes, the dough products with filling inside production have fundamental features that must be taken into account when developing automatic control systems to increase energy efficiency.

17.3 AIM OF WORK

For the above reasons, an effective SAC must solve the problem of optimizing the current mode of performing a technological process for a particular raw material directly during its processing (Groover 2009, Hraniak et al. 2018). In this case, the recommended values of mode variables taken from recipes or obtained in laboratory conditions should be considered as quasi-optimal and used in search engine optimization as initial approximations.

17.4 MATERIAL AND RESEARCH METHODS

This is due to the fact that a change in the characteristics of raw materials and ingredients modifies the values of mode variables and the properties of the controlled object (CO) for control channels. It is necessary to take into account the fact that ensuring the robustness of the system when implementing adaptive algorithms will not be as effective. After all, with sufficiently intense disturbances, these algorithms do not converge well and create a significant component of the system's motion, which, in turn, creates additional control errors. In this case, it is more expedient for us to use control algorithms (Woods 1996), which are specific to technological type objects, which enable the extension of system stability margins compared with standard algorithms. The most important control channels are the ingredient feed channels (frozen minced meat −15°C to −30°C and warm dough 20°C–25°C). In certain conditions, disturbances in raw materials and their combination, other conditions for the maintenance of TP, the process of obtaining finished products changes its properties from static to non-static. This greatly complicates the process management and can also affect the operation of the equipment and lead to negative consequences up to an emergency stop. The situation is further complicated by the fact that the most effective modes of equipment operation can be achieved when the raw materials are supplied in a mode close to sticky dough and equipment jamming mode. This contradiction can be resolved by ensuring that a predetermined compromise is maintained in real time between the probability of an emergency and the maximum allowable temperatures by guaranteeing control (Morgan & Haley 2013, Khobin 2004) (Figure 17.1).

Next, we will consider the equipment for the manufacture of dough products with filling (RP) in strict cubic form. The technological process is implemented on the equipment for producing dough products with filling.

The equipment for the production of dough products with filling in a special cubic shape (Figure 17.2) is a cylinder-type device with an RP production speed of 600–1,200 units per hour, the weight of one special form dough product with filling is 12 g, and the size is $22 \times 22 \times 22$ mm.

The new approach to the development of new products is based on the creation of targeted equipment, which is responsible for carrying out most of the work on the manufacture of new products and their storage before marketing. Parallel development of complex form products, with technologies already existing and used at the enterprise, leads to savings of not only time but also financial resources. Based on this information, this may allow using the resources related to equipment management so purposefully that the product meets the required quality to the maximum extent.

The guarantee function (Martin et al. 2014, Khobin 2008) should provide:

(a) real-time evaluation and, on its basis, sliding intervals of the current values of the probability of violating the restrictions established by requirements (tolerance fields), including the aforementioned quality characterizing parameters;
(b) correction of the current temperature modes of the incoming raw materials, in which the values of these probabilities would not exceed their specified maximum allowable, pre-set values. This can ensure compliance with product quality

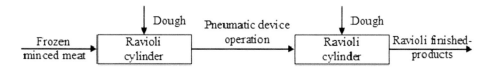

Figure 17.1 Technological diagram of the production process of dough products with filling.

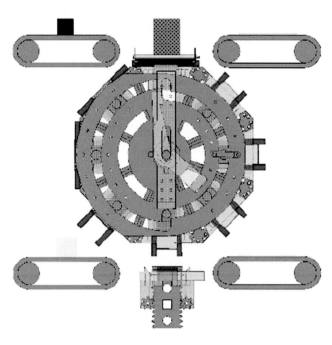

Figure 17.2 3D model of production equipment for special form dough products with filling "CUB-2015".

requirements and trouble-free faultless equipment operation, including, which is important – in the vicinity of the maximum permissible loads of pneumatic equipment. This in particular can create the necessary conditions to minimize the specific energy consumption for the TP. The function of current mode optimization should allow purposeful redistribution of process management resources to achieve the maximum economic effect with unconditional fulfillment of all technological and operational requirements. It should be emphasized that the operation of all the listed functions, which the SAC must implement in the process, is interconnected in the most essential way. At the same time, the effectiveness of the functioning of each of them largely affects the efficiency of the others and vice versa. Therefore, the development of an effective SAC for the production of dough products with filling is a holistic and fairly knowledge-intensive task. The initial stage of such development should be the construction of the most common (conceptual) production process model of RP as CO. Such a model is the first stage of concretization and formalization of the new control tasks, which were formulated above. It will form the basis for the development of an SAC with a targeted set of functions that will allow to solve the formulated control tasks and achieve the main goal of developing such an SAC – improving product quality and energy efficiency of TP RP production (Figure 17.3).

The above analysis briefly shows us the features of the RP production process and determines the composition of the backbone functions of the SAC for this process. Analysis of changes in the properties of products with changes in the properties of plant materials,

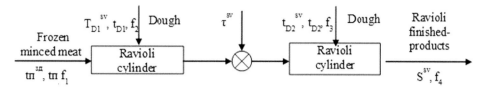

Figure 17.3 Parameterized technological scheme of the RP production process; tr^{sv} – set temperature value of dough products with filling; tr – temperature of dough products with filling; S^{sv} – set form value of dough products with filling; τ^{sv} – idle time specified by technical requirements; t_{D1}^{sv} – set value of top dough temperature (1 serve); t_{D1} – top dough temperature (1 serve); t_{D2}^{sv} – set value of bottom dough temperature (2 serve); t_{D2} – bottom dough temperature (2 serve); $f_1 - f_4$ – uncontrollable disturbances.

which significantly affect the quality indicators of the finished product, energy efficiency, and stability of TP, was carried out in the technological process tables. Complementing them with an analysis of the effects available for controlled change, i.e., having identified the composition of potential control actions, by changing which one can purposefully influence the course of the RP production process, realizing the necessary management functions, one can present a conceptual model of the RP production process as a block diagram, as shown in Figure 17.4. The arrows that indicate the variables on the diagram indicate the direction of their interaction – "from cause to effect," and the rectangles indicate the presence of non-unit operators for the transformation of these variables. The following groups of variables are highlighted on the structural diagram: variables that are indirect indicators of the RP process quality and available for measurement in real time (for example, using an intelligent channel that determines the form of video information – S_F (Yousefi-Darani et al. 2018, Khobin 2002): using the color and the degree of heterogeneity of the product surface at the exit from RP production process equipment; variables characterizing the temperature mode of operation – T^{sv} – internal temperatures T^{sv}_{D1C}, T^{sv}_{D2C}, T^{sv}_{D3C} vegetable raw materials: incoming minced meat, incoming dough in two zones, and incoming second zone of minced meat with dough; equipment performance variables: pneumatic equipment operation and time needed for auxiliary operations, τ^{id} – downtime, (variables $f_1 - f_4$ and T_{mm} indirection changing).

Changing the τ^{sv}–pasting time, according to technological standards leads to minced a meat temperature change, coming into the first zone on the conveyor belt, and minced meat coated in 50% dough upward, as well as the temperature of the dough applied to the minced meat in the direction of decreasing temperature. The amount of downtime during the execution of the technological process can be defined as the line's volumetric capacity. The productivity of the line directly depends on the time required to perform glinting of dough product with filling in the fourth zone of the production equipment and also on the waiting mode time of the equipment, which can be changed four times, according to the technological regulations, from 1 to 4 s (Khobin 2008, Khobin & Egorov 2009, Polishchuk et al. 2019).

It should also be noted that the time required for feeding the raw material to the machine, which, depending on the ingredients of the dough and minced meat, can vary by 2 times, from 3 to 6 s. The time spent on the operation by the actuator is 2 s in both directions. It is also necessary to take into account the fact that from the top to the

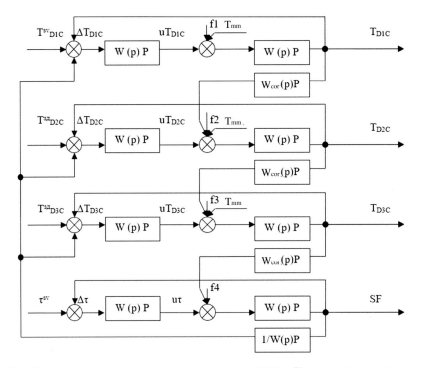

Figure 17.4 Block diagram of the conceptual model of SAC; τ^{sv} – set value of time of modeling; S^F – products with filling inside made from dough form; T^{sv}_{T1C} – set value of the temperature of dough in the first zone; T_{T1C} – temperature of dough in the first zone; T^{sv}_{T2C} – set value of the temperature of dough in the second zone; T_{T2C} – temperature of dough in the second zone; T^{sv}_{T3C} – set value of the temperature of dough in the third zone; T_{T3C} – temperature of dough in the third zone; $f_1 - f_5$ – uncontrollable disturbances; T_{mm} – uncontrolled temperature of minced meat.

lower chamber, the dough products with filling come in five cylinder cycles. As a result, the time taken to move may vary significantly for different types of products and take from 6 to 12 s to produce one unit, and the time spent in the cylinder for minced meat with 50% applied dough is from 30 to 60 s.

For a minute from extracting the dough at a temperature of 25°C, when in contact with minced meat at a temperature of (13)°C, the dough will cool down, which negatively affects the formation of seams during modeling in the lower chamber and, accordingly, may lead to increasing the required modeling time, which leads to increasing the RP production time. The values of volumetric productivity at the equipment output are denoted as τ^{sv}, $\Delta\tau^1$, $\Delta\tau^2$, τ^{TS}, and $u\tau$. Formally, the dependence of the change of these variables on $u\tau$ can be written as follows:

$$u\tau = \left(\Delta\tau^1 + \tau^{SV} + \Delta\tau^2 + \tau^{TS}\right) * 5 \tag{17.1}$$

The work of the forces (torques) necessary for the rotation of the cylinder and the extension of the grip by pressure are converted, ultimately, into heat. The energy for

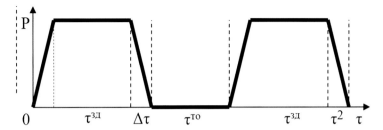

Figure 17.5 Time diagram of the pneumatic actuator operation; $\Delta\tau^1$ – rod pushing out time; τ^{id} – idle time according to technological standards; $\Delta\tau^2$ – rod pulling in time; τ^{TS} – technological shutdown time.

doing this work is taken from the electrical network and converted into mechanical energy, through the operation of the compressor and the actuator (Figure 17.5).

The power and efficiency coefficients for each specific recipe during production are nonlinear functions of the arguments of technological equipment downtime:

$$u\tau = \int_{t\,\min}^{t\,\max} \left(\Delta\tau^1 + \tau^{id} + \Delta\tau^2 + \tau^{TS}\right) * 5 \qquad (17.2)$$

where: $\Delta\tau^1$ and $\Delta\tau^2$ are the coefficient of the minimum and maximum idle time intervals in [$t1$, $t2$]. The control task realized due to changes in $\Delta\tau^1$ and $\Delta\tau^2$ should be the task of regulating (stabilizing) $u\tau n$ in a vicinity that tends toward tot_{min} subject to restrictions on other process variables. The analysis of the conceptual model of the RP production process obtained above makes it possible to concretize the principal features of the RP production process as CO (Kozlov et al. 2019, Ogorodnikov et al. 2018, Ogorodnikov & Sivak 2018).

They can be formulated the following way:

1. Incompleteness of information on the characteristics of the raw materials (minced meat and dough) during the process is not actually controlled. Laboratory measurements of these characteristics, even in the case of using express methods, introduce delays that are many times longer than the product's stay in the heating and modeling zone chambers, which does not allow this information to be used to control the current RP production process modes. In addition, the results of such measurements are represented by the lattice low-frequency function of time. Its values, due to the transposition of relatively high-frequency unfiltered factors, "measurement noise," to the low-frequency region, contain errors that cannot be estimated (Ogorodnikov et al. 2004, Dragobetskii et al. 2015, Vorobyov et al. 2017).
2. Restriction of resources on process management. Potentially, the following variables can be used as control actions in the SAC of the RP production process:
   a. changes in the supply of raw materials to the equipment;
   b. changes in power supply to the heaters (heating elements) of heating zones.
   The essence of the restrictions:

1. the number of targeted effects on regulated variables, the change of which is available during the process, is significantly less than the number of these variables;
2. heater power change range;
3. the cooling of the minced meat inside the equipment is not provided;
4. changes in the supply of raw materials during the RP production process, to a large extent, have similar consequences, and, most importantly, they affect all other regulated variables;
5. technical means for automatic resizing of dough products with filling (actuator and corresponding mechanical transmission) are not provided in the basic design of the equipment, although their installation is possible.

3. The high level of uncertainty of the control channel properties of the RP production process as CO. The recipe of products, and consequently, the composition of their raw materials, is changing quite dynamically due to the presentation of increasingly high demands on products, in particular – increasing their nutritional value, reducing production costs, expanding the raw material base, enriching with minerals and vitamins. At the same time, the characteristics of the original product, even within the same recipe, always differ from each other (due to the particulars of previous technological operations with raw materials, soils in the locations of growth, fertilizers used, precipitation, storage conditions, amount of proteins, fats and carbohydrates, etc.), and therefore, they can change significantly and unpredictably during the process.

This leads to the fact that the regulations found in the laboratory conditions for maintaining the technological process of RP production, i.e., set of values of mode variables (regulations), for production conditions should not be considered as optimal, but only, and in the best case, as quasi-optimal. Such uncertainty is caused by the inevitable differences in the characteristics of the raw materials, design, and condition of the equipment working parts, which were used in laboratory conditions and which will be used in production conditions (Vasilevskyi et al. 2018, Vasilevskyi & Didych 2018).

In addition, the listed changes to the specific conditions of the RP production affect the dynamic properties of the control channels and, therefore, the implementation quality of control functions, in particular, the stability margins of closed control loops. Since it is impossible to trace and describe the causal relationships between RP production and dynamic properties, the changes of the latter should be considered as their uncertainty and very significant. The regulation functions are implemented based on the closed principle of control of the corresponding regulators. In this case, control actions are formed based on regulation errors, in accordance with the control algorithms chosen for regulators:

$$u_d(t) = W_{I_{\text{heater}}}^p (\Delta I, t) \left(I_{\text{heater}}^{\text{svv+}}(t) - I(t) \right), \tag{17.3}$$

$$u_{\text{nom}_i}(t) = W_{\theta_i}^p (\Delta \theta_i, t) \left(\theta_i^{sv}(t) - \theta_i(t) \right), \tag{17.4}$$

$$u_d(t) = W_{\theta_d}^p (\Delta \theta_d, t) \left(\theta_d^{\text{svv+}}(t) - (\theta_e(t)) \right), \tag{17.5}$$

where: $W^p_{I_{\text{heater}}}(\Delta I,t)$, $W^p_{\theta_i}(\Delta\theta_i,t)$, $W^p_{\theta_d}(\Delta\theta_d,t)$ – control algorithms (in case of linear algorithms, these are transfer functions of regulators) of current, the temperature of heating zones, temperature in the modeling chamber; $I^{svv+}_{\text{heater}}(t)$ – set value of $I_{\text{heater}}(t)$, formulated taking into account the guarantee of compliance with the thermal mode of the third chamber and process optimization by the specific energy consumption criterion $\eta_{\text{production}}$; θ_i^{sv} – set values of $Q_i(t)$, formulated from the conditions of ensuring the required product quality; θ_d^{svv+} – set value of $\theta_d(t)$, formulated to ensure compliance with the form limits S_F, maximum proximity to which corresponds to the optimal product quality at the output.

Optimization functions are designed to ensure the achievement of optimal conditions of process energy efficiency (minimum specific energy consumption) and the quality of RP. The specific energy consumption value is calculated by the averaged sliding time interval $s\tau_{cp}$ of electrical power spent on the process:

$$\eta_{\text{producrion}}(t) = \int_{t}^{t+\tau_{av}} P_{RP}(t)dt \Big/ \int_{t}^{t+\tau_{av}} Q_m(t)dt. \tag{17.6}$$

Note the following. (1) The averaging time τ_{av} should be chosen so that the spectrum of averaged power and performance would suppress high-frequency components caused by transients in the circuit. (2) The optimal values of $\eta_{\text{production}}$ should be compared for different types of raw materials, then this would require sufficiently accurate measurements of $P_{RP}(t)$ and $Q_m(t)$. Such measurements for $Q_m(t)$ are rather complicated. But, since the optimization problem of $\eta_{\text{production}}(t)$ must be solved as the problem of current optimization, i.e., within a specific type of raw material, the value of Q_m can be taken as fairly approximate, for example, calculated from the value of u_d and the mathematical model.

Consider the structural diagram of the ACS that implements the selected functional organization. Obviously, the functional organization of the SAC discussed above defines only the most general requirements for the developed system. This commonality results in considerable freedom in the concrete realization of functions. Of course, the number of implementation options that can be attributed to the competitive group is large and can hardly be finally determined. It is important to consider not just competitive, but alternative options that have different advantages and disadvantages and therefore may be relevant for specific models of RP production equipment, with their features. Further, after the functional organization, the specification level of the SAC is its structural organization. It reflects the specific implementation of functions through the block diagram. Consider the following block diagram of the RP production process SAC, which specifies the management concept adopted here.

The principal feature of the SAC, the block diagram of which is shown in Figure 17.6, is that it uses the temperature data in the first two zones and the results of control actions in them to predict the temperature for the third zone.

It should be emphasized that this scheme involves the use of control actions for each channel in the continuous change mode. The regulation functions are implemented based on the closed principle of control of the corresponding regulators. In this case, control actions are formed based on regulation errors, according to the control

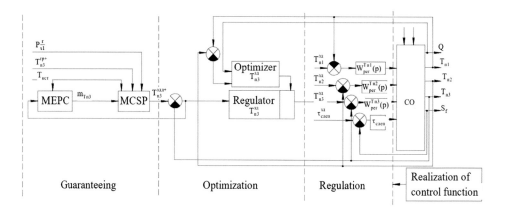

Figure 17.6 Structural scheme of the RP production process SAC in case of using the temperature predicting algorithm for the third zone.

algorithms selected for regulators. Optimization functions are designed to ensure the achievement of optimal conditions of process energy efficiency (minimum specific energy consumption) and the quality of dough products with filling.

The value of specific energy consumption is calculated from the averaged sliding time intervals τ_{av} of the electric power spent on the process and the time spent on heating the dough:

$$\eta_{\text{production}}(t) = \int_{t}^{t+\tau_{av}} P_{RP}(t)dt \,/\, \int_{t}^{t+\tau_{av}} Q_m(t)dt. \tag{17.7}$$

The task of optimizing quality in this SAC structure (Figure 17.6) is solved by changing the heater power, as well as measuring the difference between the steady-state temperatures in the first TP1 and second TP2 control loops and the transition time between zones, during which the temperature values are lowered. During the process, quality indicators stabilize at the optimal level.

17.5 RESULTS AND DISCUSSION

The developed conceptual model of the RP production process made it possible to identify the key features of the CO: (1) a high level of the information incompleteness on the state of the process and the consequences of control actions; (2) a significant limitation of process management resources; (3) a high level of uncertainty in the properties of the equipment for RP production as a CO. The structural diagram, as a graphical representation of the conceptual model, conveniently reflects the interconnection of input, internal, and output variables of the CO. At the development stage of the SAC for RP production, the features of the EA listed above are presented in the form of problems, the degree of overcoming of which, ultimately, will determine the effectiveness level of the SAC. It is obvious that a constructive approach to the development of an effective SAC should be adequate to the problems and, above all, should be systemic, in our case system-functional. Within the framework of such an approach, the methodological basis consists of the concepts of "functional system organization" and "development of the system in the direction of increasing functional integrity." It became the basis for the development of a concept of building an effective SAC of the RP production process. The implementation of such a concept, due to the specific properties of the CO, is a fairly knowledge-intensive task. In particular, because the number of regulated variables of the RP production process is very much higher than the number of control actions available for implementation, it is impossible to stabilize the process for all regulated variables based on classical approaches to SAC construction. Its solution leads to the need to expand the set of functions implemented by the system due to optimization functions, real-time measurement of indirect quality indicators of dough products with filling, and improvements to the implementation of traditional, i.e., regulatory functions. The developed block diagram of the RP production process SAC and key control algorithms specify the adopted control concept, solving the formulated problems, ensuring that all process requirements are established for the process regulation, including – restrictions on the values of variables, as well as the synergistic nature of the interrelationships of the various process control system functions.

REFERENCES

Dragobetskii, V., Shapoval, A., Mos'pan, D., Trotsko, O. & Lotous, V. 2015. Excavator bucket teeth strengthening using a plastic explosive deformation. *Metallurgical and Mining Industry* 4: 363–368.

Groover, M.P. 2009. *Automation, Production Systems, and Computer-Integrated Manufacturing. Assembly Automation*. Lehigh: Pearson Education.

Hraniak, V.F., Kukharchuk, V., Bogachuk, V.V., Vedmitskyi, Y.G., Vishtak, I.V., Piotr Popiel & Yerkeldessova, G. 2018. Phase noncontact method and procedure for measurement of axial displacement of electric machine's rotor. *Proc. SPIE 1080866*: 1825–1831.

Khobin, V.A. 2002. Guarantee function in automatic control systems. *Automation of production processes* 1(14): 145–150.

Khobin, V.A. 2004. Variable structure controller for technological type objects. *Automation. Electrotechnical complexes and systems* 1(13): 190–196.

Khobin, V.A. 2008. *Systems of Guaranteeing Control of Technological Aggregates: The Basics of Theory, Practice of Application*. Odessa: Odessa National Academy Of Food Technologies.

Khobin, V.A. & Egorov, V.B. 2009. Intellectual Video Information Channel for Process Control Systems for Extruding Plant Materials. *XVI International. conf. with an automatic machine. Management "Automation-2009"* 21: 225–226.

Kozlov, L.G., Polishchuk, L.K., Piontkevych, O.V., Korinenko, M.P., Horbatiuk, R.M., Komada, P., Orazalieva, S. & Ussatova, O. 2019. Experimental research characteristics of counter balance valve for hydraulic drive control system of mobile machine. *Przeglad Elektrotechniczny* 95(4): 104–109.

Kvaternyuk, S., Kvaternyuk, O., Petruk, R., Rakytyanska, H., Mokanyuk, O., Ławicki, T. & Kashaganova, G. 2018. Indirect measurements of the parameters of inhomogeneous natural media by a multispectral method using fuzzy logic. *Proc. SPIE 108082P*: 828–834.

Lima, J., Moreira, J.F.P. & Sousa, R.M. 2015. Remote supervision of production processes in the food industry. *2015 IEEE International Conference on Industrial Engineering and Engineering Management (IEEM)* :1123–1127.

Martin, C.K., Nicklas, T., Gunturk, B., Correa, J.B., Allen, H.R. & Champagne, C. 2014. Measuring food intake with digital photography. *Journal of Human Nutrition and Dietetics* 1(27): 72–81.

Mashkov, V., Smolarz, A. & Lytvynenko, V. 2016. Development issues in algorithms for system level self-diagnosis. *Informatyka, Automatyka, Pomiary w Gospodarce i Ochronie Środowiska* 1: 26–28.

Morgan, M.T. & Haley, T.A. 2013. Design of Food Process Controls Systems. *Handbook of Farm, Dairy, and Food Machinery* 19: 475–540.

Ogorodnikov, V.A., Dereven'ko, I.A. & Sivak, R.I. 2018. On the influence of curvature of the trajectories of deformation of a volume of the material by pressing on its plasticity under the conditions of complex loading. *Materials Science* 54(3): 326–332.

Ogorodnikov, V.A., Savchinskij, I.G. & Nakhajchuk, O.V. 2004. Stressed-strained state during forming the internal slot section by mandrel reduction. *Tyazheloe Mashinostroenie* 12: 31–33.

Ogorodnikov, V.A., Zyska, T. & Sundetov, S. 2018. The physical model of motor vehicle destruction under shock loading for analysis of road traffic accident. *Proc. SPIE* 108086C: 1–5.

Polishchuk, L., Gromaszek, K., Kozlov, L.G. & Piontkevych, O.V. 2019. Study of the dynamic stability of the belt conveyor adaptive drive. *Przegląd Elektrotechniczny* 95(4): 98–103.

Vasilevskyi, O., Didych, V., Kravchenko, A., Yakovlev, M., Andrikevych, I., Kompanets, D., Danylyuk, Y., Wójcik, W. & Nurmakhambetov, A. 2018. Method of evaluating the level of confidence based on metrological risks for determining the coverage factor in the concept of uncertainty. *Proc. SPIE* 108082C: 714–719.

Vasilevskyi, O.M., Kulakov, P., Kompanets, D., Lysenko, O.M., Prysyazhnyuk, V., Wójcik, W. & Baitussupov, D. 2018. A new approach to assessing the dynamic uncertainty of measuring devices. *Proc. SPIE 108082E*: 728–735.

Vorobyov, V., Pomazan, M., Vorobyova, L. & Shlyk, S. 2017. Simulation of dynamic fracture of the borehole bottom taking into consideration stress concentrator. *Eastern-European Journal of Enterprise Technologies* 3/1(87): 53–62.

Woods, D.D. 1996. Decomposing automation: Apparent simplicity, real complexity. *Automation and Human Performance: Theory and Applications* 1: 3–17.

Yousefi-Darani, A., Paquet-Durand, O., Zettel, V. & Hitzmann, B. 2018. Closed loop control system for dough fermentation based on image processing. *Journal of Food Process Engineering* 5(41): 1–9.

Chapter 18

Theoretical preconditions of circuit design development for the manipulator systems of actuators of special-purpose mobile robots

S. Strutynskyi, W. Wójcik, A. Kalizhanova, and M. Kozhamberdiyeva

CONTENTS

18.1 Introduction 197
18.2 Development stage analysis of constructive design of special-purpose mobile robot manipulators, substantiation of purpose, objectives, and research methods 197
18.3 Determination of required geometric, kinematic, and precision parameters of mobile robot actuators 199
18.4 Results of the development circuit and construction design of the mobile robot manipulator actuator system 205
18.5 Conclusions 208
References 209

18.1 INTRODUCTION

The development of special-purpose terrestrial robotic complexes is of great importance for the national security of Ukraine. A robotized complex includes a caterpillar chassis with a manipulator mounted on it. In this case, the required robotic complex flotation and the necessary functional capabilities for working with hazardous objects of different types are provided.

The domestic industry does not produce special-purpose serial robotic complexes. Therefore, designing robotic complexes, in particular their manipulators, is especially relevant. There is currently no theory of designing mobile robot manipulators. Therefore, the problem, in general, is to develop the theory of designing manipulators of special-purpose mobile robots.

18.2 DEVELOPMENT STAGE ANALYSIS OF CONSTRUCTIVE DESIGN OF SPECIAL-PURPOSE MOBILE ROBOT MANIPULATORS, SUBSTANTIATION OF PURPOSE, OBJECTIVES, AND RESEARCH METHODS

In recent studies and publications, there is a significant amount of materials concerning the development of terrestrial robotic systems (Li et al. 2016). The publications

DOI: 10.1201/9781003225447-18

provide the results of calculations of static and dynamic characteristics of robotic complexes (Ritzen et al. 2016). They consider the questions of course smoothness and the dynamic characteristics of the mobile robots' chassis (Strutynskyi et al. 2018). The accuracy parameters of manipulators (Jiang & Cripps 2015) have been investigated. Some publications consider the characteristics of mobile robot actuators (Kozlov et al. 2018, Marlow et al. 2016). Typically, literary sources provide limited information on the constructive execution of mobile work; however, the requirements for operating ranges and load capacity are substantiated (Strutynsky et al. 2016). Considerable attention of researchers is focused on the issues of the dynamics of mobile robots (Strutynskyi 2018). The requirements for the design of manipulator actuators in the analysis of dynamic characteristics of robotic complexes are formulated (Meoni & Carricato 2016). A number of literary sources contain information on the constructive implementation of actuators (Blanken et al. 2017). In Qian et al. (2017), the design of manipulators with precision rotary actuators is provided. Their experimental research was carried out, parameters of transient processes in drives were established. The article (Blanken et al. 2017) presents constructive designs of circuits and actuators of a mobile robot mover. The planetary motor gears for moving the drum support parts of the wheel movers are actuated. The vibration characteristics of the developed actuator system are provided. The article (Kot & Novak 2018) presents the results of research concerning a mobile robot with a lever system for changing the geometric configuration of a six-wheeled chassis. To move the levers, rotary low-speed actuators with levers are used (Kot et al. 2014). The publication (Zhao et al. 2015) shows the constructive implementation of a middle-class mobile robot. The manipulator of this robot uses rotary actuators. Some publications (Alghooneh et al. 2016) specify the types of actuators used in manipulators. They mention problems concerning their development and methods of their design (Korayem & Dehkordi 2018). The main methods recommend theoretical studies (Jeong & Cho 2016). They provide an opportunity to take into account the considerable dynamic loads that arise in mobile robots (Jeong & Cho 2016). It is noted that dynamic processes greatly impact the circuit and constructive design of robots (Mashkov et al. 2014, Rybak et al. 2013, Titov et al. 2017, Tymchyk et al. 2018). In order to improve the dynamics of mobile mechanisms, the publication (Kozlov et al. 2019) suggests the use of special drives. To clarify the constructive design of actuator system manipulators, a number of authors recommend mathematical modeling methods. Increasing the efficiency of mathematical modeling is achieved using modern information technology (Strutynskyi & Hurzhii 2017).

As a result of the review and analysis of literary sources, it has been found that they contain practically no information about the constructive performance of the actuator systems actuators of mobile robot manipulators. Therefore, as a result of the analysis of information sources, it was concluded that there is an unresolved part of the general problem, which consists in the development of circuitry and constructive design of the actuator system of special-purpose mobile robots manipulators (Kozlov et al. 2019, Polishchuk et al. 2019).

The purpose of the research described in this article is to create theoretical preconditions for the development of the constructive design of the manipulator actuator systems of special-purpose mobile robots.

The objective of the research is to analyze the geometric, kinematic, and precise parameters of the kinematic circuits of manipulators of special-purpose mobile robots, formulate the requirements for actuator systems, and on this basis, develop a constructive design of the manipulator actuator system of mobile robots (Ogorodnikov et al. 2004, 2018, Ogorodnikov & Zyska 2018).

The primary methods of research are mathematical modeling and geometrical analysis of innovative mechatronic actuator systems of manipulators for special-purpose mobile robots.

18.3 DETERMINATION OF REQUIRED GEOMETRIC, KINEMATIC, AND PRECISION PARAMETERS OF MOBILE ROBOT ACTUATORS

Special-purpose ground robotic complexes are used for conducting operations with dangerous objects. A typical terrestrial robotic complex includes a caterpillar chassis 1, which has a lever-type manipulator (Figure 18.1). The manipulator has a lower lever 2 and an upper lever 3 at the end of which a hinged rod 4 is placed in which the gripper 5 is located. At the end of the rod, a video camera 6 is installed. The gripper 5 can be rotated relative to the hinged rod 4. The levers and the bar form a hinge-lever mechanism in which the planes of movement of the levers and the bars are parallel. Three single-hinged joints 7–9 are applied in relation to the sides in which the levers are turned. The lever system of the manipulator is installed on a rotary arm 10. The manipulator has additional movable elements for moving optical devices. The camcorder 11 rotates around the axis of hinge 7, while camcorder 6 is designed to be rotated in two directions and has a hinge 12, and an additional hinge whose axis is

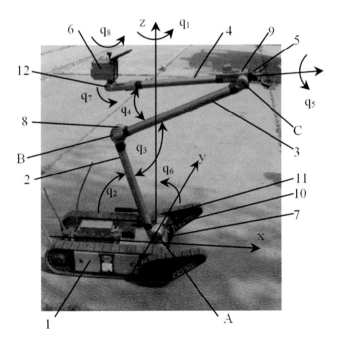

Figure 18.1 General view of a typical special-purpose mobile robot and a system of coordinates for the manipulator.

perpendicular to the axis of hinge 12. The joints 7–9, 11, 12 of the bracket, the levers, and the rods have built-in actuators, which are integrated into the locked actuator system of the robot's manipulator. According to the purpose of this article, the theoretical preconditions for developing a constructive design of the given actuator system, which provides the necessary configuration for the mobile robot's manipulator, are substantiated (Del' et al. 1975, Dragobetskii et al. 2015, Vorobyov et al. 2017).

To determine the geometric and kinematic parameters of the manipulator, the basic Cartesian coordinate system of the manipulator x, y, z is introduced, the vertical axis of which – z – is stationary and coincides with the axis of rotation of the bracket 10.

For an analytical description of the manipulator levers' position, a system of coordinates corresponding to the available degrees of freedom of the manipulator is used. The system of controlled coordinates is related to the actual position of the manipulator levers. The coordinate system determines the configuration of the manipulator, depending on the angular values q1,... q8, which are controlled coordinates. These angular values are determined by: turning the lever system relative to the z-axis (q1); angle of inclination of the lower lever 2 relative to the plane perpendicular to the axis of the bracket's rotation (q2); the angle between the axes of the lower 2 and the upper 3 levers (q3); the angle between the rod 4 and the axis of the upper lever (q4); the angle of rotation of the gripper 5 relative to the axis of the rod (q5); the turning angle of the camcorder 6 mounted on the bracket (q6); the angle of rotation of the hinge of the main video camera (q7); and the angle defining the axial rotation of the camcorder (q8).

Controlled coordinates are set up by actuators located in the manipulator hinges. The primary ones are complete rotary-type mechatronic actuators installed in the joints of the levers at points A, B, and C. For the purpose of developing constructive designs for these actuators, technical requirements have been formulated. They mainly consist of providing the necessary ranges of drive turning angles and angular rotation speeds of up to 5 rpm, the ability to perceive a significant static torque load (200 Nm), the need to compensate for gaps, and exclude the backlash in actuators. The main technical requirement is to provide the necessary sensitivity of actuators during their operation.

Its general kinematic scheme is formed based on the introduced coordinates and analysis of changes in the geometry of the manipulator (Figure 18.2) (Kukharchuk et al. 2017, Ogorodnikov et al. 2018).

The lower lever of the manipulator with actuators on the kinematic diagram corresponds to the kinematic circuit AB, the upper lever corresponds to the circuit BC, and the rod corresponds to the kinematic circuit ECD.

Rotation of the bracket at the angle q1 relative to the z-axis causes the rotation of the lever displacement planes and accordingly, the rotation of the axes of the base coordinate system x, in the associated levers. The distance between lever displacement planes is negligible. Therefore, a simplified planar geometric scheme of the manipulator that is located on a moving platform is used for calculating the geometric parameters (Figure 18.3).

To determine the necessary kinematic parameters of the actuators, an analysis of the relationship between the location of the executive body (gripper) and the controlled coordinates of the manipulator is performed (Kukharchuk et al. 2016, 2017, Vasilevskyi 2014).

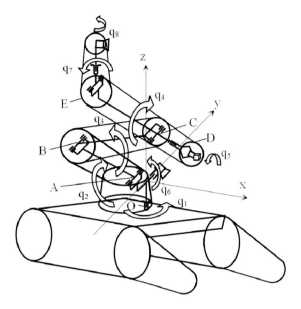

Figure 18.2 The general kinematic circuit of the mobile robot manipulator.

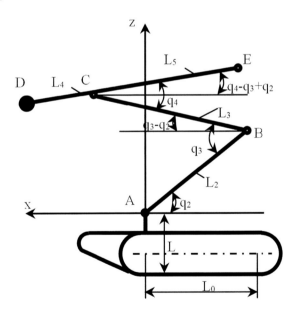

Figure 18.3 Simplified planar kinematic diagram of the mobile robot manipulator.

The main controlled coordinates for the planar kinematic diagram are the angles of leverage arrangement. $q_2,\ldots q_4$. The diagram shows the geometric dimensions of the kinematic circuits:

$AB = L_2$, $BC = L_3$, $CD = L_4$, $CE = L_5$.

Displacements in the manipulator lever system are divided into large displacements (macro-displacements) and small displacements (micro-displacements). The large displacements have the order of the length of the levers, and the small displacements are much smaller than the lengths of the levers.

Large displacements were found with methods of analytic geometry. The coordinates of the point D in the projections on the x and z axes are determined based on the geometric relations (Vasilevskyi 2013):

$$x_D = -L_2 \cdot \cos q_2 + L_3 \cos(q_3 - q_2) + L_4 \cos(q_4 - q_3 + q_2),$$

$$z_D = L_2 \cdot \sin q_2 + L_3 \sin(q_3 - q_2) - L_4 \sin(q_4 - q_3 + q_2) \tag{18.1}$$

The following formulas are used to determine the differentials of the coordinates of point D as functions of the controlled coordinates $q_2, \ldots q_4$:

$$dx_D = \frac{\partial x_D}{\partial q_2} \cdot dq_2 + \frac{\partial x_D}{\partial q_3} \cdot dq_3 + \frac{\partial x_D}{\partial q_4} \cdot dq_4$$

$$dz_D = \frac{\partial z_D}{\partial q_2} \cdot dq_2 + \frac{\partial z_D}{\partial q_3} \cdot dq_3 + \frac{\partial z_D}{\partial q_4} \cdot dq_4 \tag{18.2}$$

Partial derivatives are determined by differentiating projections of displacements of point D (1). For projection $x_D(q_2, q_3, q_4)$ we have:

$$\frac{\partial x_D}{\partial q_2} = L_2 \sin q_2 + L_3 \sin(q_3 - q_2) - L_4 \sin(q_4 - q_3 + q_2),$$

$$\frac{\partial x_D}{\partial q_3} = -L_3 \sin(q_3 - q_2) + L_4 \sin(q_4 - q_3 + q_2),$$

$$\frac{\partial x_D}{\partial q_4} = -L_4 \sin(q_4 - q_3 + q_2) \tag{18.3}$$

For projection onto the vertical axis $z_D = z_D(q_2, q_3, q_4)$ we have:

$$\frac{\partial z_D}{\partial q_2} = L_2 \cos q_2 - L_3 \cos(q_3 - q_2) - L_4 \cos(q_4 - q_3 + q_2),$$

$$\frac{\partial z_D}{\partial q_3} = L_3 \cos(q_3 - q_2) + L_4 \cos(q_4 - q_3 + q_2),$$

$$\frac{\partial z_D}{\partial q_4} = -L_4 \cos(q_4 - q_3 + q_2) \tag{18.4}$$

The partial derivatives included in the dependence data are components of the Jacobi matrix, which binds the coordinate differentials of point D to the controlled coordinate differentials. In the matrix-vector form, the relation of the differentials of the coordinates (18.2) is established by the dependence:

$$\begin{bmatrix} dx_D \\ dz_D \end{bmatrix} = I \cdot \begin{bmatrix} dq_2 \\ dq_3 \\ dq_4 \end{bmatrix}, \text{ where } I(q_2,q_3,q_4) = \begin{bmatrix} \dfrac{\partial x_D}{\partial q_2} & \dfrac{\partial x_D}{\partial q_3} & \dfrac{\partial x_D}{\partial q_4} \\ \dfrac{\partial z_D}{\partial q_2} & \dfrac{\partial z_D}{\partial q_3} & \dfrac{\partial Z_D}{\partial q_4} \end{bmatrix} \quad (18.5)$$

The components of the matrix (18.5) are determined by the dependences (18.3) and (18.4).

Moving on from the differentials in formula (18.5) to the final differentials, we obtain the matrix-vector dependence of the displacement increments of point D on the increments of the coordinates as:

$$\begin{bmatrix} \Delta x_D \\ \Delta z_D \end{bmatrix} = I(q_2,q_3,q_4) \cdot \begin{bmatrix} \Delta q_2 \\ \Delta q_3 \\ \Delta q_4 \end{bmatrix} \quad (18.6)$$

Jacobi matrices are defined based on the abovementioned formulae for the characteristic configurations of the mobile robot's manipulator. As for the position of the manipulator (Figure 18.3), the matrix has the following form:

$$I = \begin{bmatrix} 499{,}297 & -209{,}984 & -41{,}752 \\ -480{,}626 & 924{,}932 & -297{,}080 \end{bmatrix} \text{mm/Rad}$$

The increments of controlled coordinates, which depend on the precision of the actuators, determine the accuracy of the manipulator gripper output (point D) to a given position. Based on formulas (18.6), (18.3), and (18.4), the gripper output tolerances in relation to position are calculated depending on the tolerances of the actuators for different manipulator configurations. It is assumed that all the tolerances of the actuators installed in the hinges of the manipulator are identical in absolute value and are:

$$\pm\Delta q_2 = \pm\Delta q_3 = \pm\Delta q_4 = \Delta q$$

Tolerance signs for separate actuators are selected in such a way as to ensure their maximum negative impact on the output gripper tolerances in relation to position. That is, the choice of tolerance signs should provide:

$$\max \begin{bmatrix} \Delta x_D \\ \Delta z_D \end{bmatrix}$$

Different manipulator configurations are considered and the output gripper tolerances in relation to position at various values of the actuator tolerances (Table 18.1) are calculated.

Table 18.1 Calculation of the gripper positioning precision of the manipulator depending on the magnitude of the tolerances of working controlled coordinates for different manipulator configurations

№ / Manipulator configuration	Changes of controlled coordinate and Δq displacement of the point D, mm	$\Delta q = 1'$		$\Delta q = 20''$	
		Δx	Δz	Δx	Δz
1		0.2	−0.5	0.07	−0.17
2		0.81	0	0.27	0
3		0	−0.81	0	−0.27
4		−0.19	−0.77	−0.06	−0.26
5		0.08	−0.38	0.03	−0.13

The first line in the table corresponds to the typical average manipulator configuration, which is shown in Figure 18.3. The second line contains the gripper output tolerances in relation to position at its maximum elevation. The third line contains data for the maximum gripper manipulator displacement, and the fourth and fifth lines correspond to the manipulator with a vertical rod arrangement.

The general tolerance of gripper output in relation to position is defined as the geometric sum of the tolerance components according to the following dependence:

$$\Delta = \sqrt{(\Delta x)^2 + (\Delta y)^2}$$

General tolerance sets the deviation of the gripper from the nominal position, which occurs due to the tolerances of working controlled coordinates by the corresponding actuators.

From the analysis of data given in the table, it follows that the total tolerance of the gripper output in relation to position is approximately $\Delta = 0.27$ mm with an actuator tolerance of $\Delta q = 20''$. At the same time, the dependence of tolerance of the gripper output in relation to position on the tolerance of the actuators is close to the linear one:

$$\Delta = c_\Delta \cdot \Delta q$$

The proportionality coefficient in this formula depends on the geometric dimensions of the manipulator parts. For the considered manipulator these are: $c_\Delta = 0.013...0.014$ mm/angular second.

The angular second is $7.75 \cdot 10^{-5}$ of one revolution of the output actuator chain. Correspondingly, the proportionality factor is:

$$c_\Delta = (1.68...1.8) \cdot 10^4 \text{ mm/rev}$$

The input actuator shaft of the manipulator's rotary motion is turned in relation to the number of revolutions much more than the revolution of the output chain. The rotation ratio of the input shaft and the actuator output chain is determined by the transmission ratio i_p. To ensure the precision of the gripper output into the ordered position $\Delta I = 0.1$ mm, the transmission ratio of the actuator order is required:

$$i_p = c_\Delta / [\Delta] \approx (1.68...1.8) \cdot 10^5$$

The necessary transmission ratio of such an order cannot be realized by the known design of reducers. The closest to the transmission ratio is a two-stage wave transmission in which the transmission ratio can reach:

$$i_w = (1.0...1.5) \cdot 10^5$$

If discreteness of rotation angle measurements of the input shaft actuator 60° is ensured, which is carried out by the installation of a circle of six markers for angle measurement, to ensure the precision of the output into the position of $[\Delta] = 0.1$ mm, the transmission ratio of the actuator should be

$$i_p = (2.8...3.0) \cdot 10^4$$

The actuator with this transmission ratio is implemented according to the two-stage wave transmission diagram.

18.4 RESULTS OF THE DEVELOPMENT CIRCUIT AND CONSTRUCTION DESIGN OF THE MOBILE ROBOT MANIPULATOR ACTUATOR SYSTEM

The principal circuit was developed based on defined actuator parameters (Figure 18.4). The actuator has a built-in high-revving electric motor 1 whose shaft rotates to the angle φe. The shaft of the motor is connected through the coupling to the mechanical

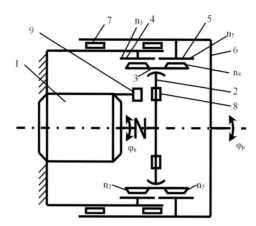

Figure 18.4 Kinematic rotary movement circuit of the lever actuator for the mobile robot's manipulator.

wave generator 2 of the wave gear. The wavelength transmission has a flexible toothed wheel 3, which engages with a fixed, rigid gear 4 and a movable hard gear 5. The mobile gear is connected to the drive unit output 6 and rotates to the angle φr. The output of the actuator is mounted on the needle roller bearings 7. There are permanent magnets on the flange of the wave generator 8, which intermittently interact with fixed mounted hermetic contacts (reeds) 9.

The transmission ratio of the drive is determined by the number of wheel teeth and is calculated according to the dependence:

$$i_p = \left(1 - \frac{n_3 n_4}{n_2 n_5}\right)^{-1}$$

where n_2, n_4 – number of teeth of the toothed crowns of the flexible wheel; n_5 – is the number of teeth of a fixed rigid wheel.

When selecting the required number of teeth, the actuator provides a transmission ratio of 2, $5·10^2...1,5·10^5$. To ensure the required precision of working out the rotation angle of the actuator output shaft, measurement of the rotation angle of the input shaft is performed. When installing six magnets, eight dimensions of rotating angle are measured with a discreteness of 60°. It is sufficient to ensure the actuator output shaft installation precision, which is not less than 20″.

According to the necessary parameters of the kinematic characteristics of mobile robots, a number of circuits and constructive design of manipulator actuators are offered. A design version of the actuators has been developed on a traditional element basis. It includes a complete unit containing a high-torque electric motor and a wave gear transmission with a cam mechanical wave generator (Figure 18.5).

The actuator has a fixed housing 1, which is mounted on the turntable base of the mobile robot chassis. There is an axle 2 mounted in the housing where a stationary stator of the electric motor 3 is located. The bearings 4 and 5 of rotor 6 are fixed on axle 2 in the axial direction. The permanent magnets 7 of the electric motor are placed

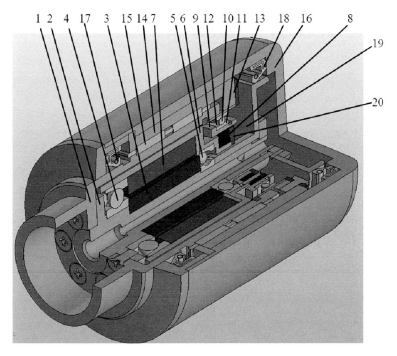

Figure 18.5 Construction version of an actuator for a mobile robot manipulator built on the traditional element basis.

on the inner surface of the rotor 6. An elastic coupling 8 with a sleeve 9 is installed on the end cylindrical surface of the rotor 6 on the outer elliptic surface of which a flexure bearing 10 is installed. The outer surface of the flexure bearing 10 interacts with the inner surface of the flexure gear wheel 11. The flexure wheel has a gear engagement with a fixed gear 12, which is made in housing 1. On the other hand, the flexure wheel has a clutch with a rotating gear wheel 13, which is made on the inner surface of the sleeve 14. Sleeve 14 is mounted on housing 1 on the needle bearings 15. Sleeve 14 has a casing 16 on the outer surface and is connected to the manipulator lever. The actuator features collar seals 17, 1 which seal the oil-filled inner surface of the actuator. A central hole in axle 2 is made for the passage of electrical communications to the manipulator's arm. To measure the total rotational speed of the motor rotor, a sealed contact 19 is used, which operates when passing by a magnet 20 placed on the rotor.

To improve the efficiency of the actuator, reduce its mass and dimensions, an advanced actuator design has been developed. A high-speed mechanical wave generator with a traditional electric motor is excluded from the proposed design. An electromagnetic wave generator that has a stationary stator with electromagnets acting on powerful permanent magnets installed on the inner surface of the wave gear transmission flexure wheel is proposed.

To increase the actuator's efficiency, an innovative variant of the actuator's constructive design is proposed where a ferromagnetic fluid is found in the gap between the flexure gear surface and the outer cylindrical stator surface (Figure 18.6).

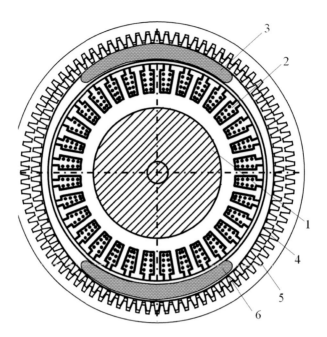

Figure 18.6 Constructive design of the innovative electrodynamic wave generator wave-tooth gear, which is the basis of the rotary of the actuator for the mobile robot manipulator: 1 – stator; 2 – electromagnet; 3 – ferromagnetic fluid; 4 – flexure wheel; 5 – a rigid gear wheel.

Ferromagnetic fluid accumulates in two diametrically opposite regions of the flexure wheel 3, 6 under the action of the electromagnets' magnetic field. An area of high pressure arises in the ferromagnetic fluid, and the pressure level is determined by the induction of the magnetic field created by the electromagnets 2. The deformation of the flexure toothed wheel 4 occurs under the influence of pressure. As a result of changes in the magnetic field along the perimeter of the stator, fluid in the region of elevated pressure moves in a circle ensuring the flexure toothed wheel deformation wave displacement.

18.5 CONCLUSIONS

1. Actuators of special-purpose mobile robot manipulators should have dimensions in the range of 100 mm with the minimum chassis. They should be self-breakable and should not contain any backlashes and gap clearances.
2. The Jacobian matrix of a manipulator, defined based on geometric relations in the manipulator lever system, allows determining the interdependence between the tolerances of the controlled coordinates which are set by the actuators and tolerances of the robot positioning in all areas of the manipulator's working space. In this case, the maximum tolerances in the gripper positioning occur at the vertical position of the levers and depend linearly on the tolerances of separate actuators.

3. To ensure positioning precision of the range of 0.1 mm, the manipulator actuators must have a transmission ratio in the range of 105. This transmission ratio can be implemented by an actuator based on a two-stage gearwheel gear. This actuator type meets the requirements for precision, mass, and dimensions. It is implemented on an existing element basis.
4. To increase the efficiency of the manipulator actuator's innovative circuit designs, actuators based on a wave gear transmission with an electromagnetic wave generator or an electro-hydrodynamic wave generator using a ferromagnetic fluid as a working medium are recommended.

REFERENCES

Alghooneh, M., Wu, C.Q. & Esfandiari, M. 2016. A passive-based physical bipedal robot with a dynamic and energy-efficient gait on the flat ground. *Journal Transactions on Mechatronics* 21(4): 1977–1984.

Blanken, L., Boeren, F., Bruijnen, D. & Oomen, T. 2017. Batch-to-batch rational feedforward control: from iterative learning to identification approaches, with application to a wafer stage, *IEEE/ASME Transactions on Mechatronics* 22(2): 826–837.

Del', G.D., Ogorodnikov, V.A. & Nakhaichuk, V.G. 1975. Criterion of deformability of pressure shaped metals. *Izv Vyssh Uchebn Zaved Mashinostr* (4): 135–140.

Dragobetskii, V., Shapoval, A., Mos'pan, D., Trotsko, O. & Lotous, V. 2015. Excavator bucket teeth strengthening using a plastic explosive deformation. *Metallurgical and Mining Industry* 4: 363–368.

Ghotbi, B., Gonzlez, F., Kovecses, J. & Angeles, J. 2015. A novel concept for analysis and performance evaluation of wheeled rovers. *Mechanism and Machine Theory* 83: 137–151.

Jeong, H.S. & Cho, J.R. 2016. Optimal design of head expander for a lightweight and high frequency vibration shaker. *International Journal of Precision Engineering and Manufacturing* 17(7): 909–916.

Jiang, X. & Cripps, R.J. 2015. A method of testing position independent geometric errors in rotary axes of a five-axis machine tool using a double ball bar. *International Journal of Machine Tools and Manufacture* 89: 151–158.

Joe, H.M. & Oh, J.H. 2018. Balance recovery through model predictive control based on capture point dynamics for biped walking robot. *Robotics and Autonomous Systems* 105: 1–10.

Korayem, M.H. & Dehkordi, S.F. 2018. Derivation of motion equation for mobile manipulator with viscoelastic links and revolute–prismatic flexible joints via recursive Gibbs–Appell formulations. *Robotics and Autonomous Systems* 103: 175–198.

Kot, T., Babijak, J., Krys, V. & Novak, P. 2014. System for automatic collisions prevention for a manipulator arm of a mobile robot. *Proceedings of the IEEE 12th International Symposium on Applied Machine Intelligence and Informatics* 1: 167–171.

Kot, T. & Novak, P. 2018. Application of virtual reality in teleoperation of the military mobile robotic system TAROS. *International Journal of Advanced Robotic Systems* 15(1): 1–6.

Kozlov, L.G., Bogachuk, V.V., Bilichenko, V.V., Tovkach, A.O., Gromaszek, K. & Sundetov, S. 2018. Determining of the optimal parameters for a mechatronic hydraulic drive. *Proc. SPIE* 1080861: 1–10.

Kozlov, L.G., Polishchuk, L.K., Piontkevych, O.V., Korinenko, M.P., Horbatiuk, R.M., Komada, P., Orazalieva, S. & Ussatova, O. 2019. Experimental research characteristics of counter balance valve for hydraulic drive control system of mobile machine. *Przeglad Elektrotechniczny* 95(4): 104–109.

Kukharchuk, V.V., Bogachuk, V.V., Hraniak, V.F., Wójcik, W., Suleimenov, B. & Karnakova, G. 2017. Method of magneto-elastic control of mechanic rigidity in assemblies of hydropower units. *Proc. SPIE* 104456A: 1–7.

Kukharchuk, V.V., Hraniak, V. F., Vedmitskyi, Y.G., Bogachuk, V.V., Zyska, T, Komada, P. & Sadikova, G. 2016. Noncontact method of temperature measurement based on the phenomenon of the luminophor temperature decreasing. *Proc. SPIE* 100312F: 1–6.

Kukharchuk, V., Kazyv, S., Bykovsky, S., Wójcik, W., Kotyra, A., Akhmetova, A., Bazarova, M. & Weryńska-Bieniasz, R. 2017. Discrete wavelet transformation in spectral analysis of vibration processes at hydropower units. *Przeglad Elektrotechniczny* 93(3): 65–68.

Li, B., Fang, Y., Hu, G. & Zhang, X. 2016. Model-free unified tracking and regulation visual servoing of wheeled mobile robots. *Journal Sensors and Actuators A: Physical, IEEE Transactions on Control Systems Technology* 24(4): 1328–1339.

Marlow, K., Isaksson, M., Dai, J.S. & Nahavandi, S. 2016. Motion force transmission analysis of parallel mechanisms with planar closed-loop subchains. *Journal of Mechanical Design* 138(6): 21–32.

Mashkov, V., Smolarz, A., Lytvynenko, V. & Gromaszek, K. 2014. The problem of system fault-tolerance. *Informatyka, Automatyka, Pomiary w Gospodarce i Ochronie Środowiska* 4: 41–44.

Meoni, F. & Carricato, M. 2016. Design of nonovercon strained energy-efficient multi-axis servo presses for deep-drawing applications. *Journal of Mechanical Design* 138(6): 1–12.

Ogorodnikov, V.A., Dereven'ko, I.A. & Sivak, R.I. 2018. On the influence of curvature of the trajectories of deformation of a volume of the material by pressing on its plasticity under the conditions of complex loading. *Materials Science* 54(3): 326–332.

Ogorodnikov, V.A., Grechanyuk, N.S. & Gubanov, A.V. 2018. Energy criterion of the reliability of structural elements in vehicles. *Materials Science*, 53(5): 645–650.

Ogorodnikov, V.A., Savchinskij, I.G. & Nakhajchuk, O.V. 2004. Stressed-strained state during forming the internal slot section by mandrel reduction. *Tyazheloe Mashinostroenie* 12: 31–33.

Ogorodnikov, V.A., Zyska, T. & Sundetov, S. 2018. The physical model of motor vehicle destruction under shock loading for analysis of road traffic accident. *Proc. SPIE* 108086C: 1–5.

Polishchuk, L., Gromaszek, K., Kozlov, L.G. & Piontkevych, O.V. 2019. Study of the dynamic stability of the belt conveyor adaptive drive. *Przeglad Elektrotechniczny* 95(4): 98–103.

Qian, J., Zi, B., Wang, D., Ma, Y. & Zhang, D. 2017. The design and development of an omnidirectional mobile robot oriented to an intelligent manufacturing system. *Sensors* 17(9): 1–15.

Ritzen, P., Roebroek, E., Van de Wouw, N. & Jiang, Z. 2016. Trailer steering control of a tractor–trailer robot. *Journal Sensors and Actuators A: Physical, IEEE Transactions on Control Systems Technology* 24(4): 1240–1252.

Rybak, L., Gaponenko, E., Chichvarin, A., Strutinsky, V., Sidorenko, R. 2013. Computer-aided modeling of dynamics of manipulator-tripod with six degree of freedom. *World Applied Sciences Journal* 25(2): 341–346.

Strutynskyi, S. 2018. Defining the dynamic accuracy of positioning of spatial drive systems through consistent analysis of processes of different range of performance. *Scientific Bulletin of the National Mining University* 3: 64–73.

Strutynskyi, S., Kravchuk, V. & Semenchuk, R. 2018. Mathematical modelling of a specialized vehicle caterpillar mover dynamic processes under condition of the distributing the parameters of the caterpillar. *International Journal of Engineering & Technology* 7(4/3): 40–46.

Strutynskyi, S.V. & Hurzhii, A.A. 2017. Definition of vibro displacements of drive systems with laser triangulation meters and setting their integral characteristics via hyper-spectral analysis methods. *Scientific Bulletin of the National Mining University* 1: 43–51.

Strutynsky, V.B., Hurzhi, A.A., Kolot, O.V. & Polunichev, V.E. 2016. Determination of development grounds and characteristics of mobile multi-coordinate robotic machines for materials machining in field conditions. *Scientific Bulletin of the National Mining University* 5(155): 43–51.

Titov, A.V., Mykhalevych, V.M., Popiel, P. & Mussabekov, K. 2017. Statement and solution of new problems of deformability theory. *Proc. SPIE* 108085E: 1611–1617.

Tymchyk, S.V., Skytsiouk, V.I., Klotchko, T.R., Ławicki, T. & Demsova, N. 2018. Distortion of geometric elements in the transition from the imaginary to the real coordinate system of technological equipment. *Proc. SPIE* 108085C: 1595–1604.

Vasilevskyi, O.M. 2013. Advanced mathematical model of measuring the starting torque motors. *Technical Electrodynamics* (6): 76–81.

Vasilevskyi, O.M. 2014. Calibration method to assess the accuracy of measurement devices using the theory of uncertainty. *International Journal of Metrology and Quality Engineering*, 5(4): 1–9.

Vorobyov, V., Pomazan, M., Vorobyova, L. & Shlyk, S. 2017. Simulation of dynamic fracture of the borehole bottom taking into consideration stress concentrator. *Eastern-European Journal of Enterprise Technologies* 3/1(87): 53–62.

Zhao, Y., Qiu, K., Wang, S. & Zhang, Z. 2015. Inverse kinematics and rigid-body dynamics for a three rotational degrees of freedom parallel manipulator. *Robotics and Computer-Integrated Manufacturing* 31(1): 40–50.

Chapter 19

Analysis of random factors in the primary motion drive of grinding machines

V. Tikhenko, O. Deribo, Z. Dusaniuk, O. Serdiuk,
A. Kotyra, S. Smailova, and Y. Amirgaliyev

CONTENTS

19.1 Introduction 213
19.2 Analysis of studies of dynamics of the primary motion drives......................214
19.3 Methods of study of random factors in the primary motion drive in
 grinding machines 214
19.4 Results and discussion 220
19.5 Conclusions 221
References..222

19.1 INTRODUCTION

The primary factors that determine machine quality are the reliability and durability of the machine parts and the quality of the surface, which is characterized by roughness, wavelength, and physical and mechanical properties of the surface layer. In most cases, it is the grinding operations that complete the long and laborious process of manufacturing the parts, which achieves a high precision of shape and size, low roughness of the treated surfaces, which determines their durability, operational reliability and, as a consequence, the quality of the machine as a whole.

The quality of the surface layer is determined by the simultaneous influence of the geometric, power and thermal factors, which are functions of the grinding process modes and the characteristics of the grinding wheels, which have a direct connection with the relative oscillations of the form-forming knots (Nikitin, 2010).

Grinding machines perform the following types of work: cylindrical external grinding, centerless external grinding, centerless internal grinding, grinding of planes by periphery of a grinding wheel, and shaping grinding, including grinding of teeth, grooves, etc., and grinding of planes by the end of a grinding wheel. Interrupting grinding wheels are becoming increasingly more widespread. A special group consists of blasting and grinding machines for abrasive cleaning of rolled metal, which belong to a group of heavy machine tools. For example, the machines of the Swedish firm "Centro Maskin" have a grinding wheel drive power of 300 kW and provide a cutting speed of 80 m/s.

To date, the following gradation of abrasive treatment has been established for cutting speed: normal grinding (up to 45 m/s), high speed (over 45–120 m/s), and super high speed (more than 120 m/s). The last range of speeds is practically unexplored today and there are very few works on this topic. Parameters of modes are usually chosen based on reference or calculation data and are accepted unchanged during the grinding operation, at best, adjusted according to the results of the processing of the workpiece surfaces. The choice of rational regimes is particularly complicated when grinding hard-working materials, such as those containing additives of tungsten, silicon, titanium, and other elements (Azarova et al., 2013).

19.2 ANALYSIS OF STUDIES OF DYNAMICS OF THE PRIMARY MOTION DRIVES

An overview of literary sources showed that the process of surface grinding is usually modeled as a system having two orthogonal degrees of freedom (Popp et al., 2008; Weidong, 2018). The cutting forces that occur when inserting into the material of the workpiece are presented in the form of constituents in the direction of feed (X) and in normal (Z), which stimulate dynamic displacements. In the simulation, the angular vibrations of the grinding wheel were not considered, that is, the angular speed of rotation ω is constant and does not change due to vibrations.

A dynamic model of the elastic system of a cylindrical grinding machine was considered in Gusev and Svetlitsky (1984), Levin (1978), and Strutynskyi et al. (2011). On the basis of analysis of kinematics and dynamics of plunge cylindrical grinding, a structural scheme of the mathematical model of the subsystem of the spindle group was considered, taking into account the transverse and twist fluctuations of the spindle and the dynamics of the subsystem of the wheel head. In the conclusions, the role of the drive of the main motion in the dynamic processes was not singled out. In addition, the differences of cylindrical grinding from other types of abrasive treatment should be taken into account. For example, the spindle rotational speed of the cylindrical grinding machine is almost two times lower than with surface grinding. The workpiece that is being processed may have a deviation from the roundness of the cross-sectional profile.

However, processes in the primary motion drive (the rotation of the grinding wheel) are an integral part (subsystem) of the dynamic system of grinding machines. The input to the subsystem is a cutting torque whose values change during the processing cycle. Fluctuations of this moment lead to a change in the load on the wheel drive and, consequently, to the excitation of the torsional oscillations of the drive elements, that is, to fluctuations in the speed of cutting. It is known that the methods of studying the dynamics of machines, which are based on deterministic perturbations of input parameters, do not always deliver results that are adequate to the reality of the physics of the oscillatory process, since the bulk of external technological factors are of a random nature. It is advisable to carry out a detailed analysis of random factors that cause the appearance of torsional oscillations, based on a dynamic model.

19.3 METHODS OF STUDY OF RANDOM FACTORS IN THE PRIMARY MOTION DRIVE IN GRINDING MACHINES

The primary motion in grinding machines is carried out from asynchronous electric motors through a V-belt drive (Figure 19.1). The transition from the kinematic

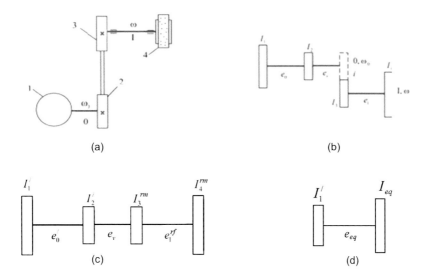

Figure 19.1 Transformation of the kinematic scheme of the drive into the calculation scheme.

(constructive) drive scheme (Figure 19.1a) to the calculation scheme is to determine the numerical values of the rotating masses, torsional flexibility, and the coefficients of damping. As the concentrated rotating masses, we accept the actuator parts that have the corresponding moments of inertia: the rotor of the electric motor I_1, the pulleys I_2 and I_3, and the face plate I_4 with the grinding wheel mounted onto it. For the compilation of the equations of the dynamical system of the wheel drive, we consider the equality of moments on each of the masses, using the principle of D'Alembert. The moment of inertia of the distributed masses of the electric motor shaft and the spindle is reduced to the moments of inertia of the concentrated masses:

$$I'_1 = I_1 + \frac{1}{6}I_0 \tag{19.1}$$

$$I'_2 = I_2 + \frac{1}{6}I_0 \tag{19.2}$$

$$I'_3 = I_3 + \frac{1}{6}I_4 \tag{19.3}$$

$$I'_4 = I_4 + \frac{1}{6}I_4 \tag{19.4}$$

In order to find the reduced flexibility, the flexibility of the shafts (e_0 and e_1) should be summed up with the flexibility of the combinations (e_2, e_3 and e_4) adjacent to the corresponding section:

$$e'_0 = e_0 + e_1 + e_2 \tag{19.5}$$

$$e'_1 = e_1 + e_3 + e_4 \tag{19.6}$$

Based on the received dependencies, it is possible to go from the structural scheme to the graduated scheme (Figure 19.1b), where e_v is the flexibility of the V-belt transmission. The presence of a V-belt transmission with a transmission coefficient i is reflected on the graduated scheme by dashed lines. To move to a linear scheme (Figure 19.1c), it is necessary to bring all rotating masses and flexibility of the shafts to one link (electric motor shaft). For this purpose, the moments of inertia of I_3 and I_4 rotating masses with angular velocity ω are brought to shaft 0 with an angular velocity ω_0.

On the basis of the equality of the kinematic energies reduced and given masses, we determine the reduced moments:

$$\frac{I'_3 \omega^2}{2} = \frac{I_3^{\ddot{i}} \omega_0^2}{2} \tag{19.7}$$

$$\frac{I'_4 \omega^2}{2} = \frac{I_4^{\ddot{i}} \omega_0^2}{2} \tag{19.8}$$

$$I_3^{\ddot{i}} = I'_3 \left(\frac{\omega}{\omega_0}\right)^2 = I'_3 i^2 \tag{19.9}$$

$$I_4^{\ddot{i}} = I'_3 \left(\frac{\omega}{\omega_0}\right)^2 = I'_4 i^2 \tag{19.10}$$

The reduced flexibility can be calculated by the expression $e_1^{\ddot{i}\partial} = \dfrac{e'_1}{i^2}$.

A linear calculation scheme can be simplified to two masses (Figure 19.1d). To do this, we store a large mass with the moment of inertia I'_1, and the moments of inertia I'_2, $I_3^{\ddot{i}\partial}$, and $I_4^{\ddot{i}\partial}$ we replace with the equivalent of $I_{\text{ек}}$. The simplified two-mass system (Fig 19.1d) has the following parameters"

$$I_{e\hat{e}} = I'_2 + I_3^{\ddot{i}\partial} + I_4^{\ddot{i}\partial} \tag{19.11}$$

$$e_{e\hat{e}} = \frac{e'_0 I'_2 + (e'_0 + e_v) I_3^{\ddot{i}} + (e'_0 + e_v + e_1^{\ddot{i}}) I_4^{\ddot{i}}}{I_{e\hat{e}}} \tag{19.12}$$

The resulting calculation scheme of the drive can be considered as a rotating shaft with the rotational speed of the electric motor, on which the masses with given moment of inertia I'_1 and $I_{\text{ек}}$ are located. Typically, processes in the load-bearing system of the machine (in contrast to the cutting process) have a weak effect on the working conditions of the primary motion drive, and they can be neglected (Ogorodnikov, Dereven'ko, & Sivak, 2018; Ogorodnikov, Grechanyuk, & Gubanov, 2018; Ogorodnikov et al., 2004;

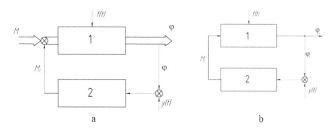

Figure 19.2 Structural scheme of the primary motion drive.

Strutynskyi et al., 2011).The study of dynamic processes is carried out by studying the structural scheme (Figure 19.2a), which is a dynamic model of the drive of the primary motion in the form of two blocks: the actual drive 1 and the grinding process 2 (Figure 19.2). These blocks are interconnected direct and reciprocal links and form a closed system.

In general, the dynamic drive system of the primary motion is multidimensional, and the input and output functions can be represented in a vector form:

$$M = \begin{bmatrix} M_1 \\ M_2 \\ \vdots \\ M_n \end{bmatrix}, \quad \phi = \begin{bmatrix} \phi_1 \\ \phi_2 \\ \vdots \\ \phi_n \end{bmatrix} \tag{19.13}$$

where $M_1, M_2, \ldots M_n$ are the components of the vector M, and $\varphi_1, \varphi_2, \ldots \varphi_n$ are the components of the vector φ.

We will assume that the random process of the initial turning angle of the grinding wheel is stationary and ergodic, as well as the input technological perturbation of the stationary, therefore the output of the dynamic system will be a stationary random process.

The stationary statement of the problem allows one to use the mathematical apparatus of spectral analysis of random processes.

In the general case, the number of components of the vector M and φ may be large, but it is expedient to accept only dominant components in the calculation. Thus, in the presence of feedback between the torsional deformation φ and torque M_z, it is possible to neglect other components of the vector, providing that they are insignificant compared with M_z.

As a result, a structural scheme for which it is easier to conduct a dynamic analysis (Figure 19.2b) will appear. External forces that do not depend on deformations and act on the drive unit are designated as $f(t)$. External factors that do not depend on deformation and affect the depth of the cutting of the grinding wheel to the surface of the part are indicated by $y(t)$. One of such factors may be a change in the allowance due to the incorrect position of the workpiece relative to the grinding wheel. The same role is played by part profile errors after the previous pass.

In order to describe such variables in the function of time in general, it is advisable to apply dependence:

$$f(\tau) = \bar{m}_f + \Delta f \cdot \xi(\tau) \tag{19.14}$$

where $\bar{m}_f$ – the mathematical expectation of the corresponding magnitude; Δf – range of deviation of value from its average value; and $\xi(\tau)$ – centered random process of fluctuation of the corresponding magnitude.

The next step is to provide a description of the structural scheme in the form of differential equations, on the basis of which, using the Laplace transform in zero initial conditions, one can obtain the transfer functions of the system through the transfer functions of its blocks.

Often a case occurs when the drive is excited by the torque applied to the spindle, the variable part of which $M_z(t)$ is considered as a function of time t. If, as the system state variables, the angular deviations $\varphi_i(t)$ and $\varphi_j(t)$ of the reduced masses I'_1 and $I_{e\kappa}$ form the rotational angle of the spindle, which is rotational with constant velocity and denote c_1 and c_2 of the rigidity of the parts of the drive chain, then the system of linear differential equations of the drive will have the form:

$$I'_1 \ddot{\varphi}_i(t) + (h_1 + h_2)\dot{\varphi}_i(t) + (c_1 + c_2)\varphi_i(t) - h_2\dot{\varphi}_j(t) - c_2\varphi_j(t) = 0 \tag{19.15}$$

$$I_{e\kappa} \ddot{\varphi}_j(t) + h_2 \dot{\varphi}_j(t) + c_2 \varphi_j(t) - h_2 \dot{\varphi}_i(t) - c_2 \varphi_i(t) = 0 \tag{19.16}$$

where c_2 – reduced stiffness of the mechanical part of the drive is given; $c_1 = 1/e_1$, where e_1 – flexibility of the electromagnetic field of the electric motor; and h_1 – coefficient of damping of the electric motor; h_2 – coefficient of damping of the mechanical part of the drive.

The greatest damping of torsional vibrations in the mechanical part of the drive provides a belt drive.

When using the Laplace operator $s = d/dt$, we can go to the operator form:

$$\left[I'_1 s^2 + (h_1 + h_2)s + (c_1 + c_2) \right] \varphi_i(s) - (h_2 s - c_2) \varphi_j = 0 \tag{19.17}$$

$$\left[I_{e\kappa} s^2 + h_2 s + c_2 \right] \varphi_j s - (h_2 s + c_2) \varphi_i(s) = M_z(s) \tag{19.18}$$

where $\varphi_i(s)$ and $\varphi_j(s)$ are the images $\varphi_i(t)$ and $\varphi_j(t)$, $M_z(s)$ is the image of $M_z(t)$.

When solving this system, the expressions $f(t)$ and $y(t)$, which characterize the angle of the twist of discs (masses) or the relative angle of the discs, are found. Then we can obtain the equation of torsional oscillations and, depending on the type of input functions $f(t)$ and $y(t)$, to analyze the transient, constant, or random processes in the system, as well as to study the stability of its motion.

For example, in the surface grinding machines, the processing process is interrupted and renewed twice in the double turn of the table with the workpiece when there is a reverse of the table. A similar phenomenon occurs when grinding a workpiece with a discontinuous surface. When interrupted grinding wheels are used, transient processes also occur (Del et al., 1975; Dragobetskii et al., 2015, Vorobyov et al., 2017).

Due to the dynamic imbalance of the electric motor rotor, there is a load that is incidental due to the action of many factors. Thus, irregularity rotation can be manifested through an imbalance of the rotor, the beating of its bearings, the variability of the gap between the rotor and the stator, as well as other causes. In Del et al. (1975), such inequalities can be approximated by the introduction of the moment from motor M_d, which is represented as a sinusoidal function, where the frequency is equal to the circular frequency ω_B of rotor rotation:

$$M_d(t) = M_d(0)\sin\omega_B \tag{19.19}$$

For technological reasons, even within one V-belt, its modulus of elasticity is not constant and can vary by 50%–60%. Therefore, when the pulley is rotated, the deformation of the V-belt periodically changes. If there are two or more V-belts in the transmission, this effect can be substantially reduced. In addition, in V-belt transmissions there are dynamic random loads that are due to transverse fluctuations of the belt. In turn, fluctuations in the V-belt, as mechanical systems with distributed parameters, lead to equivalent torsional oscillations of a spindle with a grinding wheel.

In Strutynskyi et al. (2011) and Mashkov et al. (2016), based on the consideration of transverse oscillations of the V-belt, an expression is given that allows one to determine the dependence of the equivalent turning angle on time:

$$\phi_0(t) = \frac{2\pi}{R_p l_p} C_\phi \sum_{k=1}^{N_k} \{C_{1k}\sin(\omega_{1k} + \alpha_{1k}) - C_{2k}\sin(\omega_{2k} + \alpha_{2k})\} \tag{19.20}$$

where R_p and l_p are the pulley radius and the length of the branch of the belt; C_ϕ, C_{1k}, C_{2k}, α_{1k}, α_{2k} – constant, which depends on the position of the shape and velocity of the belt; N_κ – number of forms of transverse fluctuations of belts, which are taken into account; and ω_{1k}, ω_{2k} – natural frequencies of the k-th form.

$$\omega_{1k} = \frac{\pi k}{l_p}\sqrt{\frac{F_{p1}}{m_p}\left(1 - \frac{m_p}{F_{p1}}v_p^2\right)} \tag{19.21}$$

$$\omega_{2k} = \frac{\pi k}{l_p}\sqrt{\frac{F_{p2}}{m_p}\left(1 - \frac{m_p}{F_{p2}}v_p^2\right)} \tag{19.22}$$

where F_{p1} and F_{p2} are the forces of tension of the branches of the belts in the leading and driven branch of transmission, respectively; m_p – the weight of belt distributed along the length; v_p – linear velocity of the belt.

When performing a grinding operation, the initial position and velocity of the belt are random variables. In connection with this, it can be assumed that the constant amplitudes C_ϕ, C_{1k}, C_{2k}, which determine the harmonics in expression (19.20), have random values. The same applies to the values of the harmonics α_{1k}, α_{2k}, which will also be random variables.

Assuming that the initial position and speed of the pass are random functions, the value of the correlation function can be found (Ogorodnikov et al., 2018; Strutynskyi et al., 2011; Vasilevskyi, 2013).

$$R_{\phi 0}(\tau) = \sum_{k=1}^{N_k} \frac{\bar{a}_{k1}^2}{2} \cos k\omega_{01}\tau - \sum_{k=1}^{N_k} \frac{\bar{a}_{k2}^2}{2} \cos k\omega_{02}\tau \qquad (19.23)$$

where $\bar{a}_{k1}$ and $\bar{a}_{k2}$ – mathematical expectations of the corresponding coefficients.

On the basis of expression (19.023) is the spectral density corresponding to it:

$$S_{\phi 0} = \frac{\pi}{2}\sum_{k=1}^{N_k} \bar{a}_{k1}^2 [\delta(\omega - k\omega_1) + k\omega_{01}] + \frac{\pi}{2}\sum_{k=1}^{N_k} \bar{a}_{k=1}^2 [\delta(\omega - k\omega_2) + k\omega_{02}] \qquad (19.24)$$

where δ is the Dirac's delta function.

An analysis of these dependencies has shown that the dynamic random action on the drive, due to fluctuations in the pass, can be represented by a random process with a line frequency spectrum along frequencies, multiplied by its own transverse vibration of the pass.

Reducing the intensity of transverse fluctuations in the belts, as well as the elimination of unwanted resonance modes of transmission, can be achieved with the help of specially designed damping devices. These are rollers that come into contact with the belts in a particular transmission area.

During the machining of the workpiece, the shape of the outer surface of the grinding wheel, in particular, its radius, may change randomly. In Strutynskyi et al. (2011), Vasilevskyi (2015), and Vasilevskyi et al. (2018), the dependence of the radius of a circle on its rotation in the form of deterministic and random components is presented. The correlation function and spectral density of the random component of the change in the radius of the grinding wheel are determined (Titov et al., 2017).

If the machine does not have an automatic balancing device, then at high speeds of rotation of the grinding wheel, even a small imbalance of the latter can create significant centrifugal forces that break the work of the machine and impair the quality of the surface and the correctness of its shape. The presence of the amplitude of waves on the surface of the part, which occurs when grinding "along the track," can also lead to the appearance of perturbations that affect the dynamic system of the primary motion drive (Kukharchuk, Bogachuk et al., 2017; Kukharchuk, Kazyv et al., 2017; Vedmitskyi et al., 2017; Wójcik et al., 2012).

19.4 RESULTS AND DISCUSSION

The rotational oscillations of the grinding wheel may have a random nature, but the lack of statistical characteristics (for example, the heterogeneity of the material being processed and random fluctuations of the allowance for machining) complicates the analysis. In fact, the surface coating of the polished materials contains inhomogeneities and defects of hereditary origin, having one or another degree of chance. Especially significant are stochasticity of microhomogeneities in precision magnetic-solid alloys of the ЮНДК (foreign analogues are called alnico alloys), cemented steel, and various kinds of coatings.

The absolute value of the angular velocity at a certain time τ is determined in the same way as the expression (19.14):

$$\omega(\tau) = \omega + \Delta\omega \cdot \xi(\tau) \tag{19.25}$$

where ω – the nominal value of the angular velocity of the grinding wheel, and $\Delta\omega$ is the range of angular velocity deviations.

For practical purposes, it is necessary to know the dependence of the dispersion of the angular velocity of the grinding wheel on the dispersion of the random component of the moment from the side of the electric motor. If for a zero mean, the component of the moment is characterized by a certain correlation function at zero mean values, then by using the Markov processes or statistical linearization methods, one can find the given dependence. An example of solving this problem was given in Gusev and Svetlitsky (1984), and it was concluded that the statistical linearization method yields acceptable results only for small perturbations. The range of deviation of the angular rotational speed of a grinding wheel can be determined based on priori information on the statistical characteristics of the random change of the angular velocity of the spindle, which is obtained directly from the machine.

Some issues of the dynamics of the primary motion of grinding machines require a solution or a more detailed study. For example, this applies to machine tools for super-high-speed processing. It is also worth noting that in certain models of grinding machines, multiple V-belts are already being used. Studies of the drives of the main motion of grinding machines with the use of timing belts are being conducted, but their results are not yet known. There is a need to determine the effect of fluctuations of such belts on torsional oscillations of the grinding wheel. To reduce the intensity of torsional vibrations in the drives of heavy grinding machines, additional damping devices can be used.

19.5 CONCLUSIONS

The analysis of random factors that cause the appearance of torsional oscillations in the primary motion drives of grinding machines showed that in the study of dynamic characteristics, in addition to the deterministic, a probabilistic statistical approach is required. The presence of data on the parameters of a dynamic model and the determination of dominant perturbations will allow us to calculate their own frequencies, modal coefficients of damping, and forms of oscillation in an angle, as well as the calculation of the forms of oscillations by elastic moments.

Data on dominant factors of a random nature in the primary motion drive of grinding machines can be used to develop a grinding control system and a rational control algorithm that takes into account oscillatory processes in the dynamic system of machines.

Knowing the frequency composition of the load moment calculated for the recommended grinding modes for this machine, you can also design a primary motion drive so that none of its own frequencies falls within the specified range of loading frequencies.

REFERENCES

Azarova, N. V., Osovsky E. V., Sidorov, V. A., & Tsokur, V. P. 2013. Vibrational processes during grinding hard-working materials. *Progressive Technologies and Mechanical Engineering Systems DonNTU* 3(54): 3–12.

Del, G. D. Ogorodnikov, V. A. Nakhaichuk, V. G. 1975. Criterion of deformability of pressure shaped metals. *Izv Vyssh Uchebn Zaved Mashinostr* 4: 135–140.

Dragobetskii, V., Shapoval, A., Mos'pan, D., Trotsko, O., & Lotous, V. 2015. Excavator bucket teeth strengthening using a plastic explosive deformation. *Metallurgical and Mining Industry* 4: 363–368.

Gusev, A. S., & Svetlitsky, V. A. 1984. *Calculation of Structures at Random Influences*. Moscow: Mechanical Engineering.

Kukharchuk, V., Bogachuk, V., Hraniak, V., Wójcik, W., Suleimenov, B., & Karnakova, G. 2017. Method of magneto-elastic control of mechanic rigidity in assemblies of hydropower units. *Proc. SPIE* 10445: 104456A.

Kukharchuk, V. V., Kazyv, S. S., & Bykovsky, S. A. 2017. Discrete wavelet transformation in spectral analysis of vibration processes at hydropower units. *Przeglad Elektrotechniczny* 93(5): 65–68.

Levin, A.I. 1978. *Mathematical Modeling in Researches and Designing of Machine Tools*. Moscow: Mechanical Engineering.

Mashkov, V., Smolarz, A., & Lytvynenko, V. 2016. Development issues in algorithms for system level self-diagnosis. *Informatyka, Automatyka, Pomiary w Gospodarce i Ochronie Środowiska* 1: 26–28.

Nikitin, S. P. 2010. Influence of oscillations of the dynamic system of the machine on accuracy and temperature at grinding. Vestnik Perm State Technical University. Mechanical Engineering. *Material Science* 12(3): 31–47.

Ogorodnikov, V. A., Dereven'ko, I. A., & Sivak, R. I. 2018. On the influence of curvature of the trajectories of deformation of a volume of the material by pressing on its plasticity under the conditions of complex loading. *Materials Science* 54(3): 326–332.

Ogorodnikov, V. A., Grechanyuk, N. S., & Gubanov, A. V. (2018). Energy criterion of the reliability of structural elements in vehicles. *Materials Science* 53(5), 645–650.

Ogorodnikov, V. A., Savchinskij, I. G., & Nakhajchuk, O. V. 2004. Stressed-strained state during forming the internal slot section by mandrel reduction. *Tyazheloe Mashinostroenie* (12): 31–33.

Popp, K. M., Kroger, M., Deichmueller, M., & Denkena, B. 2008. Analysis of the machine structure and dynamic response of a tool grinding machine. In *Proc. Of the 1th Int. Conf. on Process Machine Interaction*: 299–307.

Strutynskyi, V. B. & Fedorynenko, D. Y. 2011. *Statistical dynamics of spindle units on hydrostatic supports*. Nizhyn TOV Vydavnytstvo Aspekt – Poligraf.

Titov, A. V., Mykhalevych, V. M., Popiel, P., & Mussabekov, K. 2017. Statement and solution of new problems of deformability theory. *Proc. SPIE* 10808: 108085E.

Vasilevskyi, O. M. 2013. Advanced mathematical model of measuring the starting torque motors. *Technical Electrodynamics* 6: 76–81.

Vasilevskyi, O. M. 2015. A frequency method for dynamic uncertainty evaluation of measurement during modes of dynamic operation. *International Journal of Metrology and Quality Engineering* 6(2): 12–16.

Vasilevskyi, O., Kulakov, P., Kompanets, D. Lysenko, O. M., Prysyazhnyuk, V., Wójcik, W., & Baitussupov, D. 2018. A new approach to assessing the dynamic uncertainty of measuring devices. *Proc. SPIE* 10808: 108082E.

Vedmitskyi, Y. G., Kukharchuk, V. V., & Hraniak, V. F. 2017. New non-system physical quantities for vibration monitoring of transient processes at hydropower facilities, integral vibratory accelerations. *Przeglad Elektrotechniczny* 93(3): 69–72.

Vorobyov, V., Pomazan, M., Vorobyova, L., & Shlyk, S. 2017. Simulation of dynamic fracture of the borehole bottom taking into consideration stress concentrator. *Eastern-European Journal of Enterprise Technologies* 3/1(87): 53–62.

Weidong, M. 2018. *Development of the Method of Selecting the Rational Parameters of the Grinding Process on the Basis of Taking Into Account the Dynamic Characteristics of the Deformable Technological System. Diss. Candidate of Those.* Moscow: Sciences.

Wójcik, W., Gromaszek, K., Kotyra, A., & Ławicki, T. 2012. Pulverized coal combustion boiler efficient control. *Przegląd Elektrotechniczny* 88(11b): 316–319.

Chapter 20

Dynamic characteristics of "tool-workpiece" elastic system in the low stiffness parts milling process

Y. Danylchenko, A. Petryshyn, S. Repinskyi, V. Bandura,
M. Kalimoldayev, K. Gromaszek, and B. Imanbek

CONTENTS

20.1 Introduction 225
20.2 Specifics of tool and workpiece interaction consideration 226
20.3 Dynamic model of tool–workpiece system considering their contact
      interaction 227
    20.3.1 Tool–workpiece system analytical model 227
    20.3.2 Finite-element model of the tool–workpiece system 229
20.4 Results of the dynamic characteristics modeling for the
      tool–workpiece system 230
20.5 Experimental validation of modeling of tool–workpiece system's
      dynamic characteristics 233
20.6 Conclusion 235
References 235

20.1 INTRODUCTION

Vibrations are one of the few factors that hold back the growth of performance, reduce machined surface quality, and cause a decrease in tool stability during machining. Vibrations that occur due to machining are considered the result of forced vibrations or self-oscillations in the elastic technological machining system (TMS). These vibrations can cause stability loss of the machining process, and thus, it is necessary to consider their possible negative influence on the exploitation and TMS development stages.

Among the self-oscillation origins in machining, the main cause of machining instabilities is considered to be machining by chatter marks with regenerative chatter occurrence. Stability prediction of machining is performed by applying lobed stability charts, which are plotted using dynamic transfer function $G(\omega)$ of elastic TMS (Stephenson & Agapiou, 2016).

It should be noted that the "classical" method of lobed stability chart plotting is based on the representation of the elastic TMS dynamic model in a form of single mass system with one or more DOFs (two or three) (Altintas et al., 2012; Petrakov et al., 2017). Herewith, taking into account relation of vibrations and assumed values of mass, damping and specifically stiffness have a significant influence on lobed stability charts plotting (Zhang et al., 2012).

In the works, dedicated to stability prediction of high-speed milling, noted that in dynamic model it is necessary to take into account chuck, mounted in spindle or tool holder with a tool (Badrawy, 2006; Gagnol, 2007; Özşahin, 2011; Wang et al., 2012; Shuvatov et al., 2012). Herewith, a number of characteristics are being defined, which have influence on natural frequencies values of the spindle–tool holder system, which also means they has influence on dynamic transfer function $G(\omega)$:

- Stiffness of spindle–tool holder interface (Badrawy, 2006; Gagnol, 2007; Wang et al., 2012; Özşahin et al., 2011);
- Spindle bearings stiffness and its variation due to spindle (Wang et al., 2012; Jiang et al., 2010);
- Spindle, tool holder, and tool geometrical characteristics (Badrawy, 2006; Gagnol et al., 2007).

One of the ways to ensure stability of machining in real time is to use the frequency response function (FRF) tool for machining conditions control (Özşahin et al., 2011; Schmitz et al., 2001; Wang et al., 2013). For obtaining FRF, it is proposed that the Risk and Control Self-Assessment (RCSA) method be used to combine the experimental FRF of the mounted in-spindle unit tool holder with an analytical model of the tool as a body with distributed mass. Thus, the FRF on the tool end can be found for any combination of the tool holder and a tool without the necessity to repeat measurements.

Using this approach, an analytical dynamic model of the system can be simplified to a dynamic model of the tool with a known mounting condition in the tool holder (cartridge). This way, no elastic, mass, or inertia properties of the workpiece will be taken into account.

20.2 SPECIFICS OF TOOL AND WORKPIECE INTERACTION CONSIDERATION

Usually, in the dynamic models of the spindle–tool holder–tool–workpiece system, the machining process is taken into account by cutting force, applied in the point of tool and workpiece contact (Abele et al., 2010; Komada et al., 2019; Lin et al., 2013). Herewith, in determination of the relative elastic displacements of the tool and workpiece, only the elastic properties of the spindle–tool holder–tool system are considered; the elastic properties of the workpiece are ignored. This approach is also commonly used for machining stability prediction using the RCSA method.

Thus, the mechanically closed elastic system spindle–tool holder–tool–workpiece in the analytical model is presented in a form of two mechanically disconnected subsystems. Herein, the workpiece subsystem is considered to be absolutely rigid.

As a result, in stability prediction of machining with lobed stability charts usage, only the natural frequencies of the spindle–tool holder–tool subsystem are taken into account, but not of a whole mechanically closed system.

However, it is well known (Kudinov, 1967) that the action of the cutting force's normal component in the shear plane is similar to the action of the elastic interaction force of the tool and workpiece in contact. This allows one to consider this component as "elastic" and to introduce a "cutting stiffness" definition. Magnitude of this stiffness, k_p, is defined by the ratio of the cutting force's normal component to the thickness of cut variation and a change in its value relevantly to machining process laws. In fact, k_p can be determined as a partial derivative of the $P_r = P_r(H_{real})$ function:

$$k_p = \partial P_r / \partial H_{real} \tag{20.1}$$

where H_{real} – actual depth of cut.

Introduction of a "cutting stiffness" definition allows for consideration of the machining process in the analytical model of the elastic system by an additional (preloaded) elastic connection at the point of the tool and workpiece contact. Thereby, elastic properties of both subsystems can be taken into account while the relative elastic displacements of the tool and workpiece are being determined. This way one can take into account the closed-loop system when determining the dynamic characteristics of a complete elastic system (Titov et al., 2018).

In Titov et al.'s article, the results of developing a tool–workpiece closed system generalized dynamic model take into account the contact interaction in the process of machining. A simulation was performed for cantilever plate milling by end mill. Stability of milling is defined by the elastic properties of both subsystems.

20.3 DYNAMIC MODEL OF TOOL–WORKPIECE SYSTEM CONSIDERING THEIR CONTACT INTERACTION

In spindle–tool holder–tool dynamic models, the tool is considered to be an elastic uniform beam, which can be calculated using the Euler–Bernoulli beam theory (Wang et al., 2013). This approach allows one to model using modal analysis (Schmitz, 2001; Wang et al., 2013), the finite element method (FEM) (Schmitz et al., 2007), or the transfer matrices method (TMM) (Demec, 2014; Hatter, 1973; Danylchenko & Petryshyn, 2017; Starostenkov et al. 2012). This approach can be applied to the workpiece description.

20.3.1 Tool–workpiece system analytical model

A generalized analytical model of the tool–workpiece system is presented in Figure 20.1.

In Figure 20.1, the first subsystem (tool, index $s = 1$) is presented in the form of a uniform beam that consists of three sections. On the leading end of a tool (cross section $0^{(1)}$), an elastic connection with stiffness k_p, is placed, which considers the machining process. In cross sections $2^{(1)}$ and $3^{(1)}$, the tool is connected with a tool holder that is taken into account by the elastic connections with radial stiffness k_{m1}, k_{m2} and damping ratios of h_{m1}, h_{m2}. In cross section $0^{(1)}$, a dynamic component of the cutting force $P(\omega)$ has been applied.

The second subsystem (workpiece, index $s = 2$) is presented in the form of a uniform beam that consists of the three sections. In the cross section $1^{(2)}$, the workpiece has contact with a tool. In the model, the cantilever fixture of the workpiece is taken into account by two elastic connections with radial stiffness k_{b1}, k_{b2} and damping ratios h_{b1}, h_{b2}.

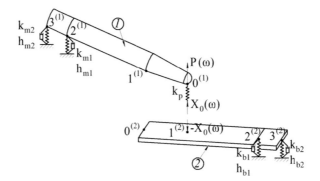

Figure 20.1 Generalized analytical model of tool–workpiece system: 1 – tool subsystem; 2 – workpiece subsystem.

According to the disconnection pattern of the tool (1) and workpiece (2) subsystems, equilibrium equations in the point of disconnection $1^{(2)}$ (Figure 20.1) will be:

$$\alpha_{00}^{(1)} \cdot P(\omega) + \left(\alpha_{00}^{(1)} + 1/k_p\right) \cdot X_0(\omega) = -\alpha_{11}^{(2)} \cdot X_0(\omega) \qquad (20.1)$$

$$\left(\alpha_{00}^{(1)} + \alpha_{11}^{(2)} + 1/k_p\right) \cdot X_0(\omega) = -\alpha_{00}^{(1)} \cdot P(\omega) \qquad (20.2)$$

where $\alpha_{ij}^{(s)}$ – terms of the receptance matrix (or influence coefficient (Hatter, 1973) or the FRF (Schmitz et al. 2001)) s – subsystem index, and the $X_0(\omega)$ – reaction of the removed connection.

Natural frequencies of the system are determined from (20.2) by:

$$\alpha_{00}^{(1)} + \alpha_{11}^{(2)} + 1/k_p = 0 \qquad (20.3)$$

Tool displacement in the point $0^{(1)}$ can be determined from equation:

$$q_0^{(1)} = \alpha_{00}^{(1)} \cdot P(\omega) + \alpha_{00}^{(1)} \cdot X_0(\omega) \qquad (20.4)$$

Workpiece displacement in point $1^{(2)}$ can be determined from equation:

$$q_1^{(2)} = -\alpha_{11}^{(2)} \cdot X_0(\omega) \qquad (20.5)$$

FRFs of the subsystems are obtained from equations:

$$G_0^{(1)}(\omega) = q_0^{(1)} / P(\omega) \qquad (20.6)$$

$$G_1^{(2)}(\omega) = q_1^{(2)} / P(\omega) \qquad (20.7)$$

FRF of the elastic system is equal to:

$$G_{el.sys}(\omega) = G_0^{(1)}(\omega) + G_1^{(2)}(\omega) \qquad (20.8)$$

20.3.2 Finite-element model of the tool–workpiece system

The milling of the cantilever plate with 120×60×4 (L×W×T) by double-tooth end mill was studied. The tool had a diameter of 16 mm, a length of 125 mm, and an overhang of 80 mm. Analytical FEM model of the system in ANSYS is shown in Figure 20.2.

The dynamic characteristics modeling of the "tool-workpiece" system in ANSYS is performed in two stages. In the first stage, the natural frequencies of the elastic system, considering the tool–workpiece contact interaction is taken into account k_p, along with the natural frequencies of the workpiece without taking into account k_p. In the second stage, the influence coefficients $\alpha_{ij}^{(s)}$ of the workpiece are determined, i.e., FRFs.

Determination of natural frequencies is performed by "modal" analysis. Cantilever fastening of the workpiece was implemented by adding "elastic support," with specifying "foundation stiffness," which is evenly distributed on the support facing (Imaoka, 2012).

The value of this stiffness is determined experimentally, $k_b = 8.1 \times 10^{11}$ N/m^3 (Figure 20.2, position A). Tool fastening is implemented in the same way (Figure 20.2, position B), and its stiffness is $k_m = 1.61 \times 10^{10}$ N/m^3. The tool and workpiece contact is taken into account in the section "connections-contact region," contact-type "frictional." "Cutting stiffness" k_p (Figurer 20.2, position C) is added in the "connections => contacts" section. The type of connection between the tool and workpiece is set to "spring," specifying and "longitudinal stiffness," which is $k_p = 4.5 \times 10^5$ N/m. k_p value is determined by (equation 20.1). The next step is to create mesh and modes calculations for the different tool and workpiece's mutual bracing.

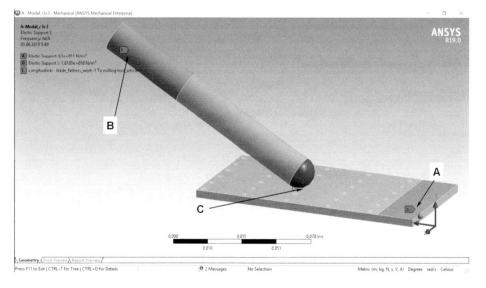

Figure 20.2 FEM model of the tool–workpiece system in ANSYS.

Determination of a workpiece influence coefficients $\alpha_{ij}^{(s)}$ is performed by a "harmonic response" study. Fastenings of the tool and workpiece, as well as adding "cutting stiffness" k_p are implemented the same way as in the "modal" analysis. Influence coefficients $\alpha_{ij}^{(s)}$ are determined as the FRFs for specific tool–workpiece contact points as a response to a uniform harmonic load. In order to do this, the calculation resolution and frequency range are specified in the "analysis settings" section. There, values are set by "solution intervals," "range minimum," and "range maximum" parameters, respectively. To determine the influence coefficient $\alpha_{ij}^{(s)}$, for a specific tool–workpiece contact point, a harmonic "force" is applied. In the "solution" section, "frequency response-deformation" results for a specific point were added, which is necessary to obtain the FRF. The resulting FRFs, obtained by the FEM modeling, can be exported into text or *.xls files.

20.4 RESULTS OF THE DYNAMIC CHARACTERISTICS MODELING FOR THE TOOL–WORKPIECE SYSTEM

The modeling of dynamic characteristics was performed for three points F, G, and H placed on the workpiece (Figure 20.3). In these points, natural frequencies and the FRFs for the "tool-workpiece system" (with k_p), and for the workpiece separately (without k_p), have been calculated. The results of the modeling are shown in Table 20.1 and in Figure 20.4).

Results of the modeling indicate that system natural frequencies, taking into consideration tool–workpiece contact interaction, differ from those for the workpiece separately; this is also displayed in the FRF's differences (Figure 20.5). In addition, the same differences can be seen in the dynamic characteristics of the system for different tool and workpiece contact points.

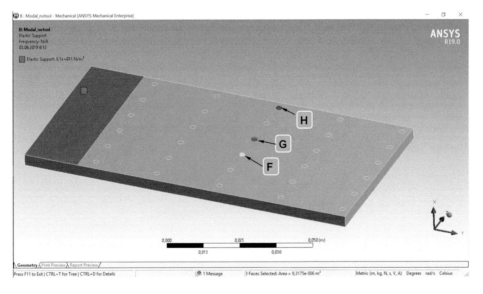

Figure 20.3 Design model of tool–workpiece system in ANSYS.

Table 20.1 System natural frequencies

No. of natural frequency	Without k_p (Point F)	With k_p	
		Point H	Point G
1	290	300	300
2	1,118	583	557
3	1,801	1,122	1,182

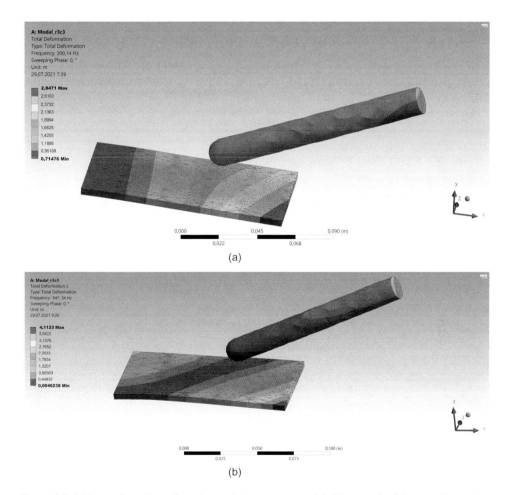

(a)

(b)

Figure 20.4 Natural modes of tool–workpiece system: (a) first mode (b) second mode.

Based on the calculated FRF and the tool analytical model, the dynamic transfer function $G(\omega)$ of elastic TMS was determined using equations (20.7 and 20.8). Following the method presented in the research by Altintas (2012), lobed stability charts were obtained using dependencies:

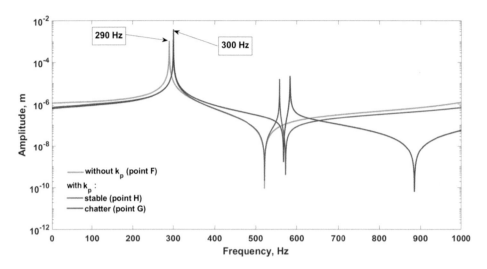

Figure 20.5 FRF of the tool–workpiece system (points G and H) and workpiece separately (point F).

$$a_{\lim} = -\frac{2\pi \Lambda_R}{NK_t}\left(1+k^2\right),\ n = \frac{60}{NT}$$

where $k = \dfrac{\Lambda_I}{\Lambda_R}$, Λ_R, Λ_I – real and imaginary parts of $G_{el.sys}(\omega)$, N – number of tool's teeth, K_t – cutting constant, which depends on the material of a tool and a workpiece, $K_t = 1{,}500$ MPa, n – spindle speed, T – tool tooth-passing period.

Lobed stability charts are shown in Figures 20.6 and 20.7.

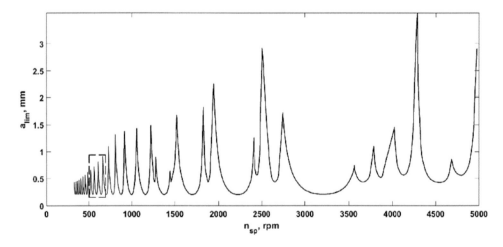

Figure 20.6 Overall lobed stability chart for plate end milling in point H.

"Tool-workpiece" elastic system 233

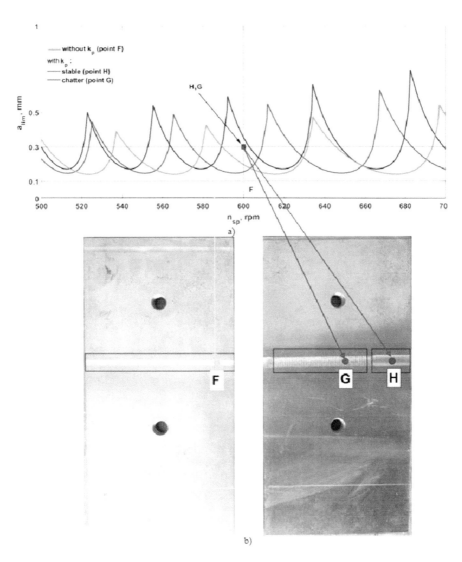

Figure 20.7 Plate end milling: (a) lobed stability charts for end milling: F – milling with a depth of cut of $t=0.1$ mm; H – milling with a depth cut of $t=0.3$ mm (stable); G – milling with a depth of cut of $t=0.3$ mm (chatter); and (b) machined workpieces.

20.5 EXPERIMENTAL VALIDATION OF MODELING OF TOOL–WORKPIECE SYSTEM'S DYNAMIC CHARACTERISTICS

The experiments were performed for plates with parameters: 120/60/4 mm (L/W/T), material – structural carbon steel Fe37B1FN; tool – ø16 mm end mill, length $L=125$ mm, overhang $l=80$ mm, number of teeth $N=2$; and carbide cutting tip RC16, material PC210F.

Cutting conditions were: spindle speed $n_{sp} = 600$ rpm; tooth loading: $s_t = 34$ mm/min. Depth of cut $t_1 = 0.1$ mm (Figure 20.7b left), $t_2 = 0.3$ mm. (Figure 20.7b right).

Vibration measurement equipment: piezo-accelerometer PCB 353B15 (Figure 20.8), amplifier PCB 480E09 and AD converter NI USB-9215 (Figure 20.9).

Experimental spectrums of plate vibrations for impact excitation (1) and machining (2) are shown in Figure 20.10.

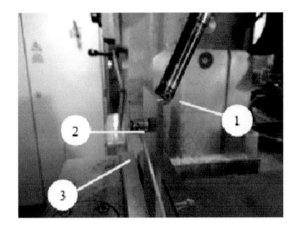

Figure 20.8 Experimental study on natural frequencies in low-stiffness parts end milling: 1 – end milling tool; 2 – workpiece; 3 – piezo-accelerometer.

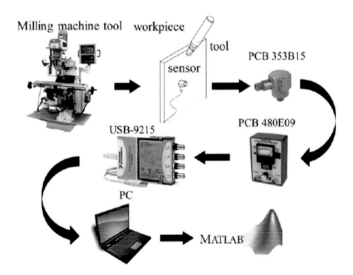

Figure 20.9 Scheme and equipment for workpiece vibration measurement.

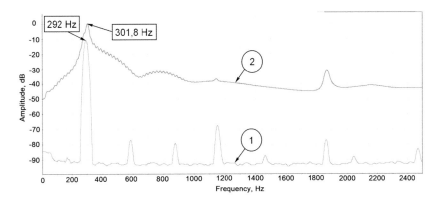

Figure 20.10 Experimental spectrums of: 1 – workpiece natural frequencies; 2 – tool–workpiece system in end milling.

20.6 CONCLUSION

The stability of the cantilever plate end milling is defined by the elastic properties of both subsystems. Experimental results revealed differences in the workpiece natural frequencies values during machining and without it. The cause was the physical contact of two bodies: the workpiece and the tool. Therefore, contact interaction of the tool and workpiece during machining must be taken into consideration when determining FRF and when plotting lobed stability charts.

FRFs were determined for the whole system, which consisted of interrelated tool and workpiece subsystems. For the low-stiffness workpieces of a geometrically complex parts, such as plates and blades, it is proposed that FRF be determined using ANSYS software, and for the tool (end mill), by the use of analytical methods. In the model, contact interaction of the tool and the workpiece was taken into consideration by introducing an additional (preloaded) elastic connection with stiffness k_p. The value of this stiffness was defined by the cutting conditions.

The adequacy of the developed procedure for FRF determination and lobed stability charts plotting was confirmed by an experiment. For this purpose, modeling of the dynamic characteristics of the "tool-workpiece" system was performed, as well as the lobed stability charts plotting for different tool and workpiece points of contact. Zones of stable machining and chatter zones were also determined. These results were validated through performing an appropriate experiment.

Further, this study is of value, because the developed model can be used to calculate relative displacements of a tool and a workpiece and to construct an accurate theoretical profile of a workpiece.

REFERENCES

Abele, E., Altintas, Y. Brecher, C. 2010. Machine tool spindle units. *CIRP Annals: Manufacturing Technology* 59(2): 781–802.

Altintas, Y. 2012. *Manufacturing automation: metal cutting mechanics, machine tool vibrations, and CNC design.*

Badrawy, S. 2006. Dynamic modeling and analysis of motorized milling spindles for optimizing the spindle cutting performance. *Moore Nanotechnology Systems*: 1–18, https://nanotechsys.com/wp-content/uploads/2019/11/DynamicModelingandAnalysis.pdf.

Danylchenko, M., Petryshyn, A. 2017. Study of contact interaction influence of workpiece and tool on lathe dynamic characteristics. *Visnyk NTUU 'KPI', Mechanical Engineering section* 77(2): 140–146.

Demec, P. 2014. Simplified dynamic analysis of grinders spindle node. *Technological Engineering* XI(1): 11–15.

Gagnol, V., Bouzgarrou, B.C., Ray, P., & Barra, C. 2007. Dynamic analyses and design optimization of high-speed spindle-bearing system. *Advances in Integrated Design and Manufacturing in Mechanical Engineering II*: 505–518.

Hatter, D.J. 1973. *Matrix Computer Methods of Vibration Analysis*. London: Butterworth. http://ansys.net/ansys/tips/EFS.pdf.

Imaoka, S. 2012. *Elastic Foundation Stiffness*. ANSY Technical Support Group. http://ansys.net/ansys/tips/EFS.pdf.

Jiang, S., Zheng, S. 2010. Dynamic design of a high-speed motorized spindle-bearing. *Journal of Mechanical Design* 132(03): 1–5.

Komada, P., Trunova, I. Miroshnyk, O. Savchenko, O. Shchur, T. 2019. The incentive scheme for maintaining or improving power supply quality. *Przegląd Elektrotechniczny* 95(5): 79–82.

Kudinov, V.A. 1967. *Dynamics of Machine-Tools*. Moskow: Mashinostroenije.

Lin, C.-W. Lin, Y.-K. Chu, C.H. 2013. Dynamic models and design of spindle-bearing systems of machine tools. *International Journal of Precision Engineering and Manufacturing* 14(3): 513–521.

Özşahin, O., Budak, E., Özgüven, H.N. 2011. Investigating dynamics of machine tool spindles under operational conditions. *Advanced Materials Research Online* 223: 610–621.

Petrakov, Y., Danylchenko, M. Petryshyn, A. 2017. Programming spindle speed variation in turning. *Eastern European Journal of Enterprise Technologies* 2(1): 86.

Schmitz, T.L. Davies, M.A. Kennedy, M.D. 2001. Tool Point Frequency Response Prediction for High-Speed Machining by RCSA. *ASME J. Manuf. Sci. Eng.* 123: 700–707.

Schmitz, T.L., Powell, K., Won, D., Duncan, G.S., Sawyer, W.G., Ziegert, J.C. 2007. Shrink fit tool holder connection stiffness/damping modeling for frequency response prediction in milling. *International Journal of Machine Tools & Manufacture* 47: 1368–1380.

Shuvatov, T., Suleimenov, B. Komada, P. 2012. Gas turbine fault diagnostic system based on fuzzy logic. *Informatyka, Automatyka, Pomiary w Gospodarce i Ochronie Środowiska* 3: 40–42.

Starostenkov, M., Demina, I. Popova, G. Denisova, N. Smolarz, A. 2012. Application of the molecular dynamics method for modelling of mass transfer on the border of Ni-Al bimetal. *Informatyka, Automatyka, Pomiary w Gospodarce i Ochronie Środowiska* 1: 36–38.

Stephenson, D.A. Agapiou J.S. 2016. *Metal cutting theory and practice*. CRC press.

Titov, A.V., Mykhalevych, V.M., Popiel, P., Mussabekov, K. 2018. Statement and solution of new problems of deformability theory. *Proc. SPIE* 108085E: 1611–1617.

Wang, E., Wu, B., Hu, Y., Yang, S., Cheng, Y. 2013. Dynamic parameter identification of tool-spindle interface based on RCSA and particle swarm optimization. *Shock and Vibration* 20(1): 69–78.

Wang, J., Wu, B., Hu, Y.M., Wang, E.H., Cheng, Y. 2012. Modeling and modal analysis of tool holder-spindle assembly on CNC milling machine using FEA. *Applied Mechanics and Materials* 157–158: 220–226.

Zhang, X.J., Xiong, C.H., Ding, Y., Feng, M.J., Xiong, Y.L. 2012. Milling stability analysis with simultaneously considering the structural mode coupling effect and regenerative effect. *International Journal of Machine Tools & Manufacture* 53: 127–140.

Chapter 21

Modeling of contact interaction of microroughnesses of treated surfaces during finishing anti-friction non-abrasive treatment FANT

I. Shepelenko, Y. Nemyrovskyi, Y. Tsekhanov, E. Posviatenko, Z. Omiotek, M. Kozhamberdiyeva, and A. Shortanbayeva

CONTENTS

21.1 The urgency of the problem 237
21.2 Analysis of publications devoted to this issue 238
21.3 Methodology of experimental studies 239
21.4 Results of the theoretical studies 241
21.5 Experimental studies 242
21.6 Results and conclusion 243
References 245

21.1 THE URGENCY OF THE PROBLEM

A promising direction in the development of mechanical engineering is the improvement of the quality of working surfaces of parts, the main indicators of which are the physicomechanical properties and the geometric characteristics of the surface layer, which are purposefully formed during the finishing operations of the technological process. An important role in the formation of physicomechanical properties is provided by the intermediate environment through which the interaction of microroughnesses occurs. Therefore, one of the ways to improve the quality of parts during the process of their manufacture and repair is the modification of their working surface, which can be achieved, for example, by creating and using anti-friction coatings.

Among the simplest, most effective, and environmentally friendly methods for producing such coatings are a group of technologies for finishing antifriction non-abrasive processing (FANT); implemented due to the frictional interaction of the machining tool with the surface of the workpiece. The result of the use of such technologies is the coating of copper and its alloys on steel and cast iron surfaces in order to improve burn-in and increase wear resistance due to the subsequent self-modification of surfaces under friction conditions during operation (Bystrov, 2011). In this case, a coating of plastic antifriction metal is formed between the contacting surfaces, creating a "third body" with a depth-positive gradient of mechanical properties (Kragel'skij et al., 1977).

The formation of AN antifriction coating FANT largely depends on the initial surface roughness, and therefore it is possible to improve the quality of the coating by

DOI: 10.1201/9781003225447-21

creating a favorable geometry of microroughness in the previous FANT operations. In this regard, studies of the influence of the shape and size of microroughness on the formation of an anti-friction coating appear to be very relevant.

21.2 ANALYSIS OF PUBLICATIONS DEVOTED TO THIS ISSUE

The existing hypotheses explaining the mechanism for the formation of an antifriction coating during FANT have been described in a number of papers (Andreeva, 1990; Garkunov, 2009; Pogonyshev & Panov, 2011; Shepelenko et al., 2013; Shepelenko et al., 2019). Despite the differences in approaches in interpreting the process of forming coatings by a friction-mechanical method, the authors of these studies agree on the need to create the following prerequisites for the implementation of the FANT process. These prerequisites include fulfillment of the conditions for micro-cutting, plastic contact, setting criterion with optimal coating parameters, and implementing the four mechanisms of activation of contact surfaces: mechanical, chemical, thermal, and ducts associated with plastic deformation.

In Shepelenko et al. (2019), it was noted that obtaining a high-quality coating on a surface that has a coarse regular micro relief is complicated by the peculiarities of filling the inter-rib hollows with an antifriction material. When the coating film is applied by a friction-mechanical method, voids appear between the individual particles of the antifriction product, which negatively affects the quality of the coating, namely the continuity and density of the antifriction film.

Based on the microscopic analysis of particles contained in glycerin after coating with FANT, the authors Pogonyshev & Panov (2011) note that when a high-quality film is formed in the contact zone "friction rod – surface to be treated," a micro-cutting process must take place. This can be achieved by fulfilling the relation (Pogonyshev & Panov, 2011):

$$\frac{h}{R} \geq \frac{1}{2}\left(1 - \frac{2\tau_n}{\sigma_s}\right) \qquad (21.1)$$

where "h" is the depth of penetration of a single unevenness of the rubbed surface into the copper-containing rod, "R" is the radius of rounding of a single microroughness, "τ_n" is the strength of adhesive bond of copper-containing bar with a rubbed surface, "σ_s" is a yield strength of the coating material.

Obviously, the creation of micro-cutting conditions requires the provision of the necessary initial surface roughness. However, the works of Andreeva (1990) and Garkunov (2009) contain rather contradictory information on the influence of the shape and size of microroughness on the formation of antifriction coating FANT.

Thus, the studies of Andreeva (1990) indicated that the implementation of this process is possible with a roughness range within $Ra = 0.16...1.25$ microns. It is also stated that with $Ra > 1.25$ microns, the diffusion process of rubbing in brass is replaced by its micro-cutting the surface microroughness, and when $Ra < 0.16$ microns, the rubbed surface is too smooth, the process is unstable, and the resulting coating is not continuous.

In studies by Garkunov (2009), it was found that to ensure a high quality of coating and high performance, the parameter Ra must have a value not lower than 1.25 μ, which allows one to obtain microroughness of the acute form, and the resulting wear products form a continuous coating. At lower values of Ra (less than 0.3 µ), penetration of tool material into cavities is complicated by the fact that the process of microcutting of a tool with a part is replaced by surface plastic deformation (SPD), resulting in tool wear rate drops, a gradual peening of the tool material to a considerable depth, and a complication in the separation of individual particles. The coating, in this case, is not continuous, which indicates its low quality.

It should be noted that nearly all researchers of the influence of the initial roughness of the treated surface on the formation of an antifriction coating use only one altitude roughness parameter – R_a (Andreeva, 1990; Pogonyshev et al., 2011; Shepelenko et al., 2019). However, there is no consensus about the optimal value of the specified parameter. In the opinion of the authors, the question of the influence of the initial roughness of a pretreated surface on the formation of antifriction coating FANT has not been studied sufficiently, and the use of only one parameter Ra does not provide a complete picture of the surface micro relief. Based on this, we can conclude that it is necessary to conduct special studies on the influence of the shape and size of microroughness on the formation of the anti-friction coating obtained by FANT. Such studies are possible, for example, by simulating the contact interaction of microroughnesses in FANT (Sedov, 1987; Tsehanov, 2015).

The *aim of this work* is to study the influence of the shape and size of microroughness on the formation of anti-friction coating FANT by conducting a simulation of the contact interaction of microroughnesses.

21.3 METHODOLOGY OF EXPERIMENTAL STUDIES

The method of the theory of similarity and dimensions was used to study the micro-cutting process (Sedov, 1987, Tsehanov, 2015, Starov, et al., 2006).

Experimental studies of the contact interaction of surfaces were carried out on special samples of gray cast iron СЧ 20 and brass Л63 according to the original method. A brass sample in the form of a plate with dimensions of $80 \times 30 \times 3$ was mounted on the working table of a milling machine, and a replaceable cast iron cutter was installed in a special device with an indicator head; the geometry of the cutting part modeled a separate microroughness of the workpiece surface (Figure 21.1). The variation of the geometric parameters of the cutting part of the cutter (front cutting angle $\gamma = +5° \div -15°$, cutting depth $h = 0.1 \div 0.5$ mm, and a cutting width of $B = 3$ mm) and, above all, the front cutting angle, due to its importance in the process of mass transfer – from micro-cutting to SPD. The load on the sample was provided by the mechanism of the vertical feed of the machine table and was controlled by the indicator head (Kozlov et al., 2019; Ogrodnikov et al., 2018).

Thus, moving under a fixed load, the cast-iron sample cuts a layer of antifriction metal (brass), thereby simulating the process of applying a brass coating during FANT at the stage of micro-cutting as shown in Figure 21.2. Two components of the force of contact interaction were measured: P_z и P_n (Figure 21.3).

Cutting depths were chosen to be small and comparable to the blunt radius r, in order to maximize compliance with the actual conditions of interaction of the micro protrusion with the anti-friction material. The efficiency of the formation of a microchip

Figure 21.1 General view of gray cast iron cutters СЧ 20 when modeling contact interaction with FANT with the value of the front cutting angle γ: (a) +5°; (b) 0°; (c) −5°; (d) −10°; (e) −15°.

Figure 21.2 Equipment working area during the simulation of the process of FANT.

was estimated by its volume V_1, cut from a unit area of $1\,mm^2$. The radius of blunting r was determined experimentally, and its initial values for all cutters were $r_0 = 0.02\,mm$ (Dragobetskii et al., 2015; Gauda & Pasierbiewicz, 2019; Ogrodnikov et al., 2018; Ogrodnikov et al., 2004).

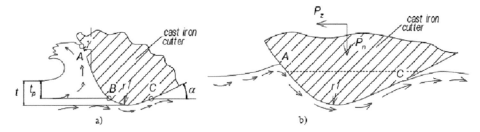

Figure 21.3 The scheme of interaction of a cast-iron tool with a brass sample during the simulation of the FANT process: (a) is a micro-cutting scheme; (b) is a micro-smoothing scheme; t is the depth of micro-cutting; r is the blunt radius of the cutter; and Pz and Pn are the components of the force of contact interaction.

21.4 RESULTS OF THE THEORETICAL STUDIES

The process of contact interaction of the surface of the workpiece with the rubbed material can be divided into two stages: (1) micro-cutting of the starting material with the tops of the micro protrusions; (2) adhesive sticking and seizure of the particles formed as a result of micro-cutting with the surface onto which the transfer and subsequent micro-smoothing take place. Since these two processes cannot be separated by precise time frames, they exist in close contact in the narrow temporal, geometric, mechanical limits of the process parameters, therefore, when studying the FANT process, we conditionally divide it into two stages. Figure 21.3 shows the contact pattern of the micro protrusion with the material.

The normal component of the force Pn, which determines the technological force of pressing the bar to the workpiece, depends on the type of material being rubbed, the depth of micro-cutting t, the width of the chip B, the geometry of the micro protrusion (the cutting angle γ and the blunting radius r); the friction conditions are determined by the friction coefficient f. The mechanical properties of the hardened material are well defined by a generalizing parameter such as hardness HV. In studies by Gauda and Pasierbiewicz (2019), HV was used to analyze the force parameters during deforming broaching, in addition to a study by Tsehanov (2015) during cutting.

Functional dependence is written as:

$$P_n = P_n(HV, t, B, r, \gamma, \alpha) \tag{21.2}$$

In accordance with the theory of dimensions (Sedov, 1987; Titov et al., 2018), we choose the main (determining) parameters: t (m), HV (N/m^2). Note that they are decisive not in terms of the degree of their influence on the cutting process, but in terms of the independence of their dimensions, since the dimension of one of them cannot be expressed in terms of the dimension of the other. The remaining parameters are expressed through these two main dimensions. The linear dimensions are expressed in units of t, and the angles and the coefficient of friction f are already dimensionless. Then, according to (Sedov, 1987):

$$P_n = HV^a \cdot t^b \cdot \overline{P_n}\left[1, 1, \frac{B}{t}, \frac{r}{t}, \gamma, \alpha, f\right] \tag{21.3}$$

Here $\overline{P_n}$ is the dimensionless (relative) force that must be determined from a model experiment.

For the dimensions of the left and right parts (21.3) to be the same, choose the exponents: $a = 1$ and $b = 2$:

$$P_n = HV \cdot t^2 \cdot \overline{P_n} \tag{21.4}$$

From the principles of super position, it is obvious that $\overline{P_n}$ should depend linearly from cut: width $\dfrac{B}{t}$

Then (4) takes the form:

$$P_n = HV \cdot t \cdot B \cdot \overline{P_n}\left[\gamma, \alpha, \frac{r}{t}, f\right] \tag{21.5}$$

In the model experiment, the angle $\alpha = 5°$ was constant as was the friction coefficient f. The blunt radius r was determined experimentally and, as shown in equation (21.6), it depends only on the front cutting angle γ. The width $B = 3$ mm was constant in all cases.

Dimensionless effort is written as:

$$\overline{P_n}\left[\frac{r}{t}, \gamma\right] = \frac{P_n}{HV \cdot t \cdot B} \tag{21.6}$$

Based on dependence (21.6), a formula has been obtained for the technological calculation of the volume of chips being removed during the interaction of a brass bar with a rubbed surface with a regular micro protrusion structure:

$$V_1 = \frac{P_{n\Sigma} \cdot S \cdot B \cdot b_1}{HV \cdot \overline{P_n}(\gamma)} \tag{21.7}$$

where $P_{n\Sigma}$ is the brass bar pressing force; S is the distance between micro protrusions on the surface of the workpiece; B is the bar width; and b_1 is the empirical coefficient taking into account filling leakage.

21.5 EXPERIMENTAL STUDIES

When conducting the experimental studies, it was noted that at the contact area AB (Figure 21.3a), there is practically no sticking of brass to the cast iron, and active sticking is observed in the area of BC, due to high contact pressures. For the mode (Figure 21.3b), only intense diffusion adhesion is characteristic due to high pressures over the entire contact micro surface due to the PSD.

It has been established that, in all cases, the cutting edge of the cast-iron cutter immediately blunted to the steady-state value r, which depends only on the front angle γ. This dependence is presented in Figure 21.4. Figure 21.5 shows the dependence of the force Pn on the process parameters. As can be seen, in all cases, the dependence of Pn on the depth of cut is linear.

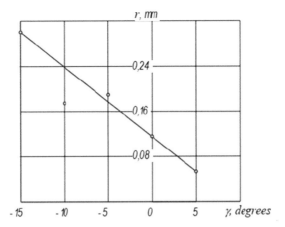

Figure 21.4 The dependence of the blunt radius of the cutting edge of the cutter r from the front cutting angle of the cutter when modeling with a cutter of cast iron сч 20 when treating brass л63.

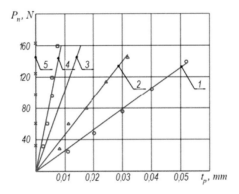

Figure 21.5 The dependence of the force Pn on the actual thickness of the slice tp in the simulation of cutting with a cutter of cast iron СЧ20 of a brass sample Л63 at the angle of the cutter: 1 – 5°; 2 – 0°; 3 – minus 5°; 4 – minus 10°.

From formula (21.6), it follows that the dimensionless force for each angle γ is a constant value. The hardness of the material of the bar, Л63 brass is, $H = 690 \text{ N/m}^2$. The dependence of Pn on γ, defined by (21.6), is shown in Figure 21.6. This dependence can be approximated by the analytical dependence $P_n(\gamma)$. The values of the volume of chips removed from 1 mm^2 of the bar surface are presented in Figure 21.7.

21.6 RESULTS AND CONCLUSION

The results of the analysis obtained from the experimental studies suggest that to effectively fill micro cavings among micro protrusions, it is necessary to create a regular micro relief with $\gamma \geq 0° \ \gamma \geq 0^0$.

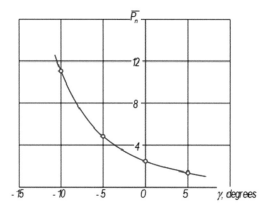

Figure 21.6 The dependence of the dimensionless force on the angle when modeling cutting with a cutter of cast iron СЧ20а sample of brass Л63.

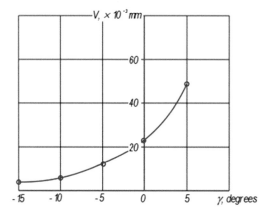

Figure 21.7 The dependence of the amount of chips removed from a unit of area V from the angle when modeling cutting with a cutter of cast iron СЧ20а brass sample Л63 at $tn = 0.4$ mm.

Note that due to the intensive blunting of the tops of the world projections of the surface of a cast iron part, the process of their interaction with the rubbed material consists of micro-cutting and the PSD of the blunted apex. The latter leads to a large strain hardening of the surface layer of the brass sample and can have a significant impact on the process of micro-cutting. This phenomenon requires additional study.

The findings also confirm the role of the micro-cutting process in the formation of a quality coating FANT. Technological recommendations have been developed for intensifying the FANT process at the first stage of applying the coating, which includes micro-cutting by filling micro-cavities. It has been established that to fill with chips micro cavities between micro protrusions, it is necessary to create a regular micro relief with a value of the front cutting angle $\gamma \geq 0°$, and the most effective micro-cutting process is carried out.

Based on the method of the theory of similarity and dimensions, model dependences were obtained for the dimensionless effort of interaction between the microroughness of the surface to be machined and the tool, as well as for the technological calculation of the volume of chips being removed when the brass bar interacts with the surface to be rubbed.

REFERENCES

Andreeva, A. G. (1990). Finishnaja antifriktsionnaja bezabrazivnaja obrabotka kak sredstvo povyshenij asrokasluzhbymashi i oborudovanija. *Dolgovechnost' truschihsja detalej mashin: Sborniknauch. statej*. Moscow: Mashinostroenie, 34–59.

Bystrov, V. N. (2011). Primenenie ustrojstv dlja friktsionno-mehanicheskogo nanesenija iznosostojkih pokrytij v uslovijah remontnogo proizvodstva. *Izobretatel'stvo, 11*(3), 29–34.

Dragobetskii, V. V., Shapoval, A. A., Mospan, D. V., Trotsko, O. V., & Lotous, V. V. (2015). Excavator bucket teeth strengthening using a plastic explosive deformation. *Metallurgical and Mining Industry, 7*(4), 363–368.

Garkunov, D. N. (2009). Finishnaja antifriktsionnaja bezabrazivnaja obrabotka (FABO) poverhnostej trenija detalej. *Remont. Vosstanovlenie. Modernizatsija, 6*, 38–42.

Gauda, K., & Pasierbiewicz, K. (2019). Zastosowanie profilometrii optycznej w analizie procesu destrukcji renowacyjnych powłok organicznych dla przemysłu motoryzacyjnego. *Informatyka, Automatyka, Pomiary w Gospodarce i Ochronie Środowiska, 9*(4), 22–25.

Kozlov, L. G., Polishchuk, L. K., Piontkevych, O. V., Korinenko, M. P., Horbatiuk, R. M., Komada, P., Orazalieva, S., & Ussatova, O. (2019). Experimental research characteristics of counter balance valve for hydraulic drive control system of mobile machine. *Przegląd Elektrotechniczny, 95*(4), 104–109.

Kragel'skij, I. V., Dobychin, M. N., & Kombalov, V. S. (1977). *Osnovy raschetovna trenie i iznos*. Moscow: Mashinostroenie.

Ogorodnikov, V. A., Dereven'ko, I. A., Sivak, R. I., (2018). On the influence of curvature of the trajectories of deformation of a volume of the material by pressing on its plasticity under the conditions of complex loading. *Materials Science, 54*(3), 326–332.

Ogorodnikov, V. A., Savchinskij, I. G., & Nakhajchuk, O. V. (2004). Stressed-strained state during forming the internal slot section by mandrel reduction. *Tyazheloe Mashinostroenie, 12*, 31–33.

Ogorodnikov, V. A., Zyska, T., & Sundetov, S. (2018, October). The physical model of motor vehicle destruction under shock loading for analysis of road traffic accident. In *Photonics Applications in Astronomy, Communications, Industry, and High-Energy Physics Experiments 2018* (Vol. 10808, p. 108086C). International Society for Optics and Photonics.

Pogonyshev, V. A., Panov, M. V., (2011). Teoreticheskie i `eksperimental'nye osnovy povyshenija iznosostojkosti detalej mashin. *Mehanika i fizika protsessov napoverhnosti i v kontakte tverdyhtel, detalej tehnologicheskogo i `energeticheskogo oborudovanija, 4*, 78–84.

Sedov, L.I. 1987. *Metodypodobija i razmernosti v mehanike*. Moscow: Nauka.

Shepelenko, I. V., & Cherkun, V. V. (2013). Obrazovanie antifriktsionnogo pokrytija finishnoj antifriktsionnoj bezabrazivnoj vibratsionnoj obrabotkoj. *Vibratsiï v tehnitsitatehnologijah, 71*(3), 99–104.

Shepelenko, I. V., Posviatenko, E. K., & Cherkun, V. V. (2019). The mechanism of formation of anti-friction coatings by employing friction-mechanical method. *Problems of Tribology, 1*, 35–39.

Starov, V. N., Tsehanov, J. A., & Eremin, M. J. (2006). *Osnovy mehaniki mikrorezanija materialov*. Voronezh: Izd-vo VGTA.

Titov, A. V., Mykhalevych, V. M., Popiel, P., & Mussabekov, K. (2018, October). Statement and solution of new problems of deformability theory. In *Photonics Applications in Astronomy, Communications, Industry, and High-Energy Physics Experiments 2018* (Vol. 10808, p. 108085E). International Society for Optics and Photonics.

Tsehanov, Ju.A. 2015. Modelirovanie 'energosilovyh parametrov rezanija metodami teorii podobija i razmernostej. *Vestnik Voronezhskogo gosudarstvennogo tehnicheskogo universiteta* *11*(2): 30–33.

Chapter 22

Practices of modernization of metal-cutting machine tool CNC systems

V. Sychuk, O. Zabolotnyi, P. Harchuk, D. Somov, A. Slabkyi, Z. Omiotek, S. Rakhmetullina, and G. Yusupova

CONTENTS

22.1 Introduction 247
22.2 Theoretical part 248
22.3 Practical part 250
22.4 Results 253
22.5 Conclusions 254
References 255

22.1 INTRODUCTION

Increasing the level of competitiveness, productivity, quality, and accuracy of product manufacturing in the machine-building sector necessarily leads to the application of equipment with computer numerical control systems (Institution of Mechanical Engineers, 2012; Ito, 2008; Kochinev et al., 2006).

Currently, there is a large park of old metal-cutting equipment in Ukraine that does not meet modern requirements in the field of control. It is worth noting that the age of these machines is not equivalent to their operating capacity. For example, let's take the difficult period of the 1990s in the post-Soviet block, where a large number of metal-working enterprises did not work at all. In this case, the mechanical portion of these machine tools has not been in operation, but their electronics are severely outdated when considering the very rapid development of computer technology.

The current state of mechanical processing is based on the use of multipurpose machines, which provide the required precision, with high-speed cutting modes (OMV technology). The cost of such equipment ranges in the hundreds of thousands of dollars. At the same time, there are many CNC machines equipped with old control systems at Ukrainian enterprises (for example, Soviet 2R22, 2C42,... systems) (Danek et al., 2000; Damodar et al., 2013; Ertruk et al., 2006; Yuchshenko & Wójcik, 2014). The productivity of such machines does not meet modern requirements.

Such a state of affairs has led to works on the modernization of such "unworn" metal-cutting machines, which were carried out by the workers and students of the Lutsk NTU.

DOI: 10.1201/9781003225447-22

22.2 THEORETICAL PART

Today, the electronics market for CNC machines is extremely diverse, in terms of both price and layout. This means that it is possible to buy an already prepared CNC solution or build, solder, and manufacture the electronics for yourself. Obviously, the second variant is extremely attractive from both an economic and an educational and scientific perspective. In general, in its most simplified form, a CNC system of the metal-cutting machine can be represented as shown in Figure 22.1.

The system's operation is as follows:

- control signals from the computer are sent to the interface board, in which they are converted into signals suitable for processing in the engine control board;
- from the motor control board, signals are sent directly to the motors, which perform the necessary operations

A number of programs can be used as operating software for a CNC machine tool: EMC2, Mach3, TurboCNC, Universal G-code Sender, and others.

The great advantage of the presented CNC system is its very flexible and user-friendly operation. A simple personal computer can be used as its operating platform. When developing control programs for the processing of parts, it is possible to use modern CAM software (FeatureCAM, PowerMill, SolidCAM, MasterCAM, and others) (Cao et al., 2006; Damodar et al., 2013; Grachev et al., 1986; Hraniak et al., 2018).

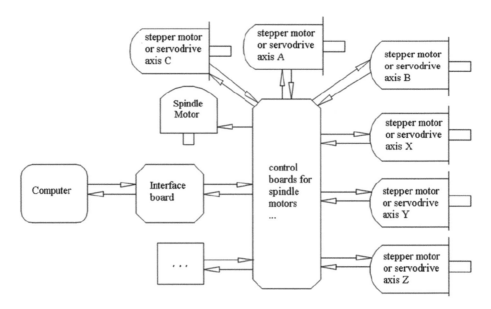

Figure 22.1 Simplified circuit diagram of a CNC metal-cutting machine.

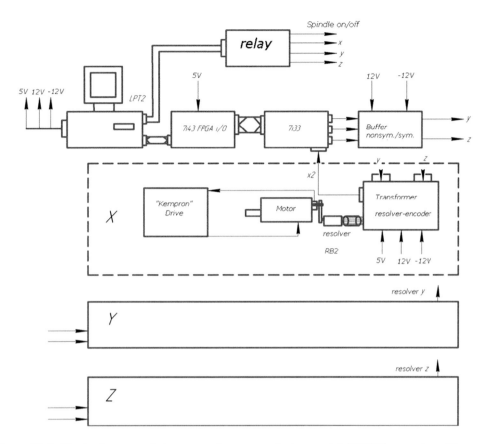

Figure 22.2 Block diagram of the control system of the model 6T13F3 console-milling machine.

In addition to the above system, another one was developed (see Figure 22.2), in which it is advisable to modernize the machines with direct current servo drives.

Using the above schemes in Lutsk, NTU successfully modernized and developed a number of metal-working equipment with CNC systems (Kyznetsov, 1991; Push, 1985):

- model 6T13F3 console-milling machine (axes X, Y, Z);
- additional turning axis A for milling machines;
- an additional two-axis turning table (axes A, C) for milling machines;
- model 6520F3 vertical milling machine (axes X, Y, Z);
- model 16A20F3 turning machine tool (axes X, Z, automatic turning of the revolving head, spindle control);
- model 16U03P turning machine tool (axes X, Z, the spindle rotation is controlled by a frequency generator).

It is worth noting that the mechanical part of the above machines is in excellent condition.

22.3 PRACTICAL PART

The method of implementing the system is shown in Figure 22.1. A four-channel stepper motor driver on the 4A DD8727T4V1 was chosen as the stepper motor driver's board (Figure 22.3).

MACH3 was chosen as the software managing the modernized 6520F3 CNC machine. As a result, it was necessary to apply a special interface board (Figure 22.4), which allows controlling the driver card from a personal computer on which the MACH3 software is installed (Dragobetskii et al., 2015; Ogrodnikov et al., 2004; Ogrodnikov et al., 2018).

All components were assembled according to the operating instructions of the aforementioned electronic circuit boards and motors, and as a result, a functional 6520F3 CNC milling machine was obtained (Figure 22.5).

The system's implementation method is presented in Figure 22.2. Additionally, the modernization of the model 6T13F3 console-milling machine was performed (Figure 22.6). The purpose of modernization was to replace the control system. Drives and the mechanical parts were in a functional condition (Kukharchuk et al., 2017; Vedmitskyi et al., 2018; Vorobyov et al., 2017).

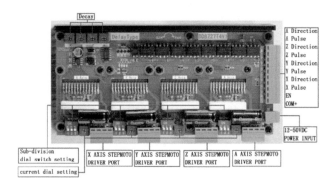

Figure 22.3 Four-channel stepper motor driver on the 4A DD8727T4V1.

Figure 22.4 MACH3 interface board.

Modernization of metal-cutting machine 251

Figure 22.5 Model 6520F3 milling machine (made in 1976) with a modernized CNC system.

Figure 22.6 Model 6T13F3 milling machine (made in 1986) with a modernized CNC system.

The chosen EMC2 CNC control system operates on a personal computer running the Linux operating system. EMC2 is an open system, which means the availability of source codes for studying and configuring the system.

The computer is connected via a parallel port to the 7i43-4 input/output card using a 24-loop plug. The computer's power supply generates voltages of 5 V, +12 V, −12 V to power all of the system's control components of the system.

The 7i43-4 I/O board contains the Xilinx X3S400 FPGA. The FPGA configuration activates when EMC2 is activated through a parallel port. To work with the board, the EMC2 program downloads the general hostmot2 driver and the hm2-7i43 driver. The path to the FPGA firmware file is transferred in the driver-enabled hm2-7i43 line. The hm2-7i43 driver loads the firmware via the parallel port in the ERP mode. Then, the driver communicates the EMC2 program with the FPGA. The SVST4_4B firmware implements the following components: 4 encoders, 4 PWM, 4 StepGen. This implementation allows simultaneous operation with four servo drives and four-step drives. The charge needed for its operation is a voltage of 5V (Kukharchuk et al. 2016, Kukharchuk et al. 2018, Vasilevskyi et al. 2018).

To the connector of the 7i43–4 board, a 7A33TA DAC board is connected using a 50-loop plug. It contains active filters on operational amplifiers, for converting a pulse-width signal into a −10 to +10 V controlling analogue signal. The board is powered via plugs from the I/O board.

The analogue control signal from the 7i33TA board reaches the buffer installed on the MS33072RG. The buffer cascade also transforms the asymmetric control signal into a symmetric one (Vasilevskyi et al. 2018).

Afterward, the control signal is sent via a twisted pair to the thyristor "Kemron" drive, located in the electrical automation cabinet. The drive controls the 4AEV6 DC motor with a feedback tachogenerator.

The motor cover contains the RB2 resolver. Connection with the shaft is carried out through the 2× multiplier gear (one revolution of the shaft equals two revolutions of the resolver). The resolver is connected to the converter block of the resolver signal to the encoder signal using a shielded cable. The unit sends a reference harmonic signal to the resolver and calculates the changes of the angle of rotation of the shaft on the *sin* and *cos* output signals. The output signal is the TTL encoder signal with 1,024 pulses per revolution. For its operation, the unit needs a power supply of 5, 12, −12 V. The encoder signal is sent to the 7i33TA board and through the loop on the 7i43–4 board (Figure 22.7).

Taking into account the multiplier, we get 2,048 imp/revolution. When moving 1cm/revolution, we get an error, about $5·10^{-3}$mm.

The rotary motion of the engine shaft, with the help of a screw-nut pair, turns into a linear movement of the table and the spindle head.

Another parallel port adapter is installed on the computer. It has a connected relay board, which operates the spindle activation and shutdown, and also engages and disengages motion drives at the startup of the program. The board requires a 12V power supply.

Figure 22.7 The developed milling machine CNC board.

Configuration of the PID controller is done manually, by direct sampling. The coefficients I and D are exposed to 0. The proportional coefficient P is increased until the cycle begins to fluctuate. Next, P is set equal to half of this value and increased and as long as the displacement will be corrected for an adequate time of the process. The coefficient D was left equal to 0 for all three coordinates. The selected coefficients provided an error of 20 μm.

22.4 RESULTS

To estimate the functionality of the modernized CNC of machine 6T13F3, control programs for two test pieces of different configurations were developed for their further processing.

In Figure 22.8, we can see a cylindrical workpiece at the end of which is a milled complex trajectory in the form of a "rat". The diameter of the cylinder is 30 mm, the milling depth is 1 mm. The material of the workpiece is aluminum. The tool used is a milling cutter with a diameter of 1 mm.

In Figure 22.9, we can see a complex spatial shape. The workpiece is rectangular with dimensions of 100 × 100 × 55 mm. The detail includes the following elements: parallel vertical and horizontal sides, convex hemisphere, hemisphere recess, half-cylinder, half-cylindrical aperture, cube, circular groove, rounding, a rectangle with sloped side, cylindrical blind hole. The material of the workpiece is aluminum. The tools used were: a milling cutter 10 mm in diameter, a final spherical milling tool with a diameter of 6 mm (replaced during processing to form the correct hemispherical recess).

Also, to check the functionality of machine 6520F3 with a modernized CNC system, a steel test piece with complicated trajectories of motion was designed and processed. The processing state is shown in Figure 22.10.

The FeatureCAM program was used to write control code for milling the above parts. In this program, all the necessary parameters were adjusted: tools, workpiece

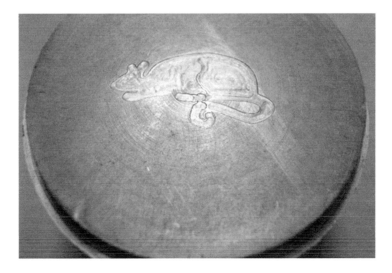

Figure 22.8 Cylindrical part at the end of which a hollow in the form of a "rat".

Figure 22.9 A part of a complex spatial form.

Figure 22.10 Processing of a test piece on the 6520F3 milling machine with a modernized CNC system.

material, milling processing strategies, configured postprocessor under the modernized CNC system of the milling machines. The generated control codes were saved in the CNC machine, on which the metal parts were manufactured.

22.5 CONCLUSIONS

This method of modernizing old CNC machines is effective and recommended for implementation. A great advantage in upgrading an old CNC system to a new one is the relatively low cost and the ability to use inexpensive computers as part of the machine control system. After such modernization, it becomes possible to use modern CAM software for manufacturing high-tech complex parts.

REFERENCES

Cao, Y. et al. 2006. Modeling of spindle-bearing and machine tool system for virtual simulation of milling operations. *International Journal of Machine Tools and Manufacture* 47(9): 1342–1350.

Damodar, A,. Kondayya, D. 2013. Static and dynamic analysis of spindle of a cnc machining centre. *International Journal of Mechanical Engineering (IJME)* 2(5): 165–170.

Danek, O. et al. 2000. *SelbsterregteSchwin- gungen an Werkzeugmaschinen*. Berlin: VerlagTechnik.

Dragobetskii V. et al. 2015. Excavator bucket teeth strengthening using a plastic explosive deformation. *Metallurgical and Mining Industry* 4:363–368.

Erturk, A. et al. 2006. Selection of design and operational parameters in spindle–holder–tool assemblies for maximum chatter stability by using a new analytical model. *International Journal of Machine Tools and Manufacture.*

Grachev, L. Kosovskij, V. Kovshov, A. 1986. *Design and Adjustment of Machine Tools with Software Control and Robotic Systems.* Moscow: Vysshajashkola.

Hraniak, V.F., Kukharchuk, V., Bogachuk, V.V., Vedmitskyi, Y.G., Vishtak, I.V., Popiel, P., Yerkeldessova, G. 2018. Phase noncontact method and procedure for measurement of axial displacement of electric machine's rotor. *Proc. SPIE* 1080866: 1825–1831.

Institution of Mechanical Engineers. 2012. 10th International Conference on Vibrations in Rotating Machinery 11–13 September 2012. Cambridge. Woodhead Publishing Limited.

Ito, Y. 2008. *Modular Design for Machine Tool.* New York. McGraw-Hill.

Kochinev, N.A., Sabirov, F.S. 2006. Quasi-static method of measuring the balance of elastic displacements of the supporting system of machine tools. *Measurement Techniques* 49(6): 572–578.

Kukharchuk, V.V. et al. 2017. Discrete wavelet transformation in spectral analysis of vibration processes at hydropower units. *Przeglad Elektrotechniczny* 93(5): 65–68.

Kyznetsov, Ju. 1991. *CNC Machines.* Kyiv: Vyshchashkola.

Ogorodnikov, V.A. et al. 2004. Stressed-strained state during forming the internal slot section by mandrel reduction. *Tyazheloe Mashinostroenie* 12: 31–33.

Ogorodnikov, V.A., Dereven'ko, I.A., Sivak, R.I. 2018. On the influence of curvature of the trajectories of deformation of a volume of the material by pressing on its plasticity under the conditions of complex loading. *Materials Science* 54(3): 326–332.

Push, V. 1985. *Metal-Cutting Machine Tool.* Moscow: Mashinostrojenije.

Vedmitskyi, Y.G. Kukharchuk, V.V. Hraniak, V.F. 2018. New non-system physical quantities for vibration monitoring of transient processes at hydropower facilities, integral vibratory accelerations. *Przeglad Elektrotechniczny* 93(3): 69–72.

Vorobyov, V. et al. 2017. Simulation of dynamic fracture of the borehole bottom taking into consideration stress concentrator. *Eastern-European Journal of Enterprise Technologies* 3/1(87): 53–62.

Yuchshenko, O. & Wójcik, W. 2014. Development of simulation model of strip pull self-regulation system in dynamic modes in a continuous hot galvanizing line. *Informatyka, Automatyka, Pomiary w Gospodarce i Ochronie Srodowiska* 1: 11–13.

Chapter 23

Improving the precision of the methods for vibration acceleration measurement using micromechanical capacitive accelerometers

V. F. Hraniak, V. V. Kukharchuk, Z. Omiotek, P. Droździel, O. Mamyrbaev, and B. Imanbek

CONTENTS

23.1 General instructions 257
23.2 Setting the task 258
23.3 Experimental investigation of vibration acceleration sensor and
      digital channel 259
23.4 Development of high-precision digital measuring channel of vibration
      acceleration 262
23.5 Conclusions 265
References 265

23.1 GENERAL INSTRUCTIONS

The rapid development of systems for control and diagnostics of high-power electric machines (including turbine- and hydro-generators) results from the increase in unit power of the latter and by the amount of equipment installed, as well as by wider opportunities for control using up-to-date measurement methods and computers (Belik, 2018a). In addition, the need for the improvement of methods and means of control and diagnostics rapidly grows due to the increase in the amount of equipment, the rated service life that has expired, while operation thereof continues. Notably, the share of such equipment among high-bower turbine- and hydro-generators has exceeded 50% in the majority of industrialized countries as of early twenty-first century (Alekseev, 2002).

Since vibration-based diagnostics is one of the most promising types of rotating electrical machines' technical state monitoring and diagnostics (Alekseev, 2002; Kukharchuk, 2015; Rao, 2007), while the overwhelming majority of existing vibration velocity and vibration displacement sensors cannot allow measuring low-frequency vibration signals (of just sporadic Hz) (Kukharchuk, 2014), the need emerges in development of brand-new approaches to the solution of this scientific-and-technical problem. Moreover, taking into account the existing trend for digitization of intermediate signals and standardization of data transmission channels between structural units of control and diagnostics systems (Belik, 2018b), which requires the introduction of intermediate programmable low-level units for digital preprocessing

DOI: 10.1201/9781003225447-23

of measured information, the possibility arises to apply analytical methods of these parameters' calculation based on the temporal implementation of vibration acceleration (Kukharchuk, 2015). This allows using, as the means for measurement of such systems' vibro-acoustic signal, exactly the measuring channels of vibration acceleration, the increase in precision of which is a crucial applied-scientific task of practical significance (Belik, 2018b; Kukharchuk, 2019; Hraniak, 2017).

23.2 SETTING THE TASK

Among known primary measuring converters of vibration acceleration, the converters based on micromechanical capacitive accelerometers (sensitive elements) have gained widespread use. The specificity that sets them apart from the sensors based on other operating principles (piezoelectric, mechanical, etc.) lies in the combination of relatively high sensitivity, the linearity of static characteristics, high overload capability, and low weight and dimensions (Kukharchuk, 2019). In such converters, under the action of linear acceleration, the inertial force

$$F = m \cdot a, \qquad (23.1)$$

is counterbalanced by spring pressure

$$F = k \cdot x, \qquad (23.2)$$

where m – weight, a – acceleration; x – weight displacement relative to the initial position.

By equating (23.1) and (23.2), we obtain

$$a = \frac{k}{m} x = S_a \cdot x \qquad (23.3)$$

where $S_a = k/m = const$ – sensitivity, the value of which depends on the sensor's structural parameters (k and m).

As it appears from (23.3), the signal at the output of vibration acceleration sensor that is based on micromechanical capacitive accelerometers will have the additive component associated with gravitational effect of the non-perpendicularity of the installation of an accelerometer on the site. And since it is extremely difficult to ensure strictly vertical position of the sensor during its on-site fastening, a significant error arises during operation of the measuring channels based on micromechanical capacitive sensors of vibration acceleration, where error impairs both the measurement channel's precision and the probability of control and diagnostics in general.

In view of the aforesaid, the objective of the chapter lies in development of a new method for deletion of additive error component during installation of capacitive micromechanical sensors of vibration acceleration, which allows the increase of the precision class of digital measurement means based thereon (Vasilevskyi, 2013; Vasilevskyi, 2015; Vasilevskyi et al., 2018).

23.3 EXPERIMENTAL INVESTIGATION OF VIBRATION ACCELERATION SENSOR AND DIGITAL CHANNEL

In order to carry out an experimental investigation of vibration acceleration sensor, which is based on capacitive micromechanical accelerometers, we used the vibration sensor based on commercially produced accelerometer ADXL322, which has two mutually perpendicular measuring axes and ensures the opportunity to measure vibration accelerations in two mutually perpendicular projections. A batch of sensors so structured currently undergoes pilot operation within vibration monitoring system of Lower Dniester HPP. The generalized structural diagram of proposed vibration acceleration sensor is shown in Figure 23.1.

The following abbreviations are used in Figure 23.1: SEA denotes the sensitive element, in which capacity accelerometer ADXL322 is used; CA-X, CA-Y denote conditioning amplifier units of X and Y measuring axes, respectively; TL – termination set.

In view of the necessity to carry out the entire range of vibration acceleration measurements, the magnification coefficients are chosen for conditioning amplifier units of measuring axes to ensure the sensor's sensitivity along measuring axes X and Y at the level of 0.08 V·s^2/m.

The essence of the experiment consisted in determination of real readings of the primary measuring converter of vibration acceleration at different turning angles of the latter in relation to the axis, which is perpendicular to measuring axes X and Y. In this case, the actual value of static acceleration for measuring axes will be theoretically defined as follows:

$$a_x = -9.81 \cdot \sin(\beta), \tag{23.4}$$

$$a_y = 9.81 \cdot \cos(\beta), \tag{23.5}$$

where a_x and a_y – static acceleration along measuring axes X and Y, respectively; β – the angle of the sensor's incremental turn.

Results of superimposition of theoretical and experimental dependence of error along measuring axes X and Y are shown in Figures 23.2 and 23.3.

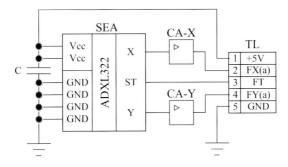

Figure 23.1 Generalized structural diagram of capacitive micromechanical sensor of vibration accelerations.

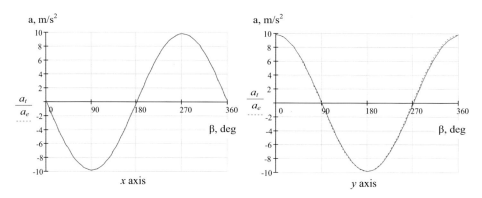

Figure 23.2 Theoretical (at) and empirical (ae) errors of capacitive micromechanical sensor of vibration accelerations associated with the influence of error during installation of vibration acceleration sensor.

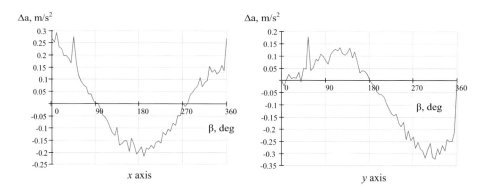

Figure 23.3 Curves of errors of capacitive micromechanical sensor of vibration accelerations.

As it is seen in Figure 23.2, a significant additive error component arises in case of deviation from installation perpendicularity. Such being the case, given the fact that the deviation of ±5 degrees of flat angle between the axis of primary measuring converter and guiding axis of coordinates system (Hraniak, 2017) is deemed normal deviation from perpendicularity of accelerometer installation in production conditions, this error component may reach 0.855 m/s², being maximal in the horizontal measuring axis (Azarov, 2011, Azarov et al., 2016).

For empirical and theoretical dependencies so obtained, we assessed the absolute error of the theoretical model:

$$\Delta a = a_t - a_e. \tag{23.6}$$

Similarly, we also experimentally established the dependence of an additive component in the absolute error of vibration acceleration's digital channel, which operates jointly with the primary measuring converter under investigation. The generalized structural diagram of vibration accelerations channel is shown in Figure 23.4.

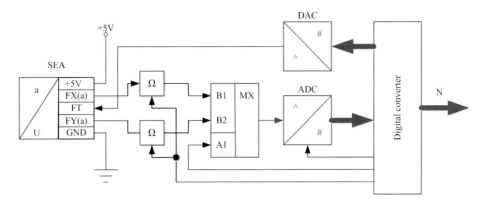

Figure 23.4 Generalized structural diagram of the digital channel of vibration accelerations.

The following abbreviations and legends are used in Figure 23.4: SEA denotes the capacitive micromechanical sensor of vibration accelerations; Ω denotes the analogue memory unit; MX denotes the analogue multiplexer; DAC denotes the digital-to-analogue converter; ADC denotes the analogue-to-digital converter.

The proposed digital channel, the pilot batch of which also undergoes commercial operation within vibration monitoring system of Lower Dniester HPP, has ten-digit ADC and quantizes vibration acceleration along measuring axes X and Y, also having a self-testing mode, which is ensured by supplying the analogue signal of +5 V voltage from the output of digital-to-analogue converter, resulting in a standard output signal of known voltage value being generated at the output of capacitive micromechanical sensor of vibration accelerations, the signal of which allows for self-testing of measuring channel in the process of its operation.

The results of experimental investigation of proposed digital measuring channel of vibrations are shown in Figure 23.5.

It follows from Figures 23.3 and 23.5 that introduction of theoretical adjustments

$$q_x = 9.81 \cdot \sin(\beta), \tag{23.7}$$

$$q_y = -9.81 \cdot \cos(\beta), \tag{23.8}$$

does not make it possible entirely to delete the additive component of error during installation of primary measuring converter. As is seen from the experimental investigations completed, even after introduction thereof the highest value of additive error component will continue to manifest itself in the horizontal axis (this may be axis X or Y depending on the type and spatial orientation of the electrical machine under investigation, and spatial arrangement of capacitive micromechanical accelerometer will also depend on this type and orientation). Yet, within normal range of deviation between the axis of primary measuring converter and the guiding axis of coordinates system not exceeding ±5 degrees of flat angle upon introduction of adjustments (7) and (8), the precision class of measuring channel with absolute error $K_\Delta = \pm 0.2 \, \text{m/s}^2$ will be achieved (Hraniak, 2017).

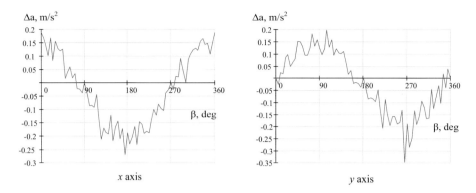

Figure 23.5 Curves of errors of digital channel of vibration accelerations.

Another weak point of using the method of errors calculated based on (23.7) and (23.8) lies in the technical complexity of high-precision determination of β angle at the measurement site, which represents another source of error's growth during the aforementioned measurement means and determines the necessity of search for other ways to delete this error component from the measurement result (Osadchuk et al., 2011a, 2011b, 2012).

23.4 DEVELOPMENT OF HIGH-PRECISION DIGITAL MEASURING CHANNEL OF VIBRATION ACCELERATION

Since the use of theoretical adjustment factors (23.7) and (23.8) does not ensure maximum improvement in precision of the methods of vibration acceleration measurement, we proposed a brand-new approach to deletion of sensor error's additive component that is based on the use of automatic self-calibration algorithm. The generalized structural diagram of intelligent self-calibrating measuring channel of vibration acceleration, which implements the said algorithm, is shown in Figure 23.6 (Osadchuk et al., 2015b).

The measuring channel that provides maximum possible number of measuring axes – three, the number of which may be adjusted depending on the technical requirement applied, is shown in Figure 23.6.

The device operates in the following way (Osadchuk et al., 2015a; Vedmitskyi et al., 2018). Measurement of signal levels at outputs of vibration acceleration sensors 1, in its essence being three-axis modification of primary measuring converter of vibration acceleration shown in Figure 23.1. From the first, second and third outputs of vibration acceleration sensors 1 to the first input of, respectively, the first 2, second 3, and third 4 conditioning amplifiers, the signals are accepted that correspond to the current level of vibration acceleration along the three coordinates axes (X, Y, Z). In the first 2, second 3, and third 4 conditioning amplifiers, the said signals are reduced to the level suitable for operation of analogue-to-digital converter 13 and supplied to the first inputs of the first 5, second 6, and third 7 analogue adder. In analogue adders 5–7, the signals from outputs of sensors of conditioning amplifiers 2–4 are complemented with adjustment

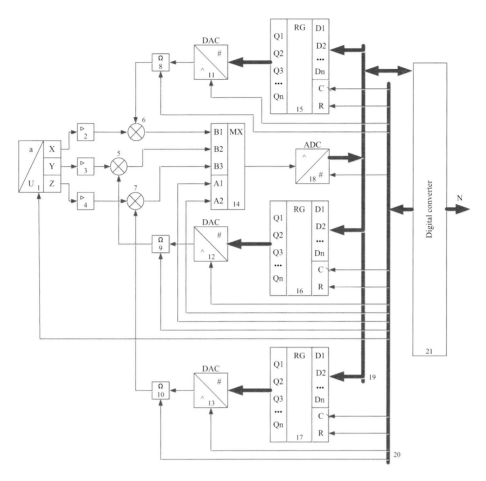

Figure 23.6 The general structural diagram of intelligent self-calibrating measuring channel of vibration acceleration.

signals accepted at the second inputs of analogue adders 5–7. From the outputs of analogue adders 5–7, the signals are supplied to, respectively, the first, second, and third informational inputs of analogue multiplexer 14. Depending on the value of digital signals supplied to the first and second address inputs of analogue multiplexer 14 from of control bus 20, the output of analogue multiplexer 14 accepts a signal from its first, second, or third informational input. From the output of analogue multiplexer 14, the signal arrives at the first input of analogue-to-digital converter 18, in which, upon arrival of triggering signal at its second input from control bus 20, the signal that arrives at its first input undergoes analogue-to-digital conversion. Upon completion of analogue-to-digital conversion, the signal of measuring conversion completion and obtained numerical code are supplied through the output of analogue-to-digital converter 18 to data bus 19, from where it is read by microcontroller 21 through its input-output. In microcontroller 21, obtained digital code is further processed and the current mode of measurement device is selected depending on software-defined algorithm. Guiding signals are supplied to control bus 20 through the first output of microcontroller 21, the signals of which regulate the operation of measurement devices (Yuchshenko & Wójcik, 2014).

The mode of compensation of the error is determined by the error during installation of vibration acceleration sensor 1. The mode of compensation of the installation error is implemented in the beginning of measurement device operation at zero value of vibration acceleration along all three coordinates axes (X, Y, Z). In this mode, levels of signals from the outputs of vibration acceleration sensor 1 are measured at zero signals at other inputs of analogue adders 5–7 according to the algorithm described above. Upon obtaining of the binary code, which is proportional to the signal at the first output of vibration acceleration sensor 1 (coordinate axis X), microcontroller 21 compares this binary code with the standardized value that corresponds to a half of the reference voltage of analogue-to-digital converter 18, and digital adjustment signal is generated. After that, one stepwise signal is supplied to the third input of register 15 through control bus 20, being in its essence the signal of the first register 15 zeroing. Further on, through the input/output of microcontroller 21, the digital adjustment code arrives at data bus 19, through which it is supplied to the first input of the first register 15. With a minimum delay after that, one stepwise signal is supplied through control bus 20 to the second input of the first register 15, this signal serving for it as the signal for memorizing of the digital adjustment signal. From the output of the first register 15, the recorded digital adjustment code constantly arrives at the first input of the first digital-to-analogue converter 11. Upon arrival of the triggering signal from control bus 2 at the second input of the first digital-to-analogue converter 11, the binary code that arrived at its first input undergoes analogue-to-digital conversion. The analogue signal obtained as a result of digital-to-analogue conversion by the first digital-to-analogue converter 11 arrives at the input of the first analogue memory unit, where, upon the signal to its second input from control bus 20, it is memorized and stored during a certain technically justified period, until the next digital-to-analogue conversion. The signal from the output of the first analogue memory unit 8 arrives at the second input of the first analogue adder 6. In such a manner, an adjusted signal is established at the output of the first analogue adder 6, the signal of which equals to a half of the reference voltage of analogue-to-digital converter 18 and contains no error associated with improper installation of vibration acceleration sensor 1.

In a similar way, using the second register 16, the second digital-to-analogue converter 12, and the second analogue memory unit 9, the bias error associated with improper installation of vibration acceleration sensor 1 is deleted from the signal at the second output of vibration acceleration sensor 1 (coordinate axis Y), and using the third register 17, the third digital-to-analogue converter 13, and the third analogue memory unit 10, the error associated with improper installation of vibration acceleration sensor 1 is deleted from the signal at the third output of vibration acceleration sensor 1 (coordinate axis Z)

Measurement mode. In this mode, instantaneous values of vibration acceleration are actually measured. This mode provides for measurement of analogue values of signals proportional to the instantaneous values of vibration acceleration along coordinate axes X, Y, Z, which arrive from the outputs of vibration acceleration sensors 1 according to the algorithm described above. Upon obtaining the binary code by microcontroller 21 according to known transformation equations, it calculates the current value of vibration acceleration. Obtained value of vibration acceleration is extracted through the second output of microcontroller 21. Upon completion of the procedure for extraction of the obtained value of vibration acceleration, vibration acceleration measurement for the next coordinate axis is launched in the current coordinates axis. Upon completion of vibration acceleration measurement in all three coordinate axes, the measurement procedure is repeated in a cyclic manner.

Self-testing mode. In this mode, the guiding signal arrives at the input of vibration acceleration sensor 1 from control bus 20, and after supply of this signal, the voltage of a priori known amplitude is established at all outputs of vibration acceleration sensor 1. After that, signals are measured at each output of vibration acceleration sensor 1 according to the algorithm described above, and the measurement result is compared with a priori known voltage value. Should these values disagree, the decision is made about system failure with a respective signal sent through the second output of microcontroller 21. Should a measured value agree with a priori known voltage value, the decision is made about the measurement device's suitability for further operation.

23.5 CONCLUSIONS

Obtained were the mathematical models of additive error components for non-perpendicularity of capacitive micromechanical accelerometer installation. It was shown that, within a normal range of deviation between the axis of the primary measuring converter and the guiding axis of coordinates system, the deviation of which should not exceed ±5 degrees of flat angle in case of correct installation, the value of its error will be maximum in the horizontal measuring axis and may reach $0.855\,m/s^2$.

Investigated was the possibility to use calculated values of errors to improve the precision of the vibration acceleration measurement methods based on capacitive micromechanical accelerometers. It was experimentally proved that, when using the calculated values of errors, an unaccounted error component remains, with its maximum value in the horizontal measuring axis to reach the values of $\pm 0.2\,m/s^2$.

Proposed was the operation algorithm and structural diagram of the intelligent self-calibrating measuring vibration channel that ensures an entire automatic deletion of the additive error component from vibration measurement results and provides for an opportunity of self-testing in the process of operation.

REFERENCES

Alekseev, B. A. 2002. *Determination of the State (Diagnostics) of Large Hydrogenerators.* Moscow: ENAS.
Azarov, O. D. 2011. *Push Pull Direct Current Amplifiers for Multidigital Self-Calibrating Information form Converters.* Vinnitsa: VNTU.
Azarov, O. D., Teplytskyi, M. Y., Bilichenko, N.O. 2016. High-speed push pull direct current amplifiers with balancing feedback. Vinnitsa: VNTU.

Belik, M. 2018a. Detection and prediction of photovoltaic panels malfunctions. *Renewable Energy and Power Quality Journal* 16: 544–548.

Belik, M. 2018b. Usage of data acquisition device NI PCI-6221 for power engineering applications. *Proceedings of the 2018 19th International Scientific Conference on Electric Power Engineering*, pp. 410–414.

Hraniak, V. F. 2017. Correlation approach to determination of weight coefficients of artificial neural network for vibration diagnostics of hydro aggregates. *Bulletin of the Engineering Academy of Ukraine* 4: 100–105.

Kukharchuk, V. V. 2014. *Monitoring, Diagnostics and Forecasting of Hydropower Units' Vibration Condition*. Vinnytsia: VNTU.

Kukharchuk, V. V. 2015. Method of analytical calculation of vibration velocity in hydropower unit acceleration mode. *Bulletin of the Engineering Academy of Ukraine* 2: 66–70.

Kukharchuk, V. V. 2019. *Measurement of the Parameters of the Rotational Motion of Electromechanical Energy Converters in Transient Operating Modes*. Vinnytsia: VNTU.

Osadchuk, A. V., Osadchuk, I. A. 2015a. Frequency transducer of the pressure on the basis of reactive properties of transistor structure with negative resistance. *Proceedings of the 2015 International Siberian Conference on Control and Communications (SIBCON)*, Omsk, 21–23 May 2015.

Osadchuk, A. V., Osadchuk, V. S. 2015b. Radio measuring microelectronic transducers of physical quantities. *Proceedings of the 2015 International Siberian Conference on Control and Communications (SIBCON)*, Omsk, 21–23 May 2015.

Osadchuk, V.S., Osadchuk, A. V. 2011a. The microelectronic radiomeasuring transducers of magnetic field with a frequency output. *Electronics and Electrical Engineering* 4(110): 67–70.

Osadchuk, V. S., Osadchuk, A. V. 2011b. The magnetic reactive effect in transistors for construction transducers of magnetic field. *Electronics and Electrical Engineering* 3(109): 119–122.

Osadchuk, V. S., Osadchuk, A. V. 2012. The microelectronic transducers of pressure with the frequency. *Electronics and Electrical Engineering* 5(121): 105–108.

Rao, S. S. 2007. *Vibration of Continuous Systems*. New York: John Wiley & Sons.

Vasilevskyi, O. M. 2013. Advanced mathematical model of measuring the starting torque motors. *Technical Electrodynamics* 6: 76–81.

Vasilevskyi, O. M. 2015. A frequency method for dynamic uncertainty evaluation of measurement during modes of dynamic operation. *International Journal of Metrology and Quality Engineering* 6 (2). doi:10.1051/ijmqe/2015008.

Vasilevskyi, O., Kulakov, P., Kompanets, D., Lysenko, O. M., Prysyazhnyuk, V., Wójcik, W., Baitussupov, D. 2018. A new approach to assessing the dynamic uncertainty of measuring devices. *Proc. SPIE* 10808. doi:10.1117/12.2501578.

Vedmitskyi, Y. G. Kukharchuk, V. V. Hraniak, V. F. 2018. New non-system physical quantities for vibration monitoring of transient processes at hydropower facilities, integral vibratory accelerations. *Przeglad Elektrotechniczny* 93(3): 69–72.

Yuchshenko, O., Wójcik, W. 2014. Development of simulation model of strip pull self regulation system in dynamic modes in a continuous hot galvanizing line. *Informatyka, Automatyka, Pomiary w Gospodarce i Ochronie Srodowiska* 1: 11–13.

Chapter 24

Modeling of the technological objects movement in metal processing on machine tools

G. S. Tymchyk, V. I. Skytsiouk, T. R. Klotchko, P. Komada, S. Smailova, and A. Kozbakova

CONTENTS

24.1 Introduction 267
24.2 Model of the object movement trajectory in working space 268
24.3 Conclusions 275
References 276

24.1 INTRODUCTION

Nowadays, there are many technological processes related to metal working. We can state that the fundamental basis of metal working is either the destruction of surplus mass or its addition to the formation of the necessary product (detail) without treating the processes related to metal working.

As a consequence, at first the research does not reveal the parameters of the cutting tool and the material's detail but the mutual movement of the cutting tool and the workpiece (Armarego & Brown 1969; Davies 2017; Kopp et al. 2016) as objects of technological process (Tymchyk et al. 2018).

If we investigate precisely from this point of view, the metal processing on the machine can be regarded as a series of steady field structures. On the one hand, it is a vector field of a tool movement and from another scalar field of a detail. It is quite natural that the observance of a properly calculated trajectory of movement of technological objects in the working space of a machining machine guarantees high quality and accuracy of details manufacturing. Therefore, modeling and calculating features of the movement of objects are necessary when designing the technological process of mechanical processing of the material (Lefeber 2012).

However, the problem of modeling the trajectory of the movement of the object to improve the accuracy of the manufacture of details is still relevant, despite existing attempts (Anton & Anton 2017; Smaoui et al. 2012; Rao & Rao2004).

Thus, the purpose of this research is to simulate the movement trajectory of the technological object during the detail processing on the machine, particularly the milling machine (Tae-Il & Myeong-Woo 1999).

At the same time, the problem is the need to consider a process of creating spatial functions that reflect the real movement of the object in the working space of the machine tool.

DOI: 10.1201/9781003225447-24

24.2 MODEL OF THE OBJECT MOVEMENT TRAJECTORY IN WORKING SPACE

Considering the task, we can assume that the detail creates in the space a scalar function $D(x, y, z)$ and a cutting tool creates a vector function $\mathbf{I}(x, y, z)$. The process of detail production is the product of these two functions. Thus, we obtain the following product in the form of a divergence of two functions, i.e.:

$$\operatorname{div} D \cdot \mathbf{I} = \frac{\partial D \cdot I_x}{\partial x} + \frac{\partial D \cdot I_y}{\partial y} + \frac{\partial D \cdot I_z}{\partial z} =$$
$$= D \cdot \left(\frac{\partial I_x}{\partial x} + \frac{\partial I_y}{\partial y} + \frac{\partial I_z}{\partial z} \right) + I_x \frac{\partial D}{\partial x} + I_y \frac{\partial D}{\partial y} + I_z \frac{\partial D}{\partial z}, \qquad (24.1)$$

or after ordering

$$\operatorname{div} D \cdot \mathbf{I} = D \cdot \operatorname{div} \mathbf{I} + \mathbf{I} \cdot \operatorname{grad} D \qquad (24.2)$$

Equation (24.2) is most suitable for describing the milling process. In addition, the first member on the right side gives a description of the tool's movement in the formation of the detail, and the second gives an idea of what is the dependent error of geometry (Koehler 2013; Tymchyk et al. 2018).

In the case of a turning, we meet a rotary product of the scalar function of the tool on the vector function of the detail

$$\operatorname{rot} I \cdot \mathbf{D} = \mathbf{i} \times \frac{\partial (I \cdot D)}{\partial x} + \mathbf{j} \times \frac{\partial (I \cdot D)}{\partial y} + \mathbf{k} \times \frac{\partial (I \cdot D)}{\partial z} =$$
$$= \left(\mathbf{i} \frac{\partial I}{\partial x} + \mathbf{j} \frac{\partial I}{\partial y} + \mathbf{k} \frac{\partial I}{\partial z} \right) \times D + I \left(\mathbf{i} \times \frac{\partial D}{\partial x} + \mathbf{j} \times \frac{\partial D}{\partial y} + \mathbf{k} \times \frac{\partial D}{\partial z} \right) \qquad (24.3)$$

or after ordering

$$\operatorname{rot} I \cdot \mathbf{D} = \operatorname{grad} I \times \mathbf{D} + I \operatorname{rot} \mathbf{D} \qquad (24.4)$$

where in the case of the turning equation (24.4) gives a description of motion along the detail and describes the detail's movement.

To begin, we simulate the instrument's movement in the technological space as a vector two-coordinate function, i.e.,

$$\mathbf{I} = \mathbf{I}(x, y) \qquad (24.5)$$

In this case, x and y are coordinates of the location of a point of the tool, which is inviolate in relation to its personal coordinate system. So, in order to set a vector $\mathbf{I}(x, y)$, it is necessary to set its projections on the coordinate axis.

Since we first consider the plane of the problem, then the expression for the vector takes the following form

$$\mathbf{I} = I_x(x, y)\mathbf{i} + I_y(x, y)\mathbf{j} \qquad (24.6)$$

If we consider the work (W), performed by the vector $\mathbf{I}$ on the vector of the path $\overline{AB}$, then this will be defined as the scalar product of the vectors

$$W = \mathbf{I} \cdot \overline{AB} \tag{24.7}$$

Since the projections of the vector $\overline{AB}$ on the axis of the coordinates are respectively equal Δx_{AB} and Δy_{AB} to that we get

$$\overline{AB} = \Delta x_{AB}\mathbf{i} + \Delta y_{AB}\mathbf{j} \tag{24.8}$$

Thus, we obtain a scalar product

$$\Delta W_{AB} = I_x(x_{AB}, y_{AB})\Delta x_{AB} + I_y(x_{AB}, y_{AB})\Delta y_{AB} \tag{24.9}$$

Consequently, if (24.9) is regarded as one of the total number of areas, then we have the opportunity to move to the integral sum

$$\lim \sum_{AB=1}^{n} I_x(x_{AB}, y_{AB})\Delta x_{AB} + I_y(x_{AB}, y_{AB})\Delta y_{AB} =$$
$$= \int_L I_x(x,y)dx + I_y(x,y)dy \tag{24.10}$$

Thus, we obtained a curvilinear integral along the trajectory of the cutting movement of the tool and the movement of the sensor and other objects (Pihnastyi 2017; Tymchyk et al. 2018). This curvilinear integral can be transformed into ordinary if imagined $x = x(t)$, $y = y(t)$.

In this case, the resulting integral is calculated at the interval of change t:

$$\int_L I_x(x,y)dx + I_y(x,y)dy =$$
$$= \int_t^{t'} \left[I_x[x(t),y(t)]x'(t) + I_y[x(t),y(t)]y' \right] dt \tag{24.11}$$

Using these theses, we consider the cutting tool's movement in the contour milling (Figure 24.1).

Consequently, we can determine the length of the path of the cutting tool by complying with the rules of integration

$$\int I_x dx + I_y dy = \int_0^{\Delta a} I_x(a+t,b)dt + \int_0^{\Delta b} I_y(a+\Delta a, b+t)dt - \int_0^{\Delta a} I_x(a+t, b+\Delta b)dt$$
$$- \int_0^{\Delta b} I_y(a,b+t)dt = -\int_0^{\Delta a} \left[I_x(a+t, b+\Delta b) - I_x(a+t,b) \right] dt$$
$$+ \int_0^{\Delta b} \left[I_y(a+\Delta a, b+t) - I_y(a, b+t) \right] dt. \tag{24.12}$$

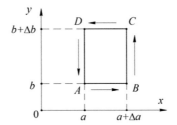

Figure 24.1 Parameters of the contour of milling.

Applying the Lagrange theorem on the mean and following a series of transformations, we obtain the following result

$$\int_L I_x \, dx + I_y \, dy = \iint_S \left(\frac{\partial I_y}{\partial x} - \frac{\partial I_x}{\partial y} \right) dxdy \qquad (24.13)$$

The resulting equation (24.13) is uncertain for the reason that in the general case the exact value of the curvilinear integral is not stored when the main condition of integration $\frac{\partial I_x}{\partial y} - \frac{\partial I_y}{\partial x}$ is not met. Nevertheless, there is a proof of equation (24.13) in (Korn & Korn 2000), which proves that it is quite possible under certain assumptions, that is, it is a method of M.V. Ostrogradsky. Equation (24.13) is a Green's formula and is a partial case of the Stokes formula, which we shall consider later.

Thus, we examined the mathematical basis that can serve to create a contour milling model. Imagine the function of the movement of an instrument in a plane as then the full differential will

$$dI(x,y) = I_x dx + I_y dy \qquad (24.14)$$

also

$$I_x = \frac{\partial I}{\partial x}, I_y = \frac{\partial I}{\partial y} \qquad (24.15)$$

In this case, there is a vector

$$\mathbf{I} = I_x \mathbf{i} + I_y \mathbf{j} = \frac{\partial I}{\partial x} \mathbf{i} + \frac{\partial I}{\partial y} \mathbf{j} = gradI(x,y) \qquad (24.16)$$

which is the gradient of the function $I(x,y)$ (24.5).

In this case, the function $I(x,y)$, the gradient of which equals the vector (24.16) is the potential of this vector. So, in this case, the curvilinear integral between two points A and B (Figure 24.1) by any trajectory L, which connects them will be equal to the potential of the vector (24.16), that is,

$$\int_A^B I_x \, dx + I_y \, dy = \int_A^B dI(x,y) = I(B) - I(A) \qquad (24.17)$$

Equation (24.17) refers to the open contour of integration, but for a closed contour (Figure 24.1), it will be zero.

$$\int_{ABCDA} I_x\,dx + I_y\,dy = 0 \tag{24.18}$$

In addition, should the condition

$$\frac{\partial I_y}{\partial x} = \frac{\partial I_x}{\partial y}; \frac{\partial I_y}{\partial x} - \frac{\partial I_x}{\partial y} = 0 \tag{24.19}$$

for which

$$\iint_S \left(\frac{\partial I_y}{\partial x} - \frac{\partial I_x}{\partial y}\right) dxy = 0 \tag{24.20}$$

All the above applies to an imaginary idealized situation. For example, the contour may not be locked during milling.

In this case, the number of vectors around the perimeter is closed by the gradient (24.16), which means that condition (24.19) is not met.

In the real situation, Equation (24.19) has an additional term in the form of the value of the Pandan zone i.e., $\lceil \mathbf{p}$, and then the expression (24.19) takes the form:

$$\frac{\partial I_y}{\partial x} - \frac{\partial I_x}{\partial y} = \frac{\partial \lceil \mathbf{p}}{\partial x} - \frac{\partial \lceil \mathbf{p}}{\partial y} \tag{24.21}$$

where $\lceil \mathbf{p}$ is the external micro Pandan zone numerically equal to the roughness of the detail's surface (Figure 24.1).

In its case, the right part in equation (24.21) is small enough but not less than [S].

We looked at the flat two-coordinate case. However, the milling tool quite often performs motion in three coordinates. In this case, the two-coordinate equation (24.14) becomes three coordinates, i.e.,

$$dI(x,y,z) = I_x dx + I_y dy + I_z dz \tag{24.22}$$

the basis of which is the motion of the vector **I**, consequently

$$\mathbf{I} = I_x(x,y,z)\mathbf{i} + I_y(x,y,z)\mathbf{j} + I_z(x,y,z)\mathbf{k} \tag{24.23}$$

In this case, the curvilinear integral

$$\int_L I_x\,dx + I_y\,dy + I_z\,dz = 0 \tag{24.24}$$

if the condition is fulfilled

$$\frac{\partial I_x}{\partial y} = \frac{\partial I_y}{\partial x}; \frac{\partial I_z}{\partial y} = \frac{\partial I_y}{\partial z}; \frac{\partial I_z}{\partial x} = \frac{\partial I_x}{\partial z} \tag{24.25}$$

So as a consequence (Figure 24.2)

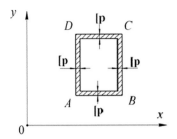

Figure 24.2 External micro-Pandan zone of contour milling.

$$\frac{\partial I_x}{\partial y} = \frac{\partial I_y}{\partial x}; \frac{\partial I_z}{\partial y} = \frac{\partial I_y}{\partial z}; \frac{\partial I_z}{\partial x} = \frac{\partial I_x}{\partial z} \qquad (24.26)$$

In this case, the gradient of the vector function (Skytsiouk & Klotchko 2013):

$$\mathrm{grad} I = \frac{\partial I}{\partial x}\mathbf{i} + \frac{\partial I}{\partial y}\mathbf{j} + \frac{\partial I}{\partial z}\mathbf{k} \qquad (24.27)$$

Thus, if we take the path of integration, then we have the opportunity to determine the next path of integration

$$I_D(x,y,z) - I_A(x_0,y_0,z_0) = \int_{(x_0,y_0,z_0)}^{(x,y,z)} I_x\,dx + I_y\,dy + I_z\,dz =$$

$$= \int_{x_0}^{x} I_x(x,y_0,z_0)\,dx + \int_{y_0}^{y} I_y(x,y,z_0)\,dy + \int_{z_0}^{z} I_z(x,y,z)\,dz \qquad (24.28)$$

It is easy to notice that we can achieve a similar result if we carry out integration and other parallelepiped's edges in Figure 24.3.

In equation (24.28), each member of the right-hand side of the equation is based on a complex mathematical expression, which not only gives a description of the milling trajectory but also the roughness of the surface. It is best considered for a parallelepiped point (Figure 24.3).

Therefore, we write equation (24.28) on the components that give an idea of the actual situation within the space of the parallelepiped in Figure 24.3, that is, (Kozlov et al. 2019; Kukharchuk et al. 2017; Vedmitskyi et al. 2017):

$$\left.\begin{array}{l} \int_{x_0}^{x} I_x(x,y_0,z_0)\,dx = l_x + R_a(x) \\ \int_{y_0}^{y} I_y(x,y,z_0)\,dy = l_y + R_a(y) \\ \int_{z_0}^{z} I_z(x,y,z)\,dz = l_z + R_a(z) \end{array}\right\} \qquad (24.29)$$

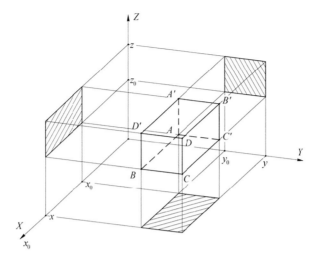

Figure 24.3 Circle of milling parallelepiped.

The reverse path during milling can be written as follows

$$\left. \begin{array}{l} \int_{x}^{x_0} I_x(x,y,z)dx = l_{\bar{x}} + R_a(\bar{x}) \\ \int_{y}^{y_0} I_y(x,y_0,z_0)dy = l_{\bar{y}} + R_a(\bar{y}) \\ \int_{z}^{z_0} I_z(x_0,y,z)dz = l_{\bar{z}} + R_a(\bar{z}) \end{array} \right\} \quad (24.30)$$

If you consider a circle of milling, then

$$l_x + l_{\bar{x}} + l_y + l_{\bar{y}} + l_z + l_{\bar{z}} + \mathrm{grad} I(x,y,z) + \\ + \mathrm{grad} R_a(x,y,z) + \mathrm{grad} R_a(\bar{x},\bar{y},\bar{z}) = 0 \quad (24.31)$$

Where $l_x, l_{\bar{x}}, l_y, l_{\bar{y}}, l_z, l_{\bar{z}}$ are the quantities of the length of the parallelepiped's edges $ABCDA'B'C'D'$. These quantities form, that is, inaccuracy of their execution.

Consequently, the gradients in (24.31) are complex functions, which are the result of the uncertainty of these quantities, i.e.,

$$\mathrm{grad} I(x,y,z) = \frac{\partial I_x}{\partial l_x} \mathrm{grad} l_x + \frac{\partial I_y}{\partial l_y} \mathrm{grad} l_y + \frac{\partial I_z}{\partial l_z} \mathrm{grad} l_z \\ + \frac{\partial I_{\bar{x}}}{\partial l_{\bar{x}}} \mathrm{grad} l_{\bar{x}} + \frac{\partial I_{\bar{y}}}{\partial l_{\bar{y}}} \mathrm{grad} l_{\bar{y}} + \frac{\partial I_{\bar{z}}}{\partial l_{\bar{z}}} \mathrm{grad} l_{\bar{z}} \quad (24.32)$$

At present, equation (24.32) indicates the basis for the contouring error of milling. The size R_a of the roughness also takes part in shaping the size during milling, since it is a direct consequence of the cyclic movement of the cutting tool surface during the metalworking (Feynman 1965; Polishchuk et al. 2019; Skytsiouk & Klotchko 2013).

Such a movement creates a micro Pandan zone of the detail with the corresponding modulation of the detail's surface, i.e.,

$$\operatorname{grad} R_a(x,y,z) = \frac{\partial R_a(x)}{\partial l_x}\operatorname{grad} l_x + \frac{\partial R_a(y)}{\partial l_y}\operatorname{grad} l_y + \\ + \frac{\partial R_a(z)}{\partial l_z}\operatorname{grad} l_z + \frac{\partial R_a(\bar{x})}{\partial l_{\bar{x}}}\operatorname{grad} l_{\bar{x}} + \frac{\partial R_a(\bar{y})}{\partial l_{\bar{y}}}\operatorname{grad} l_{\bar{y}} + \frac{\partial R_a(\bar{z})}{\partial l_{\bar{z}}}\operatorname{grad} l_{\bar{z}} \quad (24.33)$$

Focusing on the process of contour milling and the laws of the technological object's movement in space (Figures 24.1 and 24.3), we have the opportunity to make a series of analytical expressions for conventional milling.

Thus, guided by the Green's formula (24.13), we have the opportunity to obtain a number of dependencies in the contours of milling. So, the projection on the plane has the form (Feynman 1965; Korn & Korn 2000)

$$\iint_S \left(\frac{\partial I_y}{\partial x} - \frac{\partial I_x}{\partial y} \right) dx dy = \int_L I_x\, dx + I_y\, dy \quad (24.34)$$

on the plane yz

$$\iint_S \left(\frac{\partial I_z}{\partial y} - \frac{\partial I_y}{\partial z} \right) dy dz = \int_L I_z\, dz + I_y\, dy \quad (24.35)$$

on the plane xz

$$\iint_S \left(\frac{\partial I_x}{\partial z} - \frac{\partial I_z}{\partial x} \right) dx dz = \int_L I_x\, dx + I_z\, dz \quad (24.36)$$

On the mirror plane, we obtain the following result for the plane $\overline{xy}$

$$\iint_S \left(\frac{\partial I_{\bar{y}}}{\partial \bar{x}} - \frac{\partial I_{\bar{x}}}{\partial \bar{y}} \right) d\bar{x}d\bar{y} = \int_L I_{\bar{x}}\, d\bar{x} + I_{\bar{y}}\, d\bar{y} \quad (24.37)$$

on the plane $\overline{yz}$

$$\iint_S \left(\frac{\partial I_{\bar{z}}}{\partial \bar{y}} - \frac{\partial I_{\bar{y}}}{\partial \bar{z}} \right) d\bar{z}d\bar{y} = \int_L I_{\bar{z}}\, d\bar{z} + I_{\bar{y}}\, d\bar{y} \quad (24.38)$$

on the plane $\overline{xz}$

$$\iint_S \left(\frac{\partial I_{\bar{x}}}{\partial \bar{z}} - \frac{\partial I_{\bar{z}}}{\partial \bar{x}} \right) d\bar{x}d\bar{z} = \int_L I_{\bar{x}}\, d\bar{x} + I_{\bar{z}}\, d\bar{z} \quad (24.39)$$

Thus, we considered the idealized contour milling model (24.34–24.36), (24.37–24.39). Nevertheless, such an idealization does not take into account the geometry of the cutting tool.

At present, the geometry of the milling tool is usually cylindrical, and therefore equations (24.34–24.36) and (24.37–24.39) give a description of the movement of the center of mass within the borders of the technological object's Pandan zone. As a result of this movement of Pandan zones (details and cutting tools), a milling contour is formed.

Thus, the destruction of the detail's material is a consequence of the dual movement of the cutting tool.

On the one hand, it is the linear movement of the center of mass (24.34–24.36), (24.37–24.39) and from another rotational movement of the cutting tool. Both movements have a vector character. None of them in itself is a big difficulty, but in a heap, they form a complex movement of the surface of the cutter (cutting edge). So, lines of the contour are formed as a series of points.

At the moment, if we select one of the edges on the cutter, then at rotation it will form a circle of radius R. At the same time, the point will only instantly touch the imaginary surface of the milling.

If the circle moves, then this plane will form a tangentially envelope surface, which consists of individual points. The distance between the points is dependent on the speed of linear movement, the speed of rotation, and the number of cutting planes cutters. Thus, under the action of these velocity vectors, a point describes a cycloid.

If we take into account the number of teeth, then the description of such a movement will have the following implicit form

$$x + \sqrt{y(2R-y)} = R \arccos \frac{R-y}{R} \tag{24.40}$$

or in parametric form

$$\left. \begin{array}{l} x = R(t - \sin t) \\ y = R(1 - \cos t) \end{array} \right\} \tag{24.41}$$

In this case, the number of points (cuts) for one turn will be $E\left|\frac{2\pi}{n}\right|$, where the $n = 2, 3, 4, 5, 6, 7, 8, \ldots$

The presence of two vectors acting on the point leads to the fact that we need to consider the vortex function

$$rot(\mathbf{I} \times \mathbf{I}_\omega) = \mathbf{I} div \mathbf{I}_\omega - \mathbf{I}_\omega div \mathbf{I} + \frac{d\mathbf{I}}{d\mathbf{I}_\omega} - \frac{d\mathbf{I}_\omega}{d\mathbf{I}} \tag{24.42}$$

where $\mathbf{I}$ – vector of linear velocity for the chosen coordinate, $\mathbf{I}_\omega$ – vector of rotational velocity.

24.3 CONCLUSIONS

The proposed analytical dependencies can serve as a basis for determining the optimal cutting conditions, in particular, determining the value of the path length when moving the cutting tool. These dependencies take into account the type of tool movement, i.e., an integrated modeling approach is defined, which allows optimizing the machining process on a machine by various indicators.

Such a consideration makes it possible to clarify the trajectory of the cutting tool relative to the workpiece, which is processed.

Thus, the question of modeling the idealized trajectory of the movement of objects during the technological process of contour milling was considered. In this case, the modeling feature consists of the mathematical approach of the gradient distribution of the contour error during milling.

The model of the object trajectory formation taking into account the formation of micro Pandan zones of the object is presented. This allows us to consider the quality parameters of the treated surface, which increases the accuracy of manufacturing parts of the devices.

However, the prospects for further research suggest the creation of analytical dependencies that take into account the geometry of the cutting milling tool. This will provide an opportunity to improve the quality of machining details.

REFERENCES

Anton, F. D. & Anton, S. 2017. Trajectory generation for robot engraving and milling tasks. *2017 IEEE, 18th International Carpathian Control Conference (ICCC)*, 28–31 May 2017, Sinaia: Romania.

Armarego, E. J. A. & Brown, R. H. 1969. The machining of metals. Prentice-Hall. *Technology & Engineering*, 437.

Davies, B. J. 1988, *Machine tool design and research: 27th: International Conference Proceedings (Machine Tool Design and Research: International Conference Proceedings)* Hardcover – July 19, 1988, Palgrave Macmillan, 1–500.

Feynman, R. 1965. *The Character of Physical Law, A Series of Lectures Recorded by the BBC at Cornell University USA*. London: Cox and Wyman LTD, 1–173.

Koehler, D. R. 2013. Geometric-distortions and physical structure modeling. *Indian J Phys.* 87: 1029.

Kopp, T. & Stahl, J., Demmel, P., Tröber, P., Golle, R., Hoffmann, H. & and Volk, W. 2016. Experimental investigation of the lateral forces during shear cutting with an open cutting line. *Journal of Materials Processing Technology* 238: 49–54.

Korn, G. A. & Korn, T.M. 2000. *Mathematical handbook for scientists and engineers: definitions, theorems, and formulas for reference and review (Dover Civil and Mechanical Engineering).* 2 Revised Edition, 1–1152.

Kozlov, L. G., Polishchuk, L. K., Piontkevych, O. V., Korinenko, M. P., Horbatiuk, R. M., Komada, P., Orazalieva, S. & Ussatova, O. 2019. Experimental research characteristics of counter balance valve for hydraulic drive control system of mobile machine. *Przeglad Elektrotechniczny* 95(4): 104–109.

Kukharchuk, V. V., Kazyv, S. S. & Bykovsky, S. A. 2017. Discrete wavelet transformation in spectral analysis of vibration processes at hydropower units. *Przeglad Elektrotechniczny* 93(5): 65–68.

Lefeber, E. 2012. Modeling and control of manufacturing systems. Decision policies for production networks. *Springer*: 9–30. London, 1–302.

Pihnastyi, O. 2017. Analytical methods for designing technological trajectories of the object of labour in a phase space of states. *Scientific Bulletin of National Mining University* 4: 104–111.

Polishchuk, L. K., Kozlov, L. G., Piontkevych, O. V., et al. 2019. Study of the dynamic stability of the belt conveyor adaptive drive. *Przeglad Elektrotechniczny* 95(4), 2019: 98–103.

Rao, V.S. & Rao, P. V. M. 2004. Modelling of tooth trajectory and process geometry in peripheral milling of curved surfaces. *International Journal of Machine Tools & Manufacture*: 1–14.

Skytsiouk, V. I. & Klotchko, T. R. 2013. Determination of the coordinates of the pathological zones in the mass of the biological object. *Microwave & Telecommunication Technology IEEE Xplore 2*: 1083–1084.

Smaoui, M., Bouaziz, Z., Zghal, A. & Baili, M. D. 2012. Gilles Compensation of a ball end tool trajectory in complex surface milling. *International Journal of Machining and Machinability of Materials* 11(1): 51–68.

Tae-Il, S. & Myeong-Woo, C. 1999. Tool trajectory generation based on tool deflection effects in flat-end milling process(I). *KSME International Journal* 13(10): 738–751.

Tymchyk, G. S., Skytsiouk, V. I., Klotchko, T. R., Ławicki, T. & Demsova, N. 2018. Distortion of geometric elements in the transition from the imaginary to the real coordinate system of technological equipment, *Proc. SPIE 2018:* 10808.

Tymchyk, G. S., Skytsiouk, V. I., Klotchko, T. R., Popiel, P. & Begaliyeva, K. 2018. The active surface of the sensor at a contact to the technological object. *Proc. SPIE, Photonics Applications in Astronomy, Communications, Industry, and High Energy Physics Experiments* 2018 10808.

Tymchyk, G. S., Skytsiouk, V. I., Klotchko, T. R., Zyska, T. & Rakhmetullina, S. 2018. Two parameter active measuring probe for objects setting detection on CNC machines workspace. *Proc. SPIE 2018:* 10808.

Vedmitskyi, Y. G., Kukharchuk, V. V. & Hraniak, V. F. 2017. New non-system physical quantities for vibration monitoring of transient processes at hydropower facilities, integral vibratory accelerations. *Przeglad Elektrotechniczny* 93(3): 69–72.

Chapter 25

Physical bases of aggression of abstract objects existence

G. S. Tymchyk, V. I. Skytsiouk, T. R. Klotchko, W. Wójcik,
Y. Amirgaliyev, and M. Kalimoldayev

CONTENTS

25.1 Introduction...279
25.2 Model of the life cycle of abstract entities..280
25.3 Bases of law of aggression interaction objects..283
25.4 Conclusions 287
References..287

25.1 INTRODUCTION

Based on the analysis of works (Leclerc 2016, Simondon 1958, Dennen 2005), we note that there is still no mathematical and logically constructed conclusion about the life cycle of an object and their aggression. Among numerous objects, both organic and inorganic, there are very large differences in the lifetime from millionths of a second to millions of years. Against this general background, the lifetime of objects of an organic origin, and even more so technical, more than modest is look. Therefore, the most effective method is a study of short-lived objects in time, followed by extrapolation to more long-lived ones (Liu 2004).

The consumer, using only the principles known to him, has a possibility to determine his needs. The Abstract Entity (AE), first, must be necessary at this civilization level, that is, to have a social order, which causes the emergence of the primary TF and its subsequent technical, biological, chemical, and other indicators. Second, the object must have high reliability and service life without reducing its technical indicators. Third, the low cost of the object is crucial. All these three questions interest only the consumer, and he is not at all interested in what technological ways the object goes to him — this is the business of the manufacturer. Therefore, the manufacturer must decide for himself in all three questions from this list. According to the second position, AE should be eternal, but there are no examples of eternity of AE, and the solution of this question makes it possible to determine the third, that is, its product. That is why our main task in this section is to solve the main question: what is the life and death of these AE? During this time, the primary technological phantom (TF) of the highest quality must emerge, but under the influence of the destruction TF, it becomes a thing that is not needed by anyone.

Among all the properties of AE, the property of motion is such that it must be subjected to a separate consideration. From a general point of view, the ability to move with its mass, that is, controlling the external form relative to the centre of mass and

DOI: 10.1201/9781003225447-25

moving this mass in the desired direction, is a very important feature for AE. The volume, within which AE can touch the surface to any other object without disturbing its trajectory of movement, is called a Pandan zone (Tymchyk et al. 2018, Tymchyk et al. 2018).

Thus, the purpose to research is to simulate the life cycle of an abstract entity in terms of the aggression of objects interaction.

25.2 MODEL OF THE LIFE CYCLE OF ABSTRACT ENTITIES

Let us try to simulate a gradual failure of AE, immediately noting that a truly constructed AE begins to collapse from smaller particles of a higher order, that is, "n." Replacing the particles of an object with particles affected by the fracture TF, we will observe how the performance is reduced. Let's designate operational qualities as a certain indicator "K" on which they depend. Then, if we plot the dependence of "K" on the number of particles of the object affected by the fracture TF, we obtain the situation reflected in Figure 25.1. From this graphical dependence, it is clearly seen that the logical use of any object has such a period of time, until its operational qualities "K" falls to the level $K(1 - 1/e)$. This is in the first place, and second, it should be stated that the last hope for the existence of any speaker is its part at number 1 with mass as m_{AC}/e.

The rate of AE's destruction can be determined using the rate of TF destruction's change, that is, the tangent of the angle of inclination of the fracture curve $tg\alpha_i$ at point i

$$tg\alpha_i = \frac{K\left(1 - \frac{1}{e^{n-1}}\right) - K\left(1 - \frac{1}{e^{n-1-i}}\right)}{(n-i-1)-(n-i)} \qquad (25.1)$$

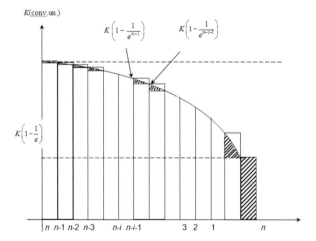

Figure 25.1 The spectral composition of the reduction of AE's properties, depending on the order of the mass of the elementary particle.

After simple mathematical transformations, we get:

$$tg\alpha_i = -\frac{K}{e^{n-i}}(e-1) \tag{25.2}$$

Taking into account that $tg\alpha_i$ is the speed of destruction on the segment from $n-1$ to $n-i-1$, it is possible to write down:

$$tg\alpha_i = \frac{S_{pi}}{t_i} = \frac{(n-i-1)-(n-i)}{t_i} = -\frac{1}{t_i} \tag{25.3}$$

Transforming expressions (25.2) and (25.3), we obtain for each part of the AE the time of existence:

$$t_i = \frac{e^{n-i}}{K_i(e-1)}, i \to 1,2,\ldots,(n-1) \tag{25.4}$$

where K_i is a qualitative indicator of i-th particle of AE, despite the fact that in the previous formula there is K of the whole AE.

The fact is that this indicator is conditional, and if the quality of the overall AE is K, then each of its shares should have the same indicator, because otherwise the quality of the speaker itself does not make sense (Feynman 1965). If we sum all the particle of AE to one whole, in order to get the lifetime of the object, we get:

$$\sum_{i=1}^{n-1} t_i = \sum_{i=1}^{n-1} \frac{e^{n-x}}{K_i(e-1)} \tag{25.5}$$

Turning to the integral form of the record (Korn & Korn 2000), we obtain the time of the logically grounded AE's life of as

$$\int_0^{t_{AE}} dt = \int_1^n \frac{e^{n-x}}{K_i(e-1)}dx \tag{25.6}$$

or

$$t_{AE} = \frac{1}{K(e-1)}(n-x)e^{n-x} \tag{25.7}$$

Summing up, we return first to $tg\alpha_i$, that is, the speed of AE's destruction. Immediately it should be noted that as soon as the AU is ready, then it is already under the influence of the fracture TF ($tg\alpha \neq 0$), and the destruction rate in this case will be $tg\alpha_n = -\frac{K}{e^n}(e-1)$.

At the final stage of AE's existence, we have:

$$tg\alpha_1 = -\frac{K}{e^{n-(n-1)}}(e-1) = -\frac{K}{e}(e-1). \tag{25.8}$$

The time of existence of an AE's particle t_i is directly proportional to its technological order through e^{n-i}, and vice versa in relation to its quality parameter $\frac{1}{K_i}$, whence it follows that there is a possibility of the existence of low-quality portions of an AE, so that their technological order is longer, and this will give durability.

This is an erroneous thesis, and this is why. The quality factor K_i of each share of AE should be equal to the quality factor of the overall speaker. But, since the quality indicators of each AE are formed under the influence of the primary TF, and that they can by no means be worse than the previous ones, the quality factor can only grow from one speaker to another speaker, i.e., we tend to $K \to \infty$. The technological order of the particle "n" in this case indicates the degree of protection of the particle "No. 1" of the AE, that is, a tendency is maintained that the loss of a particle under the number "n" should not significantly reduce the quality indicators of the object as a whole. We can give an example: a small rivet is made to the hull of a ship or aircraft and not vice versa.

A similar study, which leads to a similar result, can be carried out using the model of a power elementary electromagnetic particle (PEMP) (Tymchyk et al. 2018). In the event that a fracture TF acts on PEMP, then its reaction can be described by the following differential equation:

$$m\frac{d^2x}{dt^2} + h\frac{dx}{dt} + \gamma x = K_{TF} \tag{25.9}$$

where m – PEMP's mass, h – coefficient of elasticity; γ – coefficient of resistance; K_{TF} – TF quality of destruction.

$$P_{1,2} = -\frac{\gamma}{2m} \pm \sqrt{\left(\frac{\gamma}{2m}\right)^2 - \frac{h}{m}} \tag{25.10}$$

If, to improve the mathematical calculations, denote $\delta = \frac{\gamma}{2m}$ and $\eta = \frac{1}{\sqrt{h/m}}$, then (25.10) takes the form:

$$p_{1,2} = -\delta \pm \sqrt{\delta^2 - \eta^2} \tag{25.11}$$

Referring to the equation (25.9), it should be noted that the first term of the equation characterizes the kinetic energy of AE. The second term of the equation is its potential energy, and the third affects the rate of exchange between the kinetic and potential energies.

Assuming that their equivalent of the current quality value is K, we get the following solution:

$$k = \frac{K_{TF} - K_{AE}}{2m\sqrt{\delta^2 - \eta^2}} \left(e^{P_1 t} - e^{P_2 t}\right) \tag{25.12}$$

where K_{AE} is the maximum value of quality that AE can realize under the influence of the TF's destruction.

The maximum value of the function (25.12) reaches according to time:

$$t_n = \frac{\ln\left(\frac{P_1}{P_2}\right)}{P_2 - P_1} \tag{25.13}$$

As was shown above, the loss of quality of AE's properties occurs below the $K_{AE}\left(1-\dfrac{1}{e}\right)$ level, thus $0.632\, K_{AE}$.

In this case, referring to the thesis of active and passive life, or active or passive actions, it can be argued that this level corresponds to the distinction between demarcation of use of the quality properties of AE. Since each AE has a sufficiently wide selection of properties, it is not rational to focus on at least one of them. The object has the ability to lose properties not only of the first order but also by groups of higher orders. In this case, it is enough that the number of losses in the order be, that is, it is necessary to refer to dependencies (6–8).

25.3 BASES OF LAW OF AGGRESSION INTERACTION OBJECTS

The structure of AE is quite idealized regarding that it is possible to operate only with their abstract properties (Klotchko 2011, Skytsiouk & Klotchko 2013). In the real world, everything is much more complicated, not because the objects are complex in their primary properties, but because the properties of the second third and other higher orders are so complicated in their essence and spatial distribution that it is still not possible to determine their actions in space.

In general, if we take into account the particles in the mass of idealized AE, then it is obvious that the smaller fractions of the size cover the part with a mass $\dfrac{m_0}{e}$. Therefore, it is not surprising that AE as a self-organizing mass sacrifices its surface layer and at the same time has the property of its regeneration.

Considering a very simple question: "How is AE kept in space?", the answer is quite simple: we need some kind of support point relative to which AE can keep its position in space.

A fairly simple experience with throwing up a body proves that everything depends on the shape and weight, but all movement stops on the Earth's surface. The fact is that the AE has a constant interaction with the gravitational field of the Earth. She is trying to balance her power with the power of the gravitational field of the globe (Misner et al. 1973). In addition, air and even water interfere with this interaction if AE falls into the water. Of course, the speakers can be selected in form and weight in such a way that they are divided into three main varieties (Kittel 2004), namely: AE that keeps on the Earth's surface; AE that keeps on the surface of the water; AE contained in the air.

All these varieties are characterized by the fact that gravity and the strength of the third medium act on all objects. For the AE, contained in the air, it is characteristic that its behavior is extremely unstable, its power characteristics are identical to the power characteristics of the air and its mobility, which causes corresponding fluctuations in the coordinates of space. The same applies to the AE, which interacts with the water surface or is in the aquatic environment. However, here it is noticeable that instability in the space of AE is much less (Tymchik et al. 2017).

Compared with the two previous cases, AE on a solid surface is generally unshakable. However, this is an illusion, since you can cite of examples where a body that was on the surface eventually falls under it. Hence, the previous AE with time is under the surface. The explanation of these processes is quite simple: all AEs try to find the maximum number of resistance points in order to fix their position in space, that is, to reliably rely on some possible coordinates and even better on many coordinates. Since the AE does not have the ability to simultaneously maintain power at all coordinates simultaneously, it does this in turn. Therefore, AEs are always in a moving state, since only from time-to-time controls each of the coordinates.

If we take into account that each of the coordinates to be monitored is in itself the coordinate of the other AE, which is also in a similar state of control, then oscillations are formed at the boundary of the resultant forces. The coordinate of these oscillations and their amplitude are completely dependent on the power of both AE and the sensitivity of the sensors to the rate of change of physical laws in the touch zone. From the above, it can be argued that the following law exists when two AEs interact, namely: each entity spreads its force action in space exactly as much as neighboring entities allow it with their opposition, which confirms the action of Newton's third law.

The following conclusions can be formed from this law. The first conclusion (loss of property): AE, which does not meet resistance in space, dissolves in it and disappears. A simple example is a salt crystal that dissolves in water, smoke in the air, and the like. The second conclusion (being in space): AE, which does not have at least one equal-power interaction, cannot keep its coordinate and falls through the space.

An example is the fall of a heavy body on the ground or its immersion in water. Such technologies are widely used in modern technology.

The situation, which is considered in the static arrangement of two entities (object and sensor on Figure 25.2), is imaginary.

This is explained by the fact that as a result of the existence of special properties in any entity, it simply cannot contain a surface of equal power on the verge touch in a stationary state. An equally powerful surface is a surface where both entities have the ability to create opposing power in direction. Thus, both entities form such power Pandan zones or presence zones (Klotchko 2011, Skytsiouk & Klotchko 2013) that are intended to destroy each other. For destruction, it is necessary to fulfill a number of conditions, but if this is not possible, there is a constant test of strength zones.

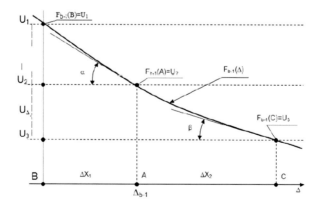

Figure 25.2 Entrance and exit from the touch of the measurement object and the sensing element according to the physical laws of F_{b-1} at their static location with neutral interaction.

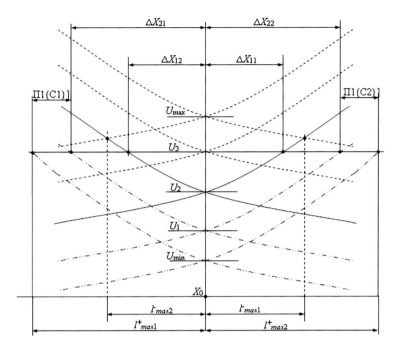

Figure 25.3 Entrance and exit from touching two entities $[\Pi_1\{C1\}]$ and $[\Pi_1\{C2\}]$, with the help of their presence zones, according to the $F_{b-1}(\Delta)$ law.

To understand how this process happens, consider what happens when two entities touch. We assume that there are two entities $[\Pi_1\{C1\}]$ and $[\Pi_1\{C2\}]$ that have the same type of Pandan zones as Π_1, formed around the entity according to the law (1.1). In the case of rest, the coordinate of the equal-power zone will be X_0, and the power level U_2 (Figure 25.3). Sensitivity to power levels U_{min}, U_1, U_2, U_3, U_{max} – for sensors of both entities is the same. In this case, none of the entities can have a stable state, since it needs to constantly have information about the neighboring entity.

Therefore, each of them forms the necessary increase in power in the zone or a step toward the neighboring entity. To obtain reliable information, each entity must either increase its power to ΔU level in order to obtain the power level in the U touch zone or make movement toward each other in ΔX_{11} and ΔX_{12} dimensions, respectively. Since AE checks for a touch, the braking process starts at $X_0 + \Delta X_{11}$ and $X_0 + \Delta X_{12}$, both for the first and for the second, when AE sensors form a touch signal (Vasilevskyi 2014, Vasilevskyi 2015, Vasilevskyi 2013).

As a result of a braking process, each of AEs deepens into the next to the depth of the second and for the first. As a consequence of inertia, the total power level in the $x = a$ coordinate rises to U_{max}. The entities, working through the reverse when they disagree through their need for "non-touch" analysis, cross the definitions of "non-touching" $X_0 - \Delta X_{21}$ and $X_0 + \Delta X_{22}$, and by inertia continue the differences until the total power in the $x = a$ coordinate falls to the minimum value U_{min}.

In the coordinates $X_1 = X_0 - l^+_{\max 1}$ and $X_2 = X_0 - l^+_{\max 2}$, they change the direction of motion, and the process repeats. Reproduction of this process again and again occurs due to those factors that will be considered in the following papers and have an endless fading relationship (Vasilevskyi et al. 2018, Polishchuk et al. 2016, Polishchuk et al. 2016).

In the general case, it is possible to write down the expression of such oscillations in a simplified way (without taking into account $U[\Pi 1]$), i.e.:

$$U_{X_0} = U[\Pi_1\{C1\}]\sin(\omega_1 t + \varphi_1) + U[\Pi_1\{C2\}]\sin(\omega_2 t + \varphi_2) \tag{25.14}$$

where U_{X_0} is the total power of AE in the coordinate $x = a$, $U[\Pi_1\{C1\}]$ Ta $U[\Pi_1\{C2\}]$ is the powers of entities from the first and second entities; $\omega_1 = \dfrac{2\pi}{T_1}$ and $\omega_2 = \dfrac{2\pi}{T_2}$ – circular oscillation frequency of entities or the power of these entities, T_1 and T_2 – their periods of oscillation; ϕ_1, ϕ_2 – phase angles.

From the equation mentioned above (14), it can be seen that, depending on the periods T_1 and T_2, and the φ_1, φ_2 phases, the zone equal to the power zones of the entities (level U_2) may either have an instantaneous coordinate X_0 or vary from ΔX_{21} to ΔX_{22} In all other cases, when its coordinates exceed the above, or it disappears completely, it has extremely undesirable consequences for entities, to catastrophic. This is especially true of entities that have a varying volume (gas, liquid, etc.), depending on the power of neighboring, on which they rely.

In the event of the loss of their opposition, they catastrophically spread until they either receive the proper resistance or lose their essence as a physical phenomenon. Both in the first and in the second case, the consequences will be negative due to the fact that an increase in volume at a constant (non-accumulating) power industry leads to a decrease in the average specific power over the entire AE's surface. In the next period of time there is its complete deformation and the restructuring of the coordinates of distancing from other entities. The inability to do it in time again leads to disastrous consequences. The higher the unit weight of an entity, the slower these processes go (Polishchuk 2018, Polishchuk 2019, Kozlov 2019).

If we take into account the overall power level of presence zone Π_1, then the equation (25.14) takes the form:

$$U_{X_0} = U_{R0} + U_{R1}\sin(\omega_R t + \varphi_R) + e^{-kt}\{U[C_1\{C1\}]\sin(\omega_R t + \varphi_R)\} \tag{25.15}$$

The stability of X_0 coordinate is also influenced by the oscillatory part of the level $U_{R1}\sin(\omega_R t + \phi_R)$, which usually has a long-period character. The general nature of this process is that entities C1 and C2 need to consider not only the actions of each other but also the fluctuations of the general level. All this, when the phases do not coincide, leads to the formation of a pulse-effect on X_0 coordinate and, as a consequence, to the formation of a conflict or catastrophic situation. Usually in this case one of the speakers is to blame: either C1 or C2, because this is a consequence of her interest in the development of the conflict. The goal of the conflict is to capture one entity of another to enhance its personal energy and manage it for personal protection (Kukharchuk 2016, Kukharchuk 2017, Vedmitskyi 2017). As a result, it is necessary to establish the fact that there is a kinematic movement of the power zones of neighboring entities, forming between them a "wall" of equal power, that is:

$$U_{X0} = U_{R0} + U_{R1}\sin(\omega_R t + \varphi_R) + \\ + 2e^{-kt} \cdot U(\Pi_1)[\sin(\omega_1 t + \varphi_1) + \sin(\omega_2 t + \varphi_2)], \quad (25.16)$$

whence it follows that in the zone of touching of two entities, the movement obeys oscillations of the general level with a common amplitude and is the sum of the amplitudes of the level and AE. In addition, there are also micro kinematic movements, which are a consequence of the φ_1, φ_2 phase lag of entities behind the phase of φ_R level (Kukharchuk 2016, Kukharchuk 2017).

25.4 CONCLUSIONS

From the point of view of the aggression of the interaction of objects, a model of the cycle of the existence of an abstract object is proposed with regard to the interaction of technical, biological, and biotechnical objects. In this case, an analysis of the spectral composition of reducing the properties of AE depending on the order of the mass of the elementary particle, taking into account an influence of the technological phantom of destruction, is given.

The situation is considered when both entities form such power Pandan zones or presence zones that are intended to destroy an object. Such cases are typical for different types of interactions and different types of objects, which take place not only in production processes but also in astronomy, medicine, and so on. Analytical dependencies that reflect the aggression of the interaction when touching objects are presented.

In further studies devoted to energy processes in AE, the overall level will be considered, which is produced by an infinite number of entities, and therefore cannot be controlled by one entity, since for this it is necessary to have an infinite energy reserve. In this case, if we take into account the quasi-stable nature of the constant level, there is a gradual shift of X_0 coordinate, due to its current variability.

REFERENCES

Dennen, J. M. G. V. D. 2005. Theories of aggression: Psychoanalytic theories of aggression. Default journal.
Feynman, R. 1965. *The Character of Physical Law, A Series Of Lectures Recorded By The BBC at Cornell University USA*. London: Cox and Wyman LTD.
Kittel, Ch. 2004. *Introduction to Solid State Physics*, Wiley: 8th Edition.
Klotchko, T. R. 2011. Formalized model of the zone presence of structures of the biological objects. *Microwave & Telecommunication Technology, IEEE Xplore* 2: 1036–1037.
Korn, G. A. & Korn, T. M. 2000. *Mathematical Handbook for Scientists and Engineers: Definitions, Theorems, and Formulas for Reference and Review (Dover Civil and Mechanical Engineering)*. 2 Revised Edition.
Kozlov, L. G., Polishchuk, L. K., Piontkevych, O. V., Korinenko, M. P., Horbatiuk, R. M., Komada, P., Orazalieva, S. & Ussatova, O. 2019. Experimental research characteristics of

counter balance valve for hydraulic drive control system of mobile machine. *Przeglad Elektrotechniczny* 95(4): 104–109.

Kukharchuk, V. V., Bogachuk, V. V., Hraniak, V. F., Wójcik, W., Suleimenov, B. & Karnakova, G. 2017. Method of magneto-elastic control of mechanic rigidity in assemblies of hydropower units. *Proc. SPIE,* 10445.

Kukharchuk, V. V., Hraniak, V. F., Vedmitskyi, Y. G., Bogachuk, V. V. et al. 2016. Noncontact method of temperature measurement based on the phenomenon of the luminophor temperature decreasing. *Proc. SPIE* 10031.

Kukharchuk, V. V., Hraniak, V. F., Vedmitskyi, Y. G., Bogachuk, V. V. et al. Noncontact method of temperature measurement based on the phenomenon of the luminophor temperature decreasing. *Proc. SPIE* 10031.

Kukharchuk, V. V., Kazyv, S. S. & Bykovsky, S.A. 2017. Discrete wavelet transformation in spectral analysis of vibration processes at hydropower units. *Przeglad Elektrotechniczny* 93(5): 65–68.

Leclerc, A. 2016. Actualism and fictional characters, *Principia* 20(1): 61–80.

Liu, J. 2004. Concept analysis: aggression. *Issues Ment Health Nurs.* 25(7): 693–714.

Misner, Ch. W., Thorne, K. S. & Wheeler, J. A. 1973. *Gravitation.* San Francisco: Freeman.

Polishchuk, L. K., Kozlov, L. G. et al. 2018. Study of the dynamic stability of the conveyor belt adaptive drive. *Proc. SPIE* 1080862.

Polishchuk, L. K., Kozlov, L. G., Piontkevych, O. V., et al. 2019. Study of the dynamic stability of the belt conveyor adaptive drive. *Przeglad Elektrotechniczny* 95(4), 2019: 98–103.

Polishchuk, L., Bilyy, O. & Kharchenko, Y. 2016. Prediction of the propagation of crack-like defects in profile elements of the boom of stack discharge conveyor. *Eastern-European Journal of Enterprise Technologies* 6(1): 44–52.

Polishchuk, L., Kharchenko, Y., Piontkevych, O. & Koval, O. 2016. The research of the dynamic processes of control system of hydraulic drive of belt conveyors with variable cargo flows. *Eastern-European Journal of Enterprise Technologies* 2(8): 22–29.

Simondon, G. 1958. *On the Mode of Existence of Technical Objects.* Paris: Aubier Editions Montaigne.

Skytsiouk, V. I. & Klotchko, T. R. 2013. Determination of the coordinates of the pathological zones in the mass of the biological object. *Microwave & Telecommunication Technology, IEEE Xplore* 2: 1083–1084.

Tymchik, G. S., Skytsiouk, V. I., Bezsmertna, H., Wójcik, W., Luganskaya, S., Orazbekov, Z. & Iskakova, A. 2017. Diagnosis abnormalities of limb movement in disorders of the nervous system. *Proc. SPIE,* 10445.

Tymchyk, G. S., Skytsiouk, V. I., Klotchko, T. R., Ławicki, T. & Demsova, N. 2018. Distortion of geometric elements in the transition from the imaginary to the real coordinate system of technological equipment, *Proc. SPIE,* 10808.

Tymchyk, G. S., Skytsiouk, V. I., Klotchko, T. R., Popiel, P. & Begaliyeva, K. 2018. The active surface of the sensor at a contact to the technological object. *Proc. SPIE,* 8 10808.

Tymchyk, G. S., Skytsiouk, V. I., Klotchko, T. R., Zyska, T. & Rakhmetullina, S. 2018. Two parameter active measuring probe for objects setting detection on CNC machines workspace. *Proc. SPIE,* 10808.

Vasilevskyi, O. M. 2013. Advanced mathematical model of measuring the starting torque motors. *Technical Electrodynamics* 6: 76–81.

Vasilevskyi, O. M. 2014. Calibration method to assess the accuracy of measurement devices using the theory of uncertainty. *International Journal of Metrology and Quality Engineering* 5(4).

Vasilevskyi, O. M. 2015. A frequency method for dynamic uncertainty evaluation of measurement during modes of dynamic operation. *International Journal of Metrology and Quality Engineering* 6(2).

Vasilevskyi, O., Kulakov, P., Kompanets, D., Lysenko, O. M., Prysyazhnyuk, V., Wójcik, W. & Baitussupov, D. 2018. A new approach to assessing the dynamic uncertainty of measuring devices. *Proc. SPIE, 10808.*

Vedmitskyi, Y. G., Kukharchuk, V. V. & Hraniak, V. F. 2017. New non-system physical quantities for vibration monitoring of transient processes at hydropower facilities, integral vibratory accelerations. *Przeglad Elektrotechniczny* 93(3): 69–72.

Chapter 26

Development and investigation of changes in the form of metal when obtaining the crankshaft's crankpin using free forging

V. Chukhlib, A. Okun, S. Gubskyi, Y. Klemeshov, R. Puzyr,
P. Komada, M. Mussabekov, D. Baitussupov, and G. Duskazaev

CONTENTS

26.1 Introduction.. 291
26.2 Research results.. 293
26.3 Conclusion and perspectives of further research.. 300
References... 301

26.1 INTRODUCTION

Nowadays, there is a tendency of the development and implementation of resource-saving technologies in enterprises while producing forged pieces. This is caused by high demands for forged pieces from alloyed and special steels. Such forged pieces require a special deformation regime to obtain high mechanical properties. That results in an increase in the cost of production of such forged pieces in comparison with pieces made of carbon steels. In addition, the process of forging is characterized by high rates of metal consumption that affects the cost of a product and associated subsequent machining.

The development of resource-saving forging technologies is an important task for today. In this area, the fifth group of forged pieces is often used, which has the most labor-intensive manufacturing process. In addition to that, it is required to apply the largest overlaps to some parts of the forged pieces, since it is impossible to produce with conventional forging methods. Moreover, this group of forged pieces includes forged pieces of crankshafts. The mass of metals, which usually are alloy steels, is often measured in tons as far as these forged pieces are concerned. An improvement of such technology has a special interest for enterprises, since when obtaining a forged piece, most of a product is covered not only by an allowance but also by an overlap in a crank portion of a shaft. Therefore, hundreds of kilograms of metal are wasted when machining. In most cases, such metal removal leads to the deterioration of mechanical properties of a product during its usage. This is caused by cutting metal fibers during machining in the most responsible and high loaded areas of a completed part.

With respect to all the abovementioned features of manufacturing, the development of resource-saving technology and definition of rational parameters for forging process of crankshaft forged pieces are a topical issue.

DOI: 10.1201/9781003225447-26

A forging technological process of crankshaft forged pieces is a labor-intensive one, which consists of a lot of forging operations and technological transitions (Kal'chenko 2014). At the same time, tracking an impact of each forging operation from the beginning on mechanical properties and quality of a completed forged piece as a whole is a difficult task. Each stage, either it is a forging of a workpiece from an ingot or a forging process of a crankshaft, requires separate consideration and study, since each stage has a certain influence on the quality of forging.

The classification of the labor intensity for creating a form of forged pieces and crankshafts, which belong to the most complex forged pieces, is given in (Ohrimenko 1976). The 40H, 40HN, 35HM, 30HN2MA, 18H2N4MA, and other steel grades are often applied for most of the crankshafts. Steels, alloyed with vanadium, chromium, molybdenum, and nickel, have an enhanced hardness, plasticity, wearing quality (30HMA, 20HN3A, 38H2MJuA, 40H2N2MA, 25H2N4MA, 38H2MJuA, etc.) and are used in manufacturing crankshafts for diesel engines with an increased output power (Bespalov 1973).

An important stage in the forging process of crankshafts, which affects mechanical properties of a forged piece, is the stage of preparing a workpiece. This involves roughing an ingot, upsetting, and broaching to sizes of a workpiece for further forging. The main parameters of the upsetting process, which influence on the stress–strain state of metal, are as follows: (1) the form factor of a workpiece (h/D); (2) the deformation degree (ε); (3) the strain rate; (4) the temperature. Main parameters of the broaching are as follows: (1) the value of the feed; (2) the degree of single compression; (3) the degree of forging reduction under deformation; (4) the form factor of a workpiece; (5) a method of applying the deformation force (canting scheme) (Grinkevich 2014).

The authors (Chukhlib et al. 2015b) carried out experiments using computer simulation; the results allowed them to determine the influence of parameters of a workpiece, such as the form factor (h/D), as well as deformation parameters such as the deformation degree (ε), on irregularity of the deformation distribution in metal during upsetting an ingot with the subsequent broaching on a round section. A canting scheme for broaching an ingot is provided in the study (Chukhlib et al. 2015a).

Various schemes of forging crankshafts, which are used in the production, have been considered. The forging process for a crankshaft may be different depending on structural features of the shaft, but forging a single-coil shaft has only one way of performance. In this case, a round blank part is marked along its length, based on the calculated volume of metal for forging-bearing journals. Then, a blacksmith operation of overpressing is carried out and the bearing journals are forged (Sokolov 2011, Polishchuk 2016, Polishchuk 2018).

The overpressing operation may be performed with the help of special tools (clips) with both round and triangular cross sections and the depth of overpressing up to 100–300mm. For overpressing to a bigger depth, the operation is only performed with trihedral clips. When forging, tightening of metal edges occurs at the point of contact with the tool. Therefore, a margin of metal is provided at the height of a crank portion of 10–25% (Polishchuk 2016). The central part of the metal forged piece (the crank portion of the shaft) remains at the same level of metal working, i.e., after the upsetting and broaching the ingot to the required diameter, the crank portion of the shaft does not deform any longer. This has a negative effect on the quality of the forged piece, due to the heterogeneity of mechanical properties in the basic part of the shaft and bearing journals. Moreover, when machining, a part of metal (overlap) is removed as a waste, which also negatively affects the quality of the forged piece, due to cutting metal fibers in the area of the crankshaft's crankpin (Polishchuk 2019a, Polishchuk 2019b, Ogorodnikov 2018).

Taking into account all the abovementioned features of the technology for manufacturing forged pieces of crankshafts, an advanced technological scheme of forging has been developed. It allows us to direct metal fibers on the form of the crank portion, as well as to reduce the consumption of metal during machining.

26.2 RESEARCH RESULTS

The objective of the study is to determine the influence of forging parameters and geometry of a deforming tool on changes in the form of the metal in a part of the crank portion of the shaft, as well as to define intervals of geometrical sizes of a forged piece, what can be obtained by applying the developed technology within the framework of PJSC Dnipropress-Stal.

Hydraulic presses are widely used for the production of forged pieces at PJSC Dnipropress-Stal. Besides, ingots of different masses are usually used in large forged pieces manufacturing. Due to the large mass of the forged pieces, it was decided to carry out computer simulation of the process to determine regularities of changes in the form. The scale of the model was chosen according to the volume of metal used to produce one forged piece by forging it from an ingot of 6.5 tons. Thus, sizes of a workpiece chosen for modeling are as follows: 80 mm is the length of the workpiece and 40 mm is the diameter. The sizes of the workpiece were selected according to the scale of the model.

To solve the main task of the study, it is necessary to consider the process of forging the crankshaft's crankpin with thin forging dies. The forging of a forged piece consists of subsequent pressing of the crankpin with a pair of thin forging dies and canting the workpiece by 90°, 45°, and 15° in order to shape it into a circle in the cross section. After forcing a thin forging die, the crankpin is forged with a pair of thin dies, while the volume of metal in the deformation zone extends in the longitudinal direction. In addition to that, it is important to establish a regularity of the influence of the tool thickness and the depth of its penetration on the length of the crankpin during forging. This is considered to be crucial especially when calculating technological transitions. It is necessary to take into account this parameter as the main one. If during the calculation of technological transitions, a larger thickness of the deforming tool is chosen, the connecting crankshaft's crankpin will receive more elongation after forging and, accordingly, will be longer than the length of the crankpin of the finished product. This is an irrevocable defect that cannot be eliminated by any kind of forging operations.

Figure 26.1 shows a schematic representation of the forged piece after forging the crankpin with a pair of thin forging dies. The dashed line in Figure 26.1 displays the crankshaft, which should be obtained as a result of the entire technological process.

The main geometrical sizes of the forged piece that has to be controlled during the forging process of the crankshaft (Figure 26.1) are:

b is the maximum achievable length of the crankpin;

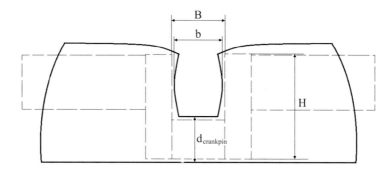

Figure 26.1 A schematic representation of the forged piece with a retracting crankpin.

B is the length of the crankpin of the completed part;
H is the height of the crank portion of the shaft;
$d_{crankpin}$ is the diameter of the crankpin.

At this stage of forging, as we see in Figure 26.1, the crankshaft forged piece is not fully forged yet, but it is still necessary to get the bearing journals of the crankshaft. However, even at this stage, the analysis of the changes in the form of the metal and determining regularities of the influence of the sizes of the tool and the depth of the penetration on the form of the received crankshaft's crankpin, and on the ratio of the basic sizes of the crank portion of the shaft (the height of the crank portion (diameter), the diameter of the crankpin, the length of the crankpin) are crucial.

Figure 26.2 depicts the general view of the forged piece of the crankshaft after forging the crankpin with a pair of the thin forging dies. In this case, forging dies with the thickness of 8 mm (0.2 $D_{general}$) were used and the depth of the pressure was 20 mm (0.5 $D_{general}$).

As can be seen Figure 26.2, the crankshaft's crankpin took the shape of a circle in a cross section (Figure 26.2b) and also extended in the longitudinal section to a certain value. In order to determine the effect of the thickness of the tool and the depth of the penetration on the elongation of the crankpin during deformation and define the ratio of the sizes of the crank portion of the shaft, the simulation of the process was carried

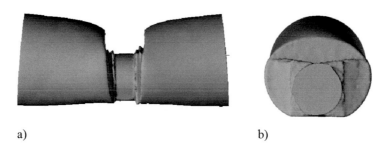

a)   b)

Figure 26.2 The general view of the forged piece of the crankshaft after forging the crankpin – (a) the side view; (b) the view of the section of the crankpin.

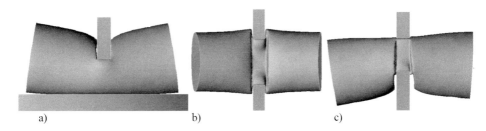

Figure 26.3 The process of the workpiece deformation with the thin forging dies when obtaining the crankshaft's crankpin using free forging.

out in accordance with the experimental plan. Each workpiece was exposed to forcing by the forging die and further forging. At the same time, the following canting scheme, which is described below, was used.

After forcing the forging die into the workpiece, it was canted by 90°, and then was pressed (Figure 26.3a). Further, the workpiece was again canted by 90° and pressed to align metal deposits, i.e., to the initial value of the forging die penetration. After that, the workpiece was canted by 90° and pressed one again, but with an increased pressure ratio. This gradual increase in the pressure ratio was made to avoid excessive barrel formation of metal in the volume of the crankpin. These metal deposits are formed similarly to the upsetting specimens with the ratio H/D > 2.5, and they are detrimental to the quality of forged pieces, because the probability of forging fold and fold formation in the volume of the crankpin increases. Canting by 90° and pressing are repeated until the crankpin has no square shape in the cross section. The sizes of square sides depend on the penetration depth of the forging die at the first stage of forging, and in this case, they are 12, 20, and 28 mm. After that, the workpiece is canted by 45°, pressed, canted by 90°, and pressed again in order to shape the crankpin into an octagon in the cross section (Figure 26.3b). Further, the crankpin is reduced to the form of a circle in the cross section (Figure 26.3c) with small pressing and canting by 15° after each pressing. In order to create a circle in the cross section, at least 10 pressings are required.

Thus, the crankshaft's crankpin is obtained. However, it is necessary to determine which typical sizes of the crank portion of the shaft are possible to obtain with the developed forging scheme. The term "typical sizes of the crank portion of the shaft" refers to the ratio of the length of the crankpin and the height of the crank portion of the shaft at a given diameter of the crankpin. To perform such a procedure, it is crucial to consider the forging of the crankshaft, after the stage of the forging of the crankpin in the longitudinal section with two thin forging dies. This requires considering the sizes of the forged piece, as shown in Figure 26.1.

After completing the second stage of the experiment with computer simulation, the results were analyzed. The second stage of the experiment was carried out in accordance with the plan of a full factorial experiment, described earlier, but after forcing the thin forging die it ensued burnishing the crankshaft's crankpin with a pair of thin forging dies according to the canting scheme, described previously. At the same time, the crank portion of the shaft was inscribed into the obtained workpiece to define the required sizes (Figure 26.1).

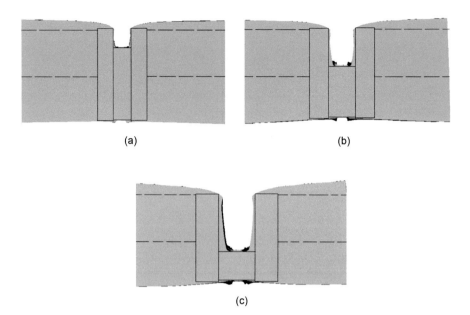

Figure 26.4 Typical sizes of the crank portion of the shaft, which can be obtained using forging dies with a thickness of 4 mm (0.1 $D_{general}$).

Figure 26.4 shows the crank portion of the shaft inscribed in the obtained workpiece under forcing the forging die and burnishing the crankpin using the forging dies with a thickness of 4 mm (0.1 $D_{general}$) for different diameters of the crankpin.

As it can be noticed, the length of the crankpin is controlled precisely by the pressing depth of the forging die at the first forging stage, which depends on the required diameter of the crankpin. Thus, when applying the developed method for forging the forged pieces of the crankshafts, the length of the crankpin and its diameter are interdependent values.

Figure 26.5 depicts the crank portion of the shaft inscribed in the obtained workpiece under forcing the forging die and burnishing the crankpin using the forging dies with a thickness of 8 mm (0.2 $D_{general}$) for different diameters of the crankpin.

In this case, it is clearly seen that during the increase of the penetration depth, the possible diameter (height) of the crank portion is being decreased. In addition, when forging the crankshaft's crankpin using the forging dies with a thickness of 8 mm (0.2 $D_{general}$), its elongation is considerably greater. Thus, the length of the crankpin with a diameter of 20 mm is equal to two thicknesses of the applied forging dies, i.e., it is 18 mm. And when the diameter of the crankpin is 12 mm, the length is 22 mm. i.e., this is almost three times more than the thickness of the forging dies.

Figure 26.6 presents the crank portion of the shaft inscribed in the obtained workpiece under forcing the forging die and burnishing the crankpin using the forging dies with a thickness of 12 mm (0.3 $D_{general}$) for different diameters of the crankpin.

The most complicated changes in the form of the work piece are these when the crankpin is forged using the forging dies with a thickness of 12 mm. This complexity

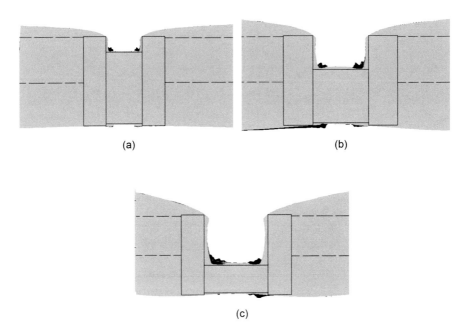

Figure 26.5 Typical sizes of the crank portion of the shaft, which can be obtained using forging dies with a thickness of 8 mm (0.2 $D_{general}$).

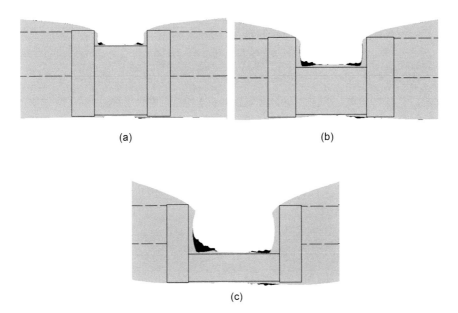

Figure 26.6 Typical sizes of the crank portion of the shaft, which can be obtained using forging dies with a thickness of 12 mm (0.3 $D_{general}$).

is due to the fact that in this case we deal with a much larger volume of deformed metal than when using the forging dies with a thickness of 4 mm. Because the volume of metal is larger, the workpiece bending is significantly increased in the process of the deformation. This is most noticeable in the case of the crankpin with a diameter of 12 mm, when the largest tightening of metal and the biggest bend of the ends of the workpiece occur.

Further, let us consider the dependence of the influence of process parameters on the sizes of the workpiece. To do this, several values were specified in relative terms. The values adopted in relative terms are the diameter of the crankpin ($d_{crankpin}$), the height of the crank portion of the shaft (H), the length of the crankpin (B), the thickness of the tool (b_{tool}), the penetration depth of the forging die in the workpiece ($h_{penetration}$). Thus, the graphical dependencies that are shown in Figures 26.7–26.10 were made. These dependencies represent changes in sizes of the workpiece as a function of the thickness of the tool and the penetration depth of the forging die.

In Figures 26.7 and 26.8, it is clearly seen that the change in the tightening value and the crankpin diameter with reference to the tightening value varies almost linearly in the case of using the forging dies with a thickness of 4 mm (0.1 $D_{general}$) and different forging parameters. Meanwhile, in the case of the forging dies with a thickness of 8 mm (0.2 $D_{general}$) and 12 mm (0.3 $D_{general}$), the dependence is nonlinear. However, it is also possible to notice almost identical curves for the case of forging with the forging dies of 8 and 12 mm and practically the same values of H\$D_{general}$ (Figure 26.7) and $d_{crankpin}$\H (Figure 26.8).

Figure 26.9 shows that the typical size of the obtained crank portion of the shaft (B\H) varies linearly for each case of applying the forging dies with different thicknesses, but each case has a specific numerical value that differs significantly for each forging case.

The change in the length of the crankshaft's crankpin, depending on the thickness of the used tool and the penetration depth of the forging die, can be seen from the

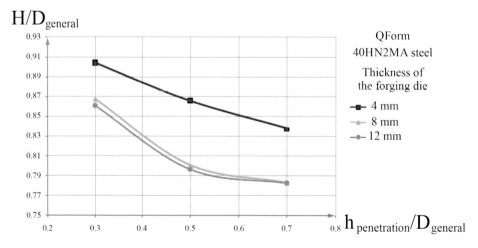

Figure 26.7 Dependence of the tightening value (H) from the forging parameters.

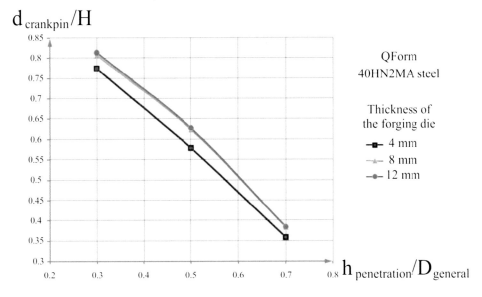

Figure 26.8 Dependence of the ratio of the crankpin diameter ($d_{crankpin}$) and the tightening value (H) from the forging parameters.

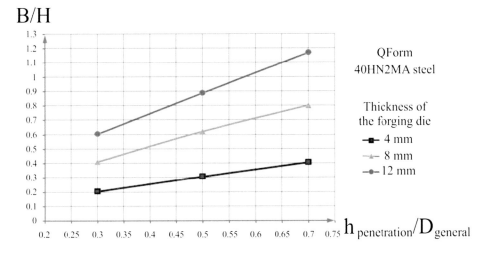

Figure 26.9 Dependence of the ratio of the crankpin length (B) and the tightening value (H) from the forging parameters.

graphical dependences shown in Figure 26.10. Therefore, we can conclude that the length of the crankpin depends almost as a linear function on the penetration depth of the forging die and insignificantly on the thickness of the deforming tool.

Based on the data of this study, practical recommendations on the use of the developed technology at PJSC Dnipropress-Stal were elaborated. The use of the developed

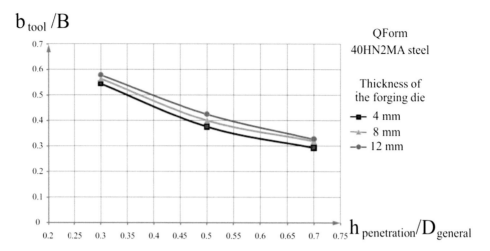

Figure 26.10 Dependence of the ratio of the deforming tool thickness (b_{tool}) and the crankpin length (B) from the forging parameters.

technology resulted in higher quality of forged pieces, saving of resources for the enterprise, and reducing the cost of production.

26.3 CONCLUSION AND PERSPECTIVES OF FURTHER RESEARCH

In this chapter, the analysis of changes in the form of metal was performed by means of computer simulation with the application of the thin deforming tool with a thickness (0.1–0.3 $D_{general}$) at the depth of forcing into the workpiece (0.3–0.7 $D_{general}$) and with further forging of the crankpin. The analysis of the computer simulation data showed that in the determined intervals of the thickness of the forging dies, as well as for the defined penetration depth of the forging die, it is possible to receive the forged piece of the crankshaft with the crankpin. With the help of the created graphical dependencies of changes in the form, the most reasonable forging schemes were defined. These schemes allowed us to obtain the crankpin with the minimal distortion of forged pieces, as well as to satisfy the required sizes in accordance with drawings of a part. Among these schemes, the most rational ones are the schemes where the forging dies with a thickness of 0.1 $D_{general}$ and under the penetration depth of 0.3–0.7 $D_{general}$ are used, as well as the forging dies with a thickness of 0.2 $D_{general}$ and under the penetration depth of 0.3–0.5 $D_{general}$ and further forging.

The developed technology was implemented into the production of PJSC Dnipropress-Stal. This made it possible to reduce metal consumption by 13.8% in the manufacture of the forged pieces of the crankshafts. The rejection of the forged pieces has decreased by 12.4% due to improving the quality of the forged pieces. The total cost of the forging pieces of metal per ton has decreased by 8.7% as a result of reducing the need for snagging forged pieces. The indicated developments allowed us to increase the efficiency of deformation redistribution in the forging of the forged pieces of the crankshafts.

REFERENCES

Bespalov, B., Glejzer, L., Kolesov, I., et al. 1973. *Tehnologija mashinostroenija (special'naja chast') (Mechanical Engineering Technology (Special Part))*. Moscow: Mashinostroenie.

Chukhlib, V., Klemeshov, J., Grynkevych, V, & Dyja, H. 2015a. Doslidzhennja napruzheno-deformovanogo stanu pry protjazhci tytanovogo splavu z metoju optymizacii' parametriv kuvannja (An investigation of the stress-strain state during the titanium alloy broaching for optimization of forging parameters). *Visnyk NTU «KhPI»* 24: 159–166.

Chukhlib, V., Klemeshov, J., Grynkevych, V. & Dyja, H. 2015b. Analiz vplyvu parametriv poperedn'oi' osadky ta i'i' vidsutnosti na nerivnomirnist' deformacii' pry protjazhci pokovok z tytanovyh splaviv (The influence analysis of the parameters of the upsetting and its absence on the inequality of deformation during the forging of titanium alloys). *Visnyk NTU «KhPI»* 47: 82–85.

Grinkevich, V., Chuhleb, V., Banashek, G. & Ashkeljanec, A. 2014. Teoreticheskie issledovanija kuznechnoj operacii protjazhki pri ispol'zovanii shemy deformacii «prohodami» (Theoretical studies of blacksmith broaching operations using the "draft" deformation scheme). *Vestnik NTU «KhPI»:* 44: 28–34.

Kal'chenko, P. & Markov, O. 2014. *Novye tehnologicheskie processy kovki krupnyh pressovyh pokovok (New technological forging processes of big forged pieces)*. Kramatorsk: DDMA.

Kozlov, L.G., Polishchuk, L.K., Piontkevych, O.V. et al. 2019. Experimental research characteristics of counter balance valve for hydraulic drive control system of mobile machine. *Przeglad Elektrotechniczny* 95(4): 104–109

Ogorodnikov, V.A., Zyska, T. & Sundetov, S. 2018. The physical model of motor vehicle destruction under shock loading for analysis of road traffic accident. *Proc. SPIE* 10808: 108086C.

Ohrimenko, J. 1976. *Tehnologija kuznechno-shtampovochnogo proizvodstva (Forging Production Technology)*. Moscow: Mashinostroenie.

Polishchuk, L., Bilyy, O. & Kharchenko, Y. 2016. Prediction of the propagation of crack-like defects in profile elements of the boom of stack discharge conveyor. *Eastern-European Journal of Enterprise Technologies* 6(1): 44–52.

Polishchuk, L.K., Kozlov, L.G., Piontkevych, O.V. et al. 2018. Study of the dynamic stability of the conveyor belt adaptive drive. *Proc. SPIE* 10808: 1080862.

Polishchuk, L.K., Kozlov, L.G., Piontkevych, O.V. et al. 2019. Study of the dynamic stability of the belt conveyor adaptive drive. *Przeglad Elektrotechniczny* 95(4): 98–103.

Sokolov, L., Aliev, I., Markov, O. & Alieva, L. 2011. *Tehnologija kuvannja (Forging Technology)*. Kramatorsk: DDMA.

Chapter 27

Approaches to automation of strength and durability analysis of crane metal structures

S. Gubskyi, V. Chukhlib, A. Okun, Y. Basova, S. Pavlov, K. Gromaszek, A. Tuleshov, and A. Toigozhinova

CONTENTS

27.1 Introduction 303
27.2 Magnetic coercive control research results 304
27.3 Automation of strength calculations 307
27.4 Practical application of automation of calculation to determine
       operational integrity of hoisting machines 309
27.5 Conclusions 313
References 314

27.1 INTRODUCTION

If hoisting machines are about to reach their normative operation life cycle, in order to determine the possibility of their further exploitation, an expert inspection must be carried out by an authorized organization (Wu 2013). One of the points of the expert inspection is to conduct calculation and analytical procedures for estimating and predicting the technical condition of the crane (DocOMD 2005; Qi 2013).

The modern market of hoisting machine inspection requires speed of work at the lowest cost. This is due to great competition in this area. Even 10–15 years ago, technical diagnostics and inspection of hoisting machines were estimated with a higher equivalent price than nowadays. At the same time, the volume and complexity of the work have not only changed, but have even increased.

Modern computer technology allows calculations of strength and predicting the remaining life cycle of crane metal structures to be automated to some extent and thus reduces the time of this stage of work (Dybała 2017; Mantic 2016). Therefore, the development of computer programs that use given calculation algorithms for strength and prediction of the residual life cycle is a topical task.

One of the tools for assessing the technical condition of metal structures of hoisting machines that are about to reach their normative life cycle is the nondestructive testing method based on the coercive force (Popov 2001; Popov 2015). However, when applying magnetic coercive control in practice, it is necessary to take into account at least the influence of some parameters on the coercive force values, namely the grain size of the controlled metal, deviations of the chemical element weight fractions in the steels, and the thicknesses of the controlled elements of the metal structures.

DOI: 10.1201/9781003225447-27

The current Ukrainian regulatory standard that specifies the use of the magnetic coercive method of nondestructive testing (hereinafter NDT) is rather limited and deficient. In 2005, methodological instructions MV 0.00–7.01–05 were approved (DocMI 2005). The permission to use methodological instructions for the expert inspection of overhead cranes is indicated in OMD 00120253.001–2005 (Wu 2013).

Today, for crane metal structures, calculation systems such as the permissible stress (Gohberg 1976), system of probabilistic calculations (Braude 1985), and the limit state design (Sokolov 2005) are used. Suitable computer programs have been developed for each calculation system. These programs automate strength calculations for metal structures of hoisting machines. However, because of the complexity of their interface, narrow application area, and limited practical usage, they have not become widespread.

It should also be mentioned that, presently, the finite element method has been widely used for calculating crane metal structures (Nemchuk 2006), but this method is considerably time-consuming and does not enable the automatic processing of calculations due to the large variety of load-carrying structures of hoisting machines.

27.2 MAGNETIC COERCIVE CONTROL RESEARCH RESULTS

For the magnetic coercive NDT of crane metal structures, applying the regulations in force is required (DocMI 2005; Wu 2013). All of the nomograms given in (DocMI 2005) were deduced for a metal thickness of 8–12 mm. Therefore, the further use of these nomograms for thicknesses other than 8–12 mm without any recalculation will lead to significant errors in the final results. In the current method, methodological instructions MV 0.00–7.01–05 propose to convert structurescope indications considering a correction factor. In other words, the value of the coercive force in the procurement state HC0, A/cm) and the value of the coercive force corresponding to the critical operation mode (HCcritical, A/cm) should be calculated according to the formula:

$$H_{C(\text{new})} = H_{C(\text{table})} + k \cdot h \tag{27.1}$$

where: $H_{C(\text{table})}$, A/sm – value of the coercive force obtained on a controlled object; h, mm – thickness of a wall of a controlled object; and k – coefficient that depends on the value of the coercive force obtained on a controlled object. In the methodological instructions MV 0.00–7.01–05, the value of the coefficient must be selected from a table. However, as practice has shown, it is more correct to give the dependence of $H_C(k)$ in the form of (27.2). This will increase the accuracy of calculation by 10%:

$$H_C = 0.5222 \cdot e^{16.082 \cdot k} \tag{27.2}$$

The use of (27.1) is proposed for all of the crane steels, for example, from St3 (with comparatively small values of $H_C^0 = 1.7$ A/cm) to 10HCND (with relatively high values of $H_C^0 = 4.0$ A/cm). At the same time, the value of the k coefficient does not change. In addition, the structure of the metal and deviations of chemical element weight fractions in steels are not taken into account. Therefore, the obtained recalculation results will not be unbiased, and this may subsequently lead to significant errors in the state evaluation of controlled metal structures using the coercive force as a parameter.

It is known that an important characteristic of metal is the grain point. An increase in the grain size is accompanied by decreasing the yield strength ($\sigma_{0.2}$, MPa), as well as the coercive force value (H_C, A/cm) (Gudinaf 1959; Mikheev 1993):

$$H_C \approx 1/d, \sigma_{0.2} \approx 1/\sqrt{d} \qquad (27.3)$$

where: d – grain size.

The research was carried out and the dependence of the grain size on the coercive force for low carbon steels was established:

$$\Delta H_C = 0.1946 \cdot e^{0.1741 \cdot b} \qquad (27.4)$$

where: b – grain point.

For instance, based on (27.4), two identical samples of low carbon steel with different grain sizes (the grain point is 6 and 10) will have a difference in the coercive forces $\Delta H_C = 0.4$ A/cm. The sample with the smaller grain size will have the bigger coercive force values. In practice, it frequently happens that the metal structures of hoisting machines with different thicknesses of their elements are made of the same metal, but the grain size in these elements of the metal structure is different. Mostly, increasing the metal thickness leads to an increase of the grain size. Consequently, if you do not take into account the grain size of the controlled metal, the results of magnetic-coercion control may be inaccurate.

Crane steels include various chemical elements (carbon (C), silicon (Si), manganese (Mn), chromium (Cr), nickel (Ni), and others). They affect the values of the coercive force differently. Table 27.1 shows the weight fraction of chemical elements in steels used in crane building.

From the earlier studies (Gudinaf 1959), the dependence of the carbon content in iron in the form of granular (1) and lamellar (2) on the coercive force value (H_C, A/cm) cementite is known (Figure 27.1).

As Figure 27.1 shows, increasing the weight of carbon leads to an increase in the coercive force. This statement was also confirmed by our study. These studies tested

Table 27.1 Results of the magnetic coercive method for samples made of 09g2c-12 rolled steel (the grain point in accordance with GOST 5639 was 9, the thickness was 6 mm)

Chemical composition (%)			H_C (A/cm)
C	Mn	Si	
0.90	1.58–1.62	0.60–0.65	3.51–3.59
0.10	1.53–1.59	0.61–0.68	3.61–3.72
0.10	1.50–1.56	0.66–0.74	3.70–3.89
0.11	1.48–1.53	0.70–0.79	3.82–3.96
0.11	1.45–1.49	0.76–0.80	3.92–4.06

Note: the metallographic characteristics and mechanical properties of the samples have a divergence of no more than 3%.

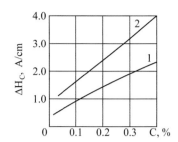

Figure 27.1 Dependence of the carbon content of iron in the form of granular (1) and lamellar (2) cementite on the coercive force value (HC, A/cm).

rolled steel that entered the crane production process with an incoming inspection not only featured chemical analysis and mechanical testing, but also metallographic analysis with initial measurements of the coercive force. During the study, the influence of the grain size on the coercive force values was excluded. Some consolidated results of the magnetic coercive method for samples made of t09G2S steel with a grain point of 9 and a thickness of 6 mm are presented in Table 27.1.

Table 27.1 shows that for 09G2C steel with a grain size of 9 and thickness of 6 mm, regulatory tolerances on chemical element weight fractions can lead to $\Delta H_C = 0.66$ A/cm.

During the study, it was also found that under the control of rolled steel from the same metal, but with different thicknesses, we received a gradual decrease in values of the coercive force with increasing the thickness of the metal to a certain value, as is given in Table 27.2. This is caused by insufficient resolution of the KRM-CK-2M structurescopes.

It can be seen from Table 27.2 that in practice, with the use of structurescopes of the KRM-CK-2M type, without recalculating the coercive force values, depending on the thickness of a controlled element of metal structures, a magnetic method of control based on the coercive force cannot be used to predict the degradation of the mechanical properties or to evaluate the stress and strain state of metal structures.

As a result of the study, it was proposed to use certificated experimental samples with variable cross sections and with known mechanical properties and chemical compositions of the metal, microstructure, and values of the coercive force in each section of a sample (Figure 27.2) (Hryhorov 2013). This eliminates the influence of the grain size, chemical composition, and thickness of the controlled metal on the results of the magnetic coercive NDT.

Table 27.2 Dependence of the coercive force values H_C (A/cm) on the thickness of the metal (d, mm)

Type of Steel	Thickness of metal d (mm)						
	6	8	10	12	16	20	30
St3	2.80–2.90	2.19–2.23	1.71–1.79	1.50–1.56	1.24–1.30	1.17–1.25	-
09G2C	3.82–3.87	3.08–3.12	-	2.05–2.11	1.62–1.69	1.51–1.60	1.40–1.50
10HCND	7.30–7.41	6.18–6.23	5.38–5.43	-	4.34–4.42	3.91–3.99	3.40–3.44

Figure 27.2 Certificated metal samples with variable cross section.

At present, the technology of making samples with variable cross sections has been devised and, according to the results of chemical analysis, mechanical tests, metallographic studies, and measurements of the coercive force, a passport for magnetic control of experimental samples with variable cross sections has also been developed. Currently, more than 100 certified experimental samples with variable cross sections and made of different steels (with different metallographic, mechanical properties and chemical compositions) have been manufactured.

To automate the processing and analysis of results of magnetic coercive NDT, the Metal v2.2 computer program (Gubskiy 2014) was developed. More than 100 units of certified samples of different crane steels have been imported into the program database. This allows us to select a desired sample from the developed database of the program.

27.3 AUTOMATION OF STRENGTH CALCULATIONS

To evaluate the technical condition of metal structures of hoisting machines with further defining their operational integrity, technical experts should distinguish such factors as static and fatigue strength, crack resistance, stability, and rigidity.

In order to assess each factor, experts should consider and study the following indicators such as safety factors and the structure of the metal, characteristic number, local and general deformations, presence of cracks, and their speed of growth. In addition to this, the construction of a crane should be taken into account. This means that a differential approach for each type of crane should be provided.

To determine the residual static strength, it is necessary to perform calculations that are similar to the calculations that are done when designing crane metal structures in accordance with regulations. However, the main difference between these calculations is that they should take into account the actual weakening of metal structures (corrosion, wear and tear, technological cuttings in the elements, etc.).

A computer program named Crane has been developed for this purpose. The program allows us to calculate the strength of metal structures of overhead cranes with their actual geometric parameters (Gubskiy 2013). We used regulations and technical documents RTM 24.090.54–79, OST 24.090.22–83, STO 24.09–5821-01–93, OST 24.090.63–87 (DocEN 1993; DocFEM 1998) in order to create the computation algorithms for the program.

The basis of the Crane program is the algorithm for calculating the construction of overhead cranes at limit states:

- condition of the first limit state is characterized by exhausting the load-carrying capacity of the material according to static characteristics:

$$\sigma_{max} \leq R \cdot \gamma \tag{27.5}$$

where: σ_{max} – maximum calculated stress (the stress intensity), which is determined by considering geometric parameters of the material, MPa; R – calculated strength of the material, MPa; and γ – coefficient of operation conditions of a calculated element of a metal structure.

- condition of the second limit state is characterized by exhausting the load-carrying capacity of the material according to the strength of repeatedly acting loadings:

$$\sigma_{max} \leq R_s \cdot \gamma_c \tag{27.6}$$

where: R_s – calculated fatigue strength, which is determined by taking into account the cyclic loading conditions, MPa.

- condition of the third limit state is characterized by the appearance of deformations, the placement of which prevents the normal operation of the crane due to reduced operation accuracy, the possibility of involuntary movement of the trolley, and leads to the occurrence of inappropriate force factors according to criteria of the first or second limit conditions:

$$\underline{Y}_{CT} \leq \left[\underline{Y}_{CT}\right]; t_{damping} < \left[t_{damping}\right] \tag{27.7}$$

As input data for the calculations (in addition to the geometrical parameters and weight characteristics of the elements of the metal structure), the operation mode of the crane, the limit calculated strength value, the safety factor depending on the load regime, the coefficients of dynamic loads, distortions from sudden failures of transmission motors, etc., are considered

According to calculations performed through the algorithm and calculation program, we obtain a scheme of forces for the main and end beams from calculated uniform loadings, from calculated concentrated loadings, from moving statically applied loadings with resulting vertical and horizontal forces, as well as torsion loadings.

On the basis of these calculations, the local stability of the vertical walls of main beams, etc., is calculated.

The main calculation results are summarized in a table with the indication of limiting and calculated loadings for the conditions of the first, second, and third limit states.

All of the performed calculations are stored in a report (database) for further printing and storage. This program database can be modified and updated from the Crane programs installed on other computers. At present, more than 110 units of overhead cranes have already been introduced in the program database.

The developed database of overhead cranes makes it possible to select the proper crane (based on similarity) and perform the required calculations. The Crane program also calculates not only new constructions but also actual constructions with all of accurate geometric parameters (for example, considering results of magnetic coercive NDT).

27.4 PRACTICAL APPLICATION OF AUTOMATION OF CALCULATION TO DETERMINE OPERATIONAL INTEGRITY OF HOISTING MACHINES

The magnetic coercive NDT of a metal structure was carried out on a casting crane with a capacity of 180/50 tons and a span length of 34 m, produced in 1976 by Sibtiazhmash (Krasnoyarsk). This crane has been operating since 1977 in severe operating conditions. Further calculations of the stress and strain state and prediction of the residual life cycle of the crane's metal structure were carried out with the help of the Metal v2.2 (Gubskiy 2014) and Crane (Gubskiy 2013) computer programs.

Figure 27.3 shows a scheme used for the coercive force measurements on the main beam of the main lift of the considered crane (the vertical walls are made of VSt3sp5 steel (GOST 380), the upper and lower belts are made of M16S steel (GOST 6713).

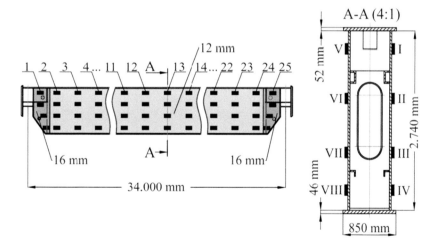

Figure 27.3 Scheme of the coercive force measurements on the main beam of the main lift of a casting crane with a carrying capacity of 180/50 t.

Magnetic control was conducted in 25 sections along the length of the main beam. In each section (along the perimeter), measurements of the coercive force were performed at 10 points. The results of the magnetic coercive NDT of the main beam of the main lift were imported into the Metal v2.2 program in the form of Table 27.3.

It can be seen in Table 27.3 that more than half of the values of the coercive force (H_C, A/cm) recorded in the upper ($\delta=52$ mm) and lower ($\delta=46$ mm) belt of the main beam are less than the coercive force value of the M16S steel in the procurement state ($H_C^0 = 1.7$ A/cm (DocMI 2005)). In addition to this, at the time of the measurements, the crane has already worked over 40 years in severe operating conditions. Therefore, the obtained results of the magnetic coercive NDT indicate a lack of resolution of the KRM-CK-2M-type structurescopes and the coercive force (H_C A/cm) of the main beam needs to be reduced to a single thickness of 8 mm.

For recalculations, two certified experimental samples (CES) were chosen from the Metal v2.2 database (No. 131 for VSt3sp5 steel, and No. 542 for M16S steel). The values of the coercive force in each section of the sample are given in Table 27.4. The functional dependence of the coercive force (HC, A/cm) on the thickness of the cross section (δ, mm) of the CES is described by formula (27.8) for No. 131 and formula (27.9) for No. 542.

Table 27.3 Results of the coercive force measurements (H_C, A/cm) on the main beam of the main lift of the casting crane with a load capacity of 180/50 t

Section no.	The coercive force values (H_C, A/cm) at a point											
	I	II	III	IV	V	VI	VII	VIII	IX	X	XI	XII
1	3.4	3.4	3.2	3.6	4.1	4.0	3.1	3.7	1.4	1.6	1.7	1.6
2	2.4	2.6	3.0	3.2	3.0	2.4	2.4	3.2	1.8	1.6	1.4	1.4
3	2.6	2.7	3.2	3.5	3.0	2.7	2.8	3.1	1.8	1.4	1.6	1.9
4	2.6	3.0	3.8	3.4	2.9	2.6	3.0	3.6	1.7	1.7	1.6	1.7
5	3.8	3.4	3.5	3.7	3.3	3.7	4.3	4.0	1.6	1.6	1.8	1.7
6	3.6	3.6	3.4	3.9	4.0	3.7	4.1	4.0	1.7	1.7	1.4	1.4
7	3.4	3.5	3.8	4.2	4.0	4.3	3.8	3.6	1.7	1.7	1.4	1.6
8	3.8	3.6	3.9	4.3	4.1	3.7	3.9	4.1	1.7	1.4	2.0	1.9
9	3.6	3.2	4.2	3.9	4.3	3.7	3.4	4.5	1.7	2.0	1.3	1.6
10	4.1	3.3	3.9	3.9	4.6	4.1	4.1	4.3	1.8	1.9	1.6	1.9
11	4.3	3.8	3.8	4.2	3.8	3.7	3.8	4.7	2.3	2.3	1.2	1.2
12	4.4	3.9	3.9	4.5	3.8	3.8	3.9	4.6	2.0	1.7	1.3	1.3
13	4.2	4.1	4.2	4.7	4.8	4.1	3.4	4.5	1.7	1.6	1.6	1.6
14	4.0	3.6	4.5	4.6	4.0	4.1	3.4	4.1	1.6	1.4	1.2	1.2
15	3.7	3.8	4.4	4.4	4.4	4.0	4.1	4.7	1.4	1.3	1.3	1.3
16	4.4	4.0	4.6	4.3	5.2	4.4	3.4	4.6	1.3	1.3	1.2	1.3
17	4.0	3.9	4.4	4.6	4.3	3.7	3.3	4.9	1.3	1.3	1.3	1.3
18	3.6	3.6	4.0	4.0	4.4	3.7	3.6	3.5	1.4	1.3	1.6	1.8
19	3.6	3.2	3.6	4.0	3.4	3.8	3.8	4.4	1.3	1.3	1.9	2.0
20	3.7	3.4	3.2	3.6	4.1	4.0	3.7	4.1	1.6	1.2	1.2	1.3
21	3.6	3.5	3.0	3.1	3.7	3.7	3.4	3.9	1.2	1.3	1.3	1.7
22	3.0	3.0	2.6	3.0	2.7	2.9	3.6	3.8	1.6	1.3	2.2	1.9
23	2.9	2.6	2.4	2.8	3.4	3.0	3.1	3.2	1.7	1.3	1.8	2.0
24	3.2	2.8	3.0	3.5	3.1	3.8	2.6	3.5	1.4	1.4	1.3	1.4
25	3.7	3.4	3.4	3.0	4.0	3.8	3.3	4.0	1.3	1.3	1.6	1.3

Note: the thickness of the metal
$\delta = 12$ mm for sections 2–24 with I-VIII points $\delta = 52$ mm for sections 1 and 25 with IX-X points.
$\delta = 16$ mm for sections 1 and 25 with I-VIII points $\delta = 46$ mm for sections 1–25 with XI-XII points.

Table 27.4 Coercive force values (H_C, A/cm) in each section of CES No. 131 and No. 542

Thickness	mm	6	8	10	12	14	16	20	24	30	34	45	52
CES No.	131	3.2	2.6	2.3	2.2	2.1	2.0	1.8					
CES No.	542		3.9	3.4	3.0		2.5	2.1	1.7	1.8	1.8	1.7	1.7

CES No.131:

$$H_C(\delta) = 1.82 + 5.33 \cdot \exp(-0.23 \cdot \delta) \tag{27.8}$$

CES No.542:

$$H_C(\delta) = 1.68 + 6.83 \cdot \exp(-0.139 \cdot \delta) \tag{27.9}$$

The results of reducing the coercive force measurements (*HC*, A/cm) on the main beam of the main lift of the casting crane to the thickness of 8 mm are presented in Figure 27.4.

Based on nomograms of the dependence of the residual number of load cycles on the coercive force values HC (A/cm) for metal structures of lifting machines (ISO 4301 (DocISO 2016)) and according to Figure 27.4, it follows that 74% of the metal of the main beam of the main lift operates in reliable mode, 25% in controlled mode, and 1% in the critical mode. An increase of the coercive force values in point V of section 16 has a local distribution, and for this reason, it is necessary to organize a permanent visual observation in this zone to prevent the appearance of cracks.

The conducted analysis of the growth rate of the coercive force in the main beam of the main lift of the casting crane with a load capacity of 180/50 *t* over the years of operation ($\Delta HC/\Delta T$, (A/cm)/year), taking into account the loading mode and according to (DocISO 2016), has shown that 88% of the metal operates in very light loading mode, 11% in light loading mode, 1% in average loading mode.

The intensity of operation for the investigated casting crane is 76 cycles per day. Subsequently, the estimated number of days (life cycle) that remains before each zone of the main beam switches into critical operating mode, while meeting the rated crane characteristics and the loading intensity, is shown in Figure 27.5. Figure 27.5 shows that, while meeting the rated casting crane characteristics and the loading intensity, 5% of the metal of the main beam of the main lift will operate in critical operating mode in 27,740 cycles (three years). Therefore, it is necessary, not later than in three years, to conduct an additional (repetitive) magnetic coercive NDT of the main beam (while meeting the rated crane characteristics), which includes recording measurement results in a magnetic control passport in order to track the dynamics of the stress and strain state of the metal.

All of the calculations were performed, and the graphics were produced with the use of the computer Metal v2.2 program (Gubskiy 2014).

Further, the metal structure of the investigated casting crane was calculated in the Crane program (Gubskiy 2013). Through this, the results of the magnetic coercive NDT were taken into account and a trolley rail and crane track were found to be in an unsatisfactory state.

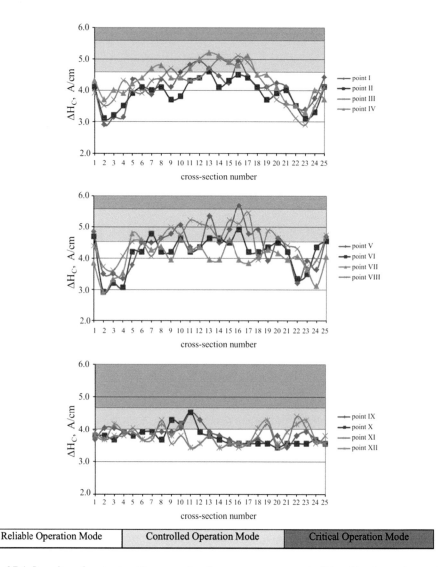

Figure 27.4 Results of reducing the coercive force measurements (H_C, A/cm) on the main beam of the main lift of the casting crane with a load capacity of 180/50 t to the thickness of 8 mm (according to Fig. 3)

Based on the magnetic coercive control results, processed with the Metal v2.2 program, and subsequent calculations of the metal structure using the Crane program, the life cycle of the casting crane with a load capacity of 180/50 t was extended for three years (while meeting the rated crane characteristics). It is necessary to conduct a second magnetic coercive NDT of the metal structure of the crane in three years, which includes recording the results of the measurements in the magnetic control passport in order to monitor the dynamics of the stress and strain state of the metal.

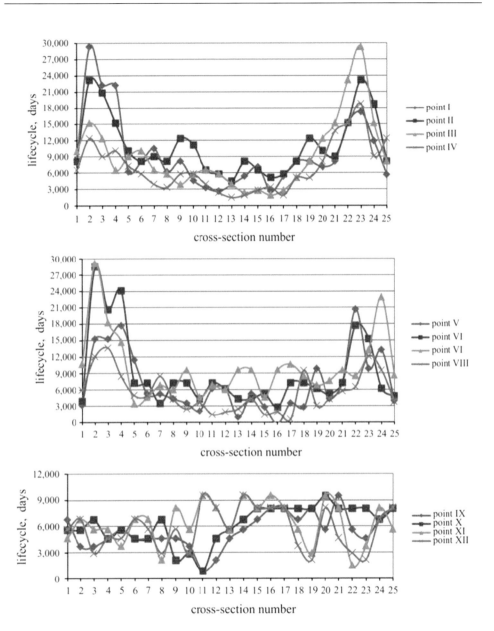

Figure 27.5 Estimated number of days (life cycle) that remain before each zone of the main beam switches into critical operating mode while meeting the rated crane characteristics and the loading intensity.

27.5 CONCLUSIONS

In the paper, we propose approaches to the automation of calculation and analytical procedures for estimating and predicting the technical condition of the metal structures of cranes. These calculations will enable an authorized organization, as a result of an inspection, to determine the operational integrity of the metal structures of hoisting machines that are about to reach their normative operating life cycle more efficiently. This is achieved by combining the processing automation of the magnetic coercive NDT results with the subsequent use of these results in the calculation algorithm of the limit states, which is implemented in a program. At present, we are conducting theoretical and practical studies of the influence of the loading intensity of the metal structure of the crane on the growth of the coercive force values as well as the application of the finite element method for the limit state calculation, followed by their automation.

REFERENCES

Braude, V., Ter-Mhitarov, M. 1985. *Sistemnye metody rascheta gruzopodemnyh mashin (System Methods for Calculating Lifting Machines)*: 1–232. Moscow: Mashinostroenie.
Dybała, J., Nadulicz, K. 2017. Diagnostics of welded joints using passive and active magnetic methods. *Diagnostyka* 18(4): 51–59.
EN 1993-1-9: Eurocode 3: Design of steel structures – Part 1–9: Fatigue.
F.E.M. 1.001 3rd edition revised 1998.10.0. Rules for the design of hoisting appliances. Booklet 2. Classification and loading on structures and mechanisms, 1998: 1–60.
Gohberg, M. 1976. *Metallicheskie konstrukcii podemno-transportnyh mashin (Metal Structures of Hoisting Machines)*: 1–456. Moscow: Mashinostroenie.
Gubskiy, S., Okun, A. 2013. *Komp'yuterna programa "Rozrahunok metalokonstrukciyi mostovogo krana ('Crane')" (Computer program "Calculation of overhead crane metal structure ('Crane')")*. Ukraine, Patent number 47890.
Gubskiy, S., Okun, A. 2014. *Metal v2.2 "Avtomatizaciya obrobki ta analizu rezul'tativ magnitnokoercitivnogo kontrolyu" (Metal v2.2 "Automation of processing and analysis of the results of magnetic coercive control")*. Ukraine, Patent number 54238.
Gudinaf, D. 1959. Teoriya vozniknoveniya oblastej samoproizvolnoj namagnichesvosti i koercitivnoj sily v polikristallicheskih ferromagnetikah (Theory of the occurrence of spontaneous magnetization and coercive force regions in polycrystalline ferromagnets). *Magnitnaya struktura ferromagnetikov* 4: 19–57.
Hryhorov, O., Okun, A., Gubskiy, S., Popov, V., Horlo, M. 2013. *Eksperimental'nij zrazok dlya kalibruvannya strukturoskopa (Experimental sample for calibrating the structuroscope)*, Ukraine, Patent number 77319.
ISO 4301:2016 Parts 1...5. Cranes. Classification.
Mantič, M., Kul'ka, J., Faltinová, E., Kopas, M. 2016. Autonomous online system for evaluating steel structure durability. *Diagnostyka* 17(3): 15–20.
MI 0.00–7.01–05. Metodychni vkazivky z provedennya magnitnogo kontrolyu napruzheno–deformovanogo stanu metalokonstrukcij pidjomnyh sporud ta vyznachennya yix zalyshkovogo resursu (Methodical instructions on the magnetic control of the stress-strain state of metal structures of lifting structures and determination of their residual lifecycle), 2005: 1–58.
Mikheev, M., Gorkunov, E. 1993. *Magnitnye metody stukturnogo analiza i nerazrushayushego kontrolya (Magnetic methods of structural analysis and non-destructive testing)*: 1–252. Moscow: Nauka.
Nemchuk, A. Starikov, M. 2006. Ocenka rabotosposobnosti kranovyh metalokonstrukcij na osnove chislennyh metodov (Evaluation of the performance of crane metal structures based on numerical methods). Podyomnye sooruzheniya. *Specialnaya tehnika*, 2006, vol. 7: 30–31.
OMD 20253.001–2005. Metodyka provedennya ekspertnogo obstezhennya (tehnichnogo diagnostuvannya) kraniv mostovogo typu (Methods of conducting expert certification (technical diagnostics) overhead type cranes). Kyiv, Derzhnaglyadohoronpraci Ukrayiny, 2005: 1–157.

Popov, V., Kotel'nikov V. 2001. Magnitnaya diagnostika i ostatochnyj resurs podyomnyh sooruzhenij (Magnetic diagnostics and residual lifecycle of lifting machines). *Bezopasnost truda v promyshlennosti* vol. 2: 44–49.

Popov, V., Rudnev, A., Gudoshnyk, V. 2015. Algoritm otvetstvennosti (Responsibility algorithm). Podyomnye sooruzheniya. *Specialnaya tehnika* vol. 10–12: 12–16.

Qi, K., Wang, W., Wang, X., Jiang, A., Liu, B., Guo, Z., Liu, J. 2013. Safety Assessment and Fatigue Life analysis of Aged Crane Structures. *Proc. of 13th International Conference on Fracture June 16–21*, 1–4, Beijing.

Sokolov, S. 2005. *Metallicheskie konstrukcii podemno-transportnyh mashin (Metal Structures of Hoisting Machines)*: 1–423. St.Petersburg: Politehnika.

Wu, F.Q., Zhang, J, Yao, W.Q. 2013. Fatigue life analysis of metallurgical bridge crane structure. *Applied Mechanics and Materials* vol. 437: 181–185.